WITHDRAWN
University of
Illinois Library
at Urbana-Champaign

INTRODUCTION TO THE THEORY OF SWITCHING CIRCUITS

McGRAW-HILL ELECTRICAL AND ELECTRONIC ENGINEERING SERIES

FREDERICK EMMONS TERMAN, *Consulting Editor*
W. W. HARMAN AND J. G. TRUXAL,
Associated Consulting Editors

AHRENDT AND SAVANT · Servomechanism Practice
ANGELO · Electronic Circuits
ASELTINE · Transform Method in Linear System Analysis
ATWATER · Introduction to Microwave Theory
BAILEY AND GAULT · Alternating-current Machinery
BERANEK · Acoustics
BRACEWELL · The Fourier Transform and Its Application
BRENNER AND JAVID · Analysis of Electric Circuits
BROWN · Analysis of Linear Time-invariant Systems
BRUNS AND SAUNDERS · Analysis of Feedback Control Systems
CAGE · Theory and Application of Industrial Electronics
CAUER · Synthesis of Linear Communication Networks
CHEN · The Analysis of Linear Systems
CHEN · Linear Network Design and Synthesis
CHIRLIAN · Analysis and Design of Electronic Circuits
CHIRLIAN AND ZEMANIAN · Electronics
CLEMENT AND JOHNSON · Electrical Engineering Science
COTE AND OAKES · Linear Vacuum-tube and Transistor Circuits
CUCCIA · Harmonics, Sidebands, and Transients in Communication Engineering
CUNNINGHAM · Introduction to Nonlinear Analysis
D'AZZO AND HOUPIS · Feedback Control System Analysis and Synthesis
EASTMAN · Fundamentals of Vacuum Tubes
ELGERD · Control Systems Theory
EVELEIGH · Adaptive Control and Optimization Techniques
FEINSTEIN · Foundations of Information Theory
FITZGERALD, HIGGINBOTHAN, AND GRABEL · Basic Electrical Engineering
FITZGERALD AND KINGSLEY · Electric Machinery
FRANK · Electrical Measurement Analysis
FRIEDLAND, WING, AND ASH · Principles of Linear Networks
GEHMLICH AND HAMMOND · Electromechanical Systems
GHAUSI · Principles and Design of Linear Active Circuits
GHOSE · Microwave Circuit Theory and Analysis
GREINER · Semiconductor Devices and Applications
HAMMOND · Electrical Engineering
HANCOCK · An Introduction to the Principles of Communication Theory
HAPPELL AND HESSELBERTH · Engineering Electronics
HARMAN · Fundamentals of Electronic Motion
HARMAN · Principles of the Statistical Theory of Communication
HARMAN AND LYTLE · Electrical and Mechanical Networks
HARRINGTON · Introduction to Electromagnetic Engineering
HARRINGTON · Time-harmonic Electromagnetic Fields
HAYASHI · Nonlinear Oscillations in Physical Systems
HAYT · Engineering Electromagnetics
HAYT AND KEMMERLY · Engineering Circuit Analysis
HILL · Electronics in Engineering
JAVID AND BRENNER · Analysis, Transmission, and Filtering of Signals
JAVID AND BROWN · Field Analysis and Electromagnetics
JOHNSON · Transmission Lines and Networks
KOENIG AND BLACKWELL · Electromechanical System Theory
KOENIG, TOKAD, AND KESAVAN · Analysis of Discrete Physical Systems
KRAUS · Antennas
KRAUS · Electromagnetics
KUH AND PEDERSON · Principles of Circuit Synthesis
KUO · Linear Networks and Systems
LEDLEY · Digital Computer and Control Engineering

LePage · Analysis of Alternating-current Circuits
LePage · Complex Variables and the Laplace Transform for Engineering
LePage and Seely · General Network Analy is
Levi and Panzer · Electromechanical Power Conversion
Ley, Lutz, and Rehberg · Linear Circuit Analysis
Linvill and Gibbons · Transistors and Active Circuits
Littauer · Pulse Electronics
Lynch and Truxal · Introductory System Analysis
Lynch and Truxal · Principles of Electronic Instrumentation
Lynch and Truxal · Signals and Systems in Electrical Engineering
Manning · Electrical Circuits
McCluskey · Introduction to the Theory of Switching Circuits
Meisel · Principles of Electromechanical-energy Conversion
Millman · Vacuum-tube and Semiconductor Electronics
Millman and Halkias · Electronic Devices and Circuits
Millman and Seely · Electronics
Mishkin and Braun · Adaptive Control Systems
Moore · Traveling-wave Engineering
Nanavati · An Introduction to Semiconductor Electronics
Pettit · Electronic Switching, Timing, and Pulse Circuits
Pettit and McWhorter · Electronic Amplifier Circuits
Pfeiffer · Concepts of Probability Theory
Pfeiffer · Linear Systems Analysis
Reza · An Introduction to Information Theory
Reza and Seely · Modern Network Analysis
Rogers · Introduction to Electric Fields
Ruston and Bordogna · Electric Networks: Functions, Filters, Analysis
Ryder · Engineering Electronics
Schwartz · Information Transmission, Modulation, and Noise
Schwarz and Friedland · Linear Systems
Seely · Electromechanical Energy Conversion
Seely · Electron-tube Circuits
Seely · Electronic Engineering
Seely · Introduction to Electromagnetic Fields
Seely · Radio Electronics
Seifert and Steeg · Control Systems Engineering
Siskind · Direct-current Machinery
Skilling · Electric Transmission Lines
Skilling · Transient Electric Currents
Spangenberg · Fundamentals of Electron Devices
Spangenberg · Vacuum Tubes
Stevenson · Elements of Power System Analysis
Stewart · Fundamentals of Signal Theory
Storer · Passive Network Synthesis
Strauss · Wave Generation and Shaping
Su · Active Network Synthesis
Terman · Electronic and Radio Engineering
Terman and Pettit · Electronic Measurements
Thaler · Elements of Servomechanism Theory
Thaler and Brown · Analysis and Design of Feedback Control Systems
Thaler and Pastel · Analysis and Design of Nonlinear Feedback Control Systems
Thompson · Alternating-current and Transient Circuit Analysis
Tou · Digital and Sampled-data Control Systems
Tou · Modern Control Theory
Truxal · Automatic Feedback Control System Synthesis
Tuttle · Electric Networks: Analysis and Synthesis
Valdes · The Physical Theory of Transistors
Van Bladel · Electromagnetic Fields
Weinberg · Network Analysis and Synthesis
Williams and Young · Electrical Engineering Problems

Princeton University Series

Clement and Johnson Electrical Engineering Science
Johnson Mathematical and Physical Principles of Engineering Analysis
Johnson Transmission Lines and Networks
McCluskey Introduction to the Theory of Switching Circuits

E. J. McCluskey

PROFESSOR OF ELECTRICAL ENGINEERING
PRINCETON UNIVERSITY

INTRODUCTION TO THE THEORY OF SWITCHING CIRCUITS

McGraw-Hill Book Company

NEW YORK   ST. LOUIS   SAN FRANCISCO   TORONTO   LONDON   SYDNEY

INTRODUCTION TO THE THEORY OF SWITCHING CIRCUITS

Copyright © 1965 by McGraw-Hill, Inc. All Rights Reserved. Printed in the United States of America. This book, or parts thereof, may not be reproduced in any form without permission of the publishers. Library of Congress Catalog Card Number 65-17394

44843

4567890 MP 72106987

PREFACE

This book is intended as a text for a first course in "classical" switching theory for engineers. A vigorous attempt has been made to include only topics that will be of lasting general importance. In keeping with this objective, and considering the increasing use of digital-computer programs for design, the emphasis of this book is on algorithms and general theory. Design techniques that are useful only for particular components or that depend on the skill of the user are not treated in detail. I feel that such techniques are neither sufficiently general nor of permanent enough importance to justify inclusion in an introductory course. Since this book is planned mainly as a text for engineers, a criterion of practical importance rather than mathematical beauty has been used in selecting topics and methods of exposition. However, many students in mathematics and the physical sciences should find the material presented here of value. Specific applications are usually treated in the problems rather than in the text, and the problems are also used as a vehicle for introducing some of the more specialized concepts.

It has been my experience that the material in this book can be covered in a one-term, advanced-undergraduate course or a first-year graduate course. I have been able to cover this material in a three-hour-per-week course by excluding Chap. 1 (Switching Devices) from the formal lectures and adding a weekly laboratory in which the material of Chap. 1 was discussed and illustrated by experiments performed by the students. It is my belief that a basic knowledge of switching devices is essential to the students' understanding of the mathematical models and that this knowledge should be supplied either via a laboratory or a separate course on digital devices and circuits.

Since 1956 this text material, in its evolutionary form, has been used each year in my teaching, first at the City College of New York and later at Princeton University. My acknowledgment of the influence of this classroom use on the final form of the text is somewhat superfluous. Major reorganizations were carried out as a result of student reaction. In preparing a text for a one-term course, it was necessary to exclude a number of interesting topics. I feel strongly that a second course should be available for students having a specific interest in the field of switching theory. A student who has completed a course using this book should be well prepared to make use of the current literature in switching theory.

The organization of the book is as follows. There are two introductory chapters—Chap. 1 on switching devices and Chap. 2 on number systems and codes. These are followed by two chapters (3 and 4) on combinational switching circuits, two chapters (5 and 6) on sequential switching circuits, and a final chapter (7) in which the effects of spurious delays are considered.

Chapter 1 is a discussion of some of the basic circuits used in the construction of digital systems, with major attention devoted to transistor circuits. Several other types of elements are also considered, either because of their importance in the early literature on switching theory, as in the case of relays and vacuum tubes, or to illustrate the wide range of applicability of switching theory to such diverse devices as magnetic cores and cryotrons. Chapter 2 includes material necessary for some of the algorithms discussed in the later chapters. This material also makes possible the inclusion of more meaningful problems and examples throughout the remainder of the book.

In Chapter 3, Boolean algebra is first introduced rather informally as a model for switching circuits, and then at the end of the chapter it is reconsidered as an abstract mathematical system. Elementary analysis and synthesis of networks of contacts, AND gates and OR gates, and NOR gates (AND-NOT or OR-NOT gates) are presented along with the Boolean-algebra theorems and a discussion of complete sets of basic elements.

The problem of simplifying switching functions is covered in Chap. 4. The emphasis in this chapter is almost exclusively on two-stage forms, since I feel that the results in this area will probably be of more lasting importance than the techniques developed to date for handling multistage problems. Karnaugh map methods and tabular minimization procedures for both single-output and multiple-output networks are treated in considerable detail. The often-neglected technique of iterative consensus is also covered thoroughly with a careful proof of the basic theorem as well as detailed examples of the algorithm.

The analysis of sequential circuits with either pulse or level inputs is treated in Chap. 5. This chapter introduces the flow-table technique and demonstrates its applicability to a wide variety of sequential circuits, including those synchronized by an external clock input. The state-diagram representation of sequential circuits is also included.

Chapter 6 presents synthesis techniques for sequential circuits, including a thorough treatment of state minimization for incompletely specified flow tables. Of particular importance is the integrated treatment of both pulse-input (pulse-mode) and level-input (fundamental-mode) sequential circuits.

The consequences of spurious delays being present in switching circuits are considered in Chap. 7. Static, dynamic, and essential hazards are treated in detail.

I am indebted to many individuals for their assistance in the writing of this book. Mrs. Hannah Kresse typed most of the final manuscript. Others who helped with the typing were Mrs. Jayne De Micheli, Mrs. Joycie Miura, Miss Florence Armstrong, Mrs. Frances Turner, and Miss Elizabeth Stetson. Many of my collegues and graduate students contributed helpful suggestions. Particular thanks are due to A. Lo, E. Eichleberger, A. Grasselli, H. Schorr, S. Gaines, J. Brzozowski, and A. Hall. My wife Roberta supplied understanding and patience throughout many years of "work in progress."

E. J. McCluskey

CONTENTS

Preface   *vii*

1. Switching Devices   1

1.1 *Electromechanical Devices*   *1*
Switches, or Keys *1*,   Relays *3*

1.2 *Vacuum Tubes*   *4*
Flip-flops *4*,   Logic Circuits *5*

1.3 *Diodes*   *8*
Circuit Operation *10*,   Circuit Losses *10*

1.4 *Transistors*   *11*
Transistor Characteristics *11*,   Basic Transistor Switch *13*,   Transistor Amplifier *15*,   Transistor-diode Logic *16*,   Transistor-resistor Logic *18*,   Direct-coupled Transistor Logic (DCTL) *19*,   Transistor Current-switching Gates *20*

1.5 *Cryotrons*   *22*
Basic Physical Principles *22*,   The Wire-wound Cryotron *23*,   A Flip-flop Using Wire-wound Cryotrons *23*,   The Crossed-film Cryotron *26*,   Crossed-film-cryotron Circuits *27*

1.6 *Magnetic Cores*   *28*
Equivalent Circuit *30*,   Combining Input Signals *31*,   Mirror Symbols *32*,   Output Connections *33*

1.7 *Symbols for Electronic Gates*   *35*

Problems   *39*
References   *40*

2. Number Systems and Codes — 42

- 2.1 Positional Notation — 42
  Conversion of Base 43
- 2.2 Binary Arithmetic — 46
  Binary Addition 47, Binary Subtraction 47, Complements 48, Shifting, Binary Multiplication 51, Binary Division 51
- 2.3 Binary Codes — 53
  Binary-coded-decimal Numbers 53
- 2.4 Geometric Representation of Binary Numbers — 55
  Distance 57, Unit-distance Codes 59, Symmetries of the n Cube 61, Error-detecting and -correcting Codes 62

Problems — 63
References — 65

3. Switching Algebra — 66

- 3.1 Postulates — 67
- 3.2 Analysis of Switching Circuits — 70
  Contact Networks 70, AND Gates and OR Gates 74, Table of Combinations 75
- 3.3 Synthesis — 76
  Canonical Expressions 78, Networks 81, Number of Functions 83
- 3.4 Theorems — 84
  Single-variable Theorems 85, Two- and Three-variable Theorems 86, n-Variable Theorems 87
- 3.5 General Gate Networks — 89
  Complete Sets 89, Analysis of AND-NOT Networks and OR-NOT Networks 93
- 3.6 Boolean Algebra — 97
  Propositional Logic 99, Algebra of Sets 103, Boolean Rings 104, Summary of Postulates and Theorems for Switching Algebra 105

	Problems	106
	References	113

4. Simplification of Switching Functions — 114

4.1 The Map Method — 114
Maps for Two, Three, and Four Variables *116*, Prime Implicants *117*, Maps for Five and Six Variables *119*, Formation of Minimal Sums *123*, Incompletely Specified Functions *126*, Minimal Products *127*

4.2 General Properties of Minimal Sums and Products — 128
Prime Implicant Theorem *129*, Generalized Prime Implicant Theorem *131*

4.3 Multiple-output Networks — 131
Multiple-output Prime Implicants *132*, Essential Multiple-output Prime Implicants and Maps *136*

4.4 Tabular Determination of Prime Implicants — 140
Binary-character Method *140*, Use of Octal Numbers *143*

4.5 Prime Implicant Tables — 146
Essential Rows *146*, Dominance *147*, Cyclic Prime Implicant Tables *152*

4.6 Tabular Methods for Multiple-output Circuits — 157
Multiple-output Prime Implicant Tables *160*

4.7 Iterative Consensus — 165
The Consensus Operation *166*, Complete Sums *167*, An Algorithm *171*

Problems — 174
References — 178

5. Sequential-circuits Analysis — 180

5.1 Introduction — 180
Illustration *180*

5.2 Formal Analysis of Circuits Containing S-R Flip-flops   184

    Transition Table *186*,   The Transition Diagram and State Table *187*,   The Excitation Table *188*

5.3 Various "Memory" Devices for Sequential Circuits   190

    Physical Requirements *191*,   Analysis of Sequential Circuits Constructed of Diode Gates *192*,   Analysis of Sequential Circuits Constructed of AND-NOT Gates *194*

5.4 Races in Sequential Circuits   195

    The Flow Table *199*

5.5 Pulse-Mode Operation   199

    Operating Points *207*,   Races in Pulse-mode Operation *208*

5.6 Clocked Sequential Circuits   210

    Fundamental-mode Clocked Circuits *215*

5.7 State Diagrams   216

Problems   218
References   223

6. Sequential-circuit Synthesis   224

6.1 Formation of the Flow Table   224

    Output Specifications–Pulse Mode *227*,   Fundamental Mode–Primitive-form Flow Tables *230*,   Output Specifications–Fundamental Mode *233*,   Initial States *235*,   Impossible Specifications *235*

6.2 Simplification of Completely Specified Flow Tables   237

    Inaccessible States *237*,   Indistinguishable Circuits *239*,   Indistinguishable States *241*,   Indistinguishability Classes *243*,   Minimum-state Flow Tables *244*,   Distinguishable States *246*,   Determination of Indistinguishability Sets *246*

6.3 Simplification of Incompletely Specified Flow Tables   248

    Unspecified Outputs *249*,   Unspecified Next States *251*,   Covering of Flow Tables *252*,   Maximum Compatibility

Classes *258*, Bounds on Minimum Number of States *260*, Forming Closed Collections of Compatibility Classes *260*, Type A Flow Tables *261*

6.4 *Formation of Transition and Excitation Tables*   *261*

Formation of Flip-flop Excitation Tables *262*, Internal Variable Assignments–Symmetries *263*, Critical Races *268*, Standard Assignments for General Flow Tables *272*

Problems   *277*
References   *282*

7. *Transient Behavior of Switching Circuits*   **284**

7.1 *Combinational Networks*   *284*

7.2 *Analysis*   *285*

Contact Networks *285*, Gate Networks *287*

7.3 *Static Hazards*   *288*

Removal of Subscripts *292*, An Example of the Use of Theorem 7.3.3 *293*, A Test Procedure *295*, Alternative Conditions for Static Hazards *295*, Analysis-procedure Example *297*

7.4 *Dynamic Hazards*   *299*

Analysis Example *302*

7.5 *Synthesis of Hazard-free Combinational Networks*   *303*

7.6 *Essential Hazards*   *306*

Problems   *310*
References   *313*
Additional References   *314*

Index   **315**

1 SWITCHING DEVICES

The types of circuits to be discussed in this book have a very important distinguishing characteristic—there are two possible values for the signals, and each signal is equal to one or the other of these values. Such two-valued signals are called *binary signals*. For example, a switching circuit constructed of transistors might have all voltages equal to either 0 or $+4$ volts. This type of circuit operation is quite different from that of a circuit such as a radio receiver, in which it is very important that the signals vary over all possible values between some upper and lower limits. In spite of this difference in operation there are similarities between these two classes of circuits. A radio receiver is designed by interconnecting circuits which perform specific tasks such as amplification, modulation, etc. Similarly a digital system is made up of circuits which carry out specific jobs of combining, generating and storing binary signals. The object of this chapter is to describe some typical circuits which are used to perform these functions. Since entire books [1,2]† are devoted to this topic, only the most important of such circuits will be described and general features rather than detailed design information will be stressed.

1.1 ELECTROMECHANICAL DEVICES

Switches, or Keys

Probably the most familiar switching device is the switch which is used to control the lights in our homes. This device usually consists of two metal strips called *springs* on which are mounted metal contacts. A lever or "push button" controls the contacts so that they either touch and form a closed electrical circuit (short circuit) or are separated and therefore constitute an open circuit.

† Numbers in brackets throughout the book correspond to references at the ends of chapters.

Switching Devices

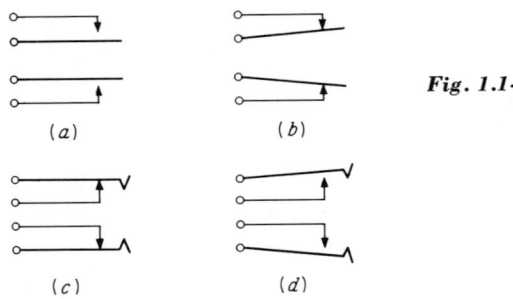

Fig. 1.1-1. *Symbols for various types of contacts. (a) Two nonlocking make contacts (released); (b) the same (operated); (c) two locking break contacts (released); (d) the same (operated).*

There are many different types of switches, or keys, but they fall into two general classes:

1. *Nonlocking*—those which return to a standard position when released
2. *Locking*—those which retain their position after the controlling lever or button is released

Usually either the contacts are normally open, or "make," contacts, or they are normally closed, or "break," contacts (Fig. 1.1-1). There are two types of *transfer contact* in which a break and a make contact are combined. The "break-before-make," or transfer, contact contains three springs. There is a common spring, which is normally connected to one of the other springs. When the switch is operated, these two springs separate and the common spring moves over to make contact with the third spring (Fig. 1.1-2a). When this type of contact is in the process of operating or releasing, there is a certain amount of time when the common spring is not touching either of the other springs. There is also a "make-before-break," or continuity transfer, which is identical to the transfer contact except that when the contact is operating or releasing there is a time when the common spring is touching *both* the other springs.

The foregoing remarks assumed that switches had two positions—operated and released. Such switches are called *single-throw*. *Double-throw switches*, which have a released position and two operated positions, are also common (Fig. 1.1-3).

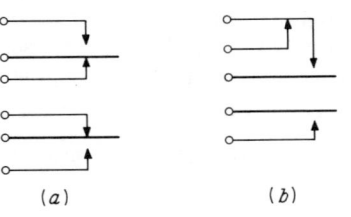

Fig. 1.1-2. *Symbols for transfer contacts. (a) Two break-before-make, or transfer, contacts (nonlocking); (b) a make-before-break, or continuity transfer and a make contact (nonlocking).*

1.1 electromechanical devices

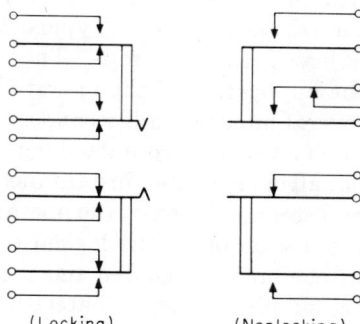

Fig. 1.1-3. Double-throw switch. (Locking)   (Nonlocking)

Relays

A switch whose lever is operated by an electromagnet is called a *relay*. In this case the lever is called the *armature*. The contact arrangements are the same as for switches, and the same symbols are often used. For logical design it is more convenient to use the *detached-contact* symbols shown in Fig. 1.1-4. The same symbol is used for both make and break contacts—they are distinguished by the labels x for a make contact of relay X and x' for a break contact of relay X. Transfer and continuity transfer contacts are represented by the appropriate combination of make and break contact symbols, and usually no distinction is made between the two types of transfer contacts.

Many different types of special-purpose relays such as polarized relays, a-c relays, sensitive relays exist; but this book will be concerned only with the general-purpose d-c relay. This type of relay operates when the current through its coil exceeds a certain value (called the *operate current*) and

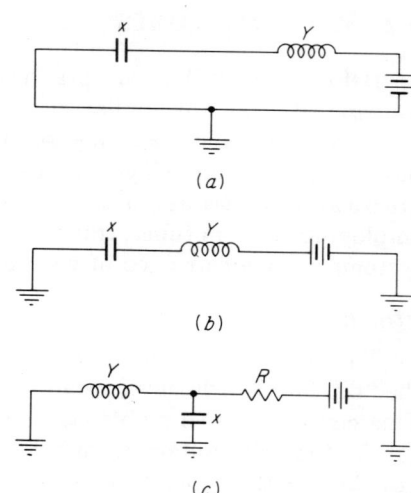

Fig. 1.1-4. Methods of controlling relays. (a) Circuit for series control of relay Y by contact x; (b) form of (a) to be used in this book; (c) shunt control of relay Y by contact x.

Switching Devices

remains operated until the current diminishes below another value (called the *release current*). Typical values for these currents range from a few milliamperes to 1 amp. The time it takes for the contacts to close after the application of the operate current ranges from a few milliseconds to several hundred milliseconds. An important characteristic of the operation of a relay is contact stagger. When a relay operates, the contacts do not all open or close instantaneously, and so there may be a delay of several milliseconds between the operation of two contacts of the same relay. In the discussion of the logical design of relay networks, this phenomenon will be initially ignored and the assumption will be made that all contacts act simultaneously. Later the effects of contact stagger will be investigated, and techniques for eliminating undesirable consequences will be presented. These relays are ordinarily used with voltages of a few volts to approximately 100 volts.

The symbol used for a relay coil is the "coil" symbol shown in Fig. 1.1-4. The most common method for controlling the operation of a relay is shown in Figs. 1.1-4a and b. Here a contact is placed in series with the relay coil and a battery. When the contact is open, no current flows and the relay does not operate. When the contact is closed, current flows through the relay coil and the relay operates. Occasionally, a relay is controlled by a shunt contact as in Fig. 1.1-4c. When this contact is closed, it places a very low-impedance path in parallel with the relay coil and the current from the battery flows through this shunt path rather than through the relay coil, preventing the relay from operating. When the contacts are open, the current flows through the relay and it operates (provided that the value of R is chosen properly) [3,4].

1.2 VACUUM TUBES

Before World War II, practically all switching systems were constructed of relays or similar electromechanical components. However, the relatively slow operating speeds of these components constitute a basic limitation on relay systems, and the modern digital computer is practical only because of the development of high-speed digital circuits employing vacuum tubes, transistors, etc. The first high-speed switching systems were constructed of vacuum tubes [5].

Flip-flops

The oldest and perhaps the most basic vacuum-tube digital circuit is the "flip-flop," or Eccles-Jordan trigger [1, chap. 5;6], shown in Fig. 1.2-1. This circuit has two stable states in which either tube 1 or tube 2 is at cutoff (the grid voltage is sufficiently negative so that no plate current flows); and the other tube is in clamp (grid current flows, the grid-to-

1.2 vacuum tubes

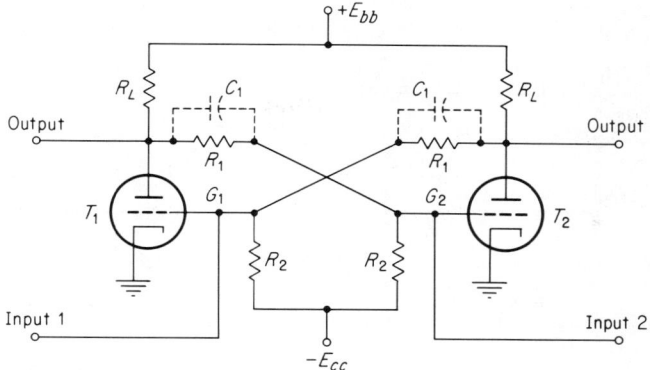

Fig. 1.2-1. Flip-flop circuit.

cathode voltage is approximately zero, and the tube is conducting). There are only two possible values of plate voltage: that corresponding to a conducting and that corresponding to a cutoff tube. These plate voltages are commonly used as the outputs from this circuit. In a sense a flip-flop performs the function of generating binary output signals.

It is possible to place the flip-flop in either of its two stable states by applying a signal to one of the input leads. Thus the flip-flop can be used to "remember" which of its input leads was pulsed last.

The symbol for a flip-flop is shown in Fig. 1.2-2. As indicated in this figure, if the S input lead was last pulsed, then the y output lead will have a high voltage and the y' output lead will have a low voltage. If the last-pulsed lead was R, y' has a high voltage and y has a low voltage.

It is also possible to construct a flip-flop which changes state whenever its single input lead is pulsed. A circuit for such a flip-flop, which is sometimes called a *binary-counter*, or *symmetrical-triggering flip-flop*, is shown in Fig. 1.2-3 along with the symbol to be used to represent it.

Logic Circuits

The term *logic circuits* is usually applied to the circuits which perform operations on binary signals [1, chap. 12;7,8]. One of the most important

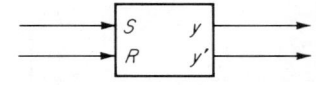

Lead pulsed last	y	y'
S	High	Low
R	Low	High

Fig. 1.2-2. Flip-flop symbol.

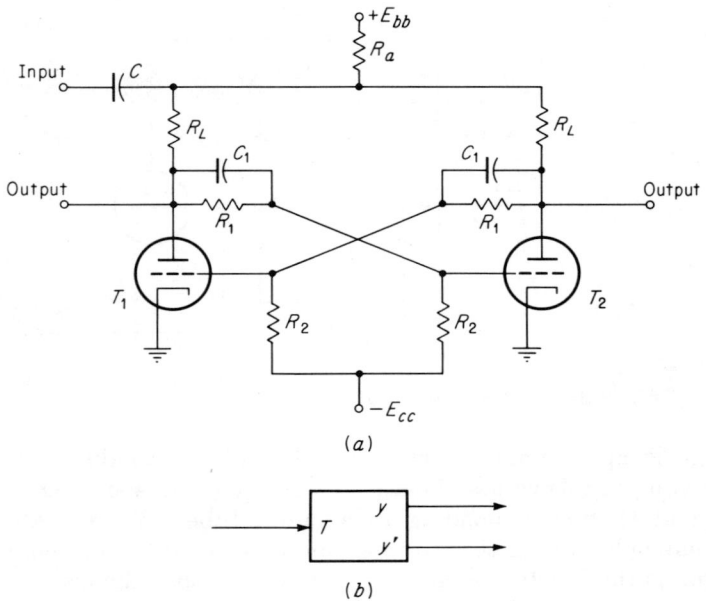

Fig. 1.2-3. *Symmetrical-triggering flip-flop.* (a) *Circuit;* (b) *symbol.*

types of logic circuit is that which has a high output signal when the input signal is low, and vice versa. Such a circuit is called an *inverter*. An example of an inverter employing a triode is shown in Fig. 1.2-4a. In this circuit, the input voltage e_1 is equal either to E_L—a voltage sufficiently negative to prevent the tube from conducting—or to E_H—a voltage of approximately zero volts. When e_1 equals E_L, the tube is cut off and the values of R_L, R_1, and R_2 are chosen so that $e_o = E_H$. When e_1 equals E_H, the tube conducts and the voltage at point 1 drops so that e_o becomes

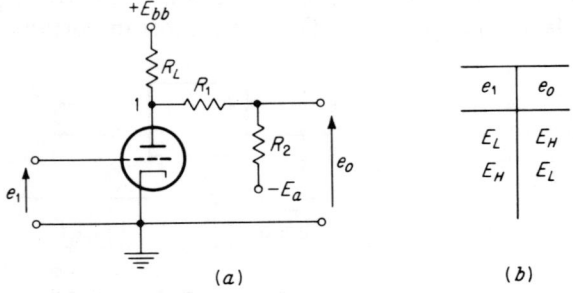

Fig. 1.2-4. *A triode inverter circuit.*

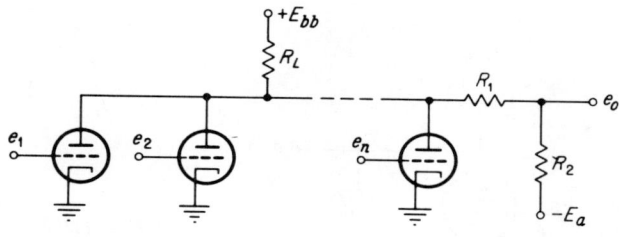

$e_o = E_H$ only if e_1 and e_2 and ... e_n all equal E_L

Fig. 1.2-5. *A circuit with n triodes sharing a common plate resistor.*

equal to E_L. This performance is summarized in Fig. 1.2-4b. The purpose of the resistances R_1 and R_2 is to shift the voltage level of point 1 so that the output voltage ranges over the same values as the input voltage. This is necessary since the output from this circuit will be used as the input to a similar circuit.

When several of these circuits are connected with a common plate resistance as in Fig. 1.2-5, a circuit which can be used for combining several binary signals results. In this circuit, when all the input voltages equal E_L, all tubes are cut off and the output voltages are equal to E_H. If any of the input voltages are equal to E_H, the corresponding tubes will conduct and the output voltage will drop to E_L. The output voltage will actually vary according to the number of tubes which are conducting, but this variation can be minimized by choosing R_L much larger than the plate resistances of the triodes. A similar circuit with the common load resistance in the cathode circuit instead of the plate circuit is shown in Fig. 1.2-6. When all the input voltages are equal to E_L, all the tubes are cut off and e_o is equal to E_L (R_1 and R_2 are chosen to ensure this). If one or more of the input voltages are equal to E_H, the corresponding tubes will

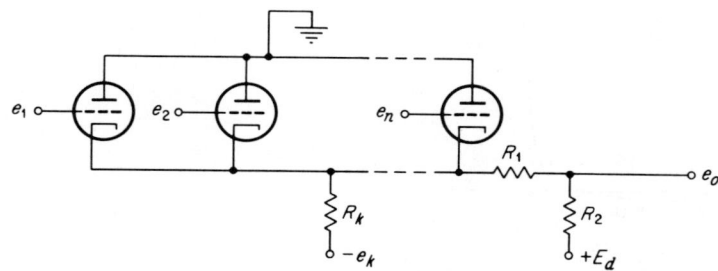

$e_o = E_H$ only if one or more of the input voltages $e_1, e_2, ..., e_n$ are equal to E_H

Fig. 1.2-6. *A circuit with n triodes sharing a common cathode resistor.*

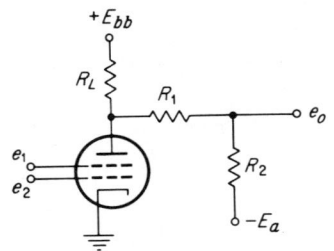

Fig. 1.2-7. *A circuit employing a tube with two control grids.*

$e_o = E_H$ if e_1 or e_2 or e_1 and e_2 are equal to E_L

conduct and the output voltage will rise to E_H. The variation in e_o due to different numbers of tubes conducting is minimized by making R_K much larger than the tube plate resistance.

Similar circuits can be constructed using tubes with more than one control grid. The basic form of such circuits is shown in Fig. 1.2-7. When both e_1 and e_2 are equal to E_H, the tube conducts and e_o is equal to E_L. If either e_1 or e_2 or both equal E_L, the tube is cut off and e_o is equal to E_H. Tubes with more than one control grid can be combined in the same manner in which triodes are combined in Figs. 1.2-5 and 1.2-6.

At the present time, semiconductor diodes and transistors are more commonly used for digital systems than are vacuum tubes.

1.3 DIODES

It is possible to construct switching systems in which the logic or combining of signals is done by means of circuits constructed of diodes; electronic tubes or transistors are used only for flip-flops, amplifiers, or inverters. An ideal diode is a two-terminal device which presents zero resistance to current flowing through it in the positive, or forward,

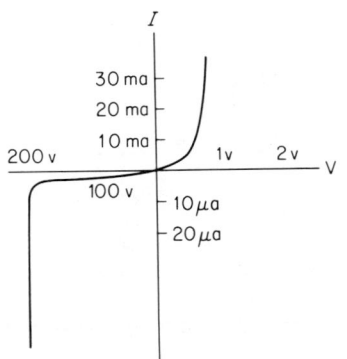

Fig. 1.3-1. *Typical junction-diode characteristics.*

direction and infinite resistance to current flowing in the negative, or backward, direction. The actual junction diodes which are used in switching circuits have a volt-ampere characteristic similar to that shown in Fig. 1.3-1.

The two basic diode logic circuits are shown in Fig. 1.3-2. Their performance will be analyzed on the assumption that ideal diodes are used, and then the effect of the departure of the physical diode characteristic from the ideal will be considered. Only three inputs are shown in the

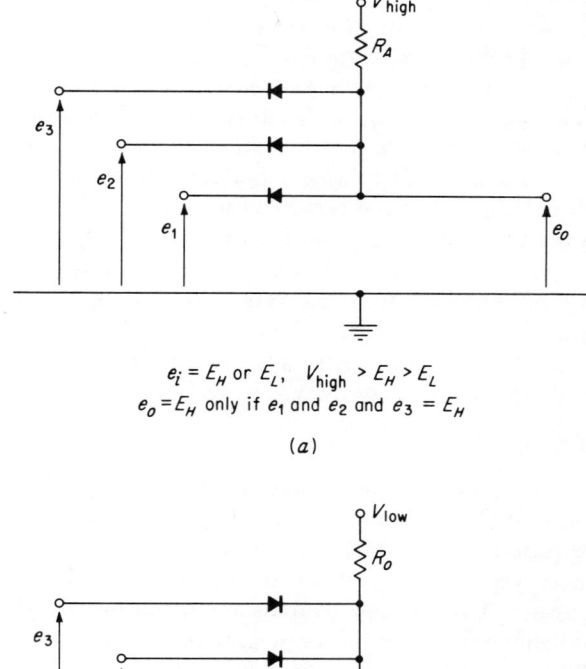

Fig. 1.3-2. *Diode logic circuits.* (a) **AND gate;** (b) **OR gate.**

figure, but the operation is essentially unchanged when more inputs are added.

Circuit Operation

If any of the inputs to the circuit of Fig. 1.3-2a are equal to E_L, the output e_o must also be equal to E_L. If the output voltage were greater than E_L, the diode connected to E_L would be conducting forward current. By definition the resistance of this diode would equal zero so that there could not be any voltage drop across it and e_o could not be greater than E_L. The output voltage cannot be less than E_L since none of the input, or supply, voltages are less than E_L. When all the input voltages are equal to E_H, the output voltage e_o must also be equal to E_H. For the reasons just discussed, e_o cannot be greater than E_H nor can e_o be less than E_H, because in this case none of the input, or supply, voltages are less than E_H. This circuit is frequently called an AND *gate* because the output voltage is at the high level only when e_1 *and* e_2 *and* e_3 are at the high level.

In the circuit of Fig. 1.3-2b, e_o will equal E_H when any of the inputs equal E_H. It is not possible for e_o to be greater than E_H since no voltages greater than E_H are connected to the circuit. If the output were less than E_H, the diodes connected to it would be forward-biased and would therefore have zero resistance, forcing e_o to equal E_H. When all the inputs are equal to E_L, the output e_o must also be equal to E_L, by similar reasoning. This circuit is frequently called an OR *gate* because the output voltage is at the high level when e_1 *or* e_2 *or* e_3 is at the high level.†

Circuit Losses

With ideal diodes, there would be no degradation of the voltage levels E_L and E_H in a circuit composed of interconnected diode gates. However, this does not mean that there would be no need for amplification in such a network. Even with ideal diodes there would be two types of current losses. One of these sources of current loss is called *fan-out* and is present whenever the output of one gate is used as an input to several other gates. The other loss, *current-transfer loss*, causes the output current from a series connection of an AND gate and an OR gate to be less than the input current [9].

The fact that a physical diode does not have an infinite back resistance introduces another current loss due to the leakage currents in reversed-biased diodes. In addition, the nonzero forward resistance of an actual diode causes voltage losses because of the potential difference across a forward-biased diode.

In a circuit having several of these diode gates connected in series, it is not possible to use the same value of resistance for all the OR-gate resistors

† The or in this expression is an inclusive or—e_1 and e_2, e_1 and e_3, e_1 and e_2 and e_3, etc., being at the high level will also cause e_o to be at the high level.

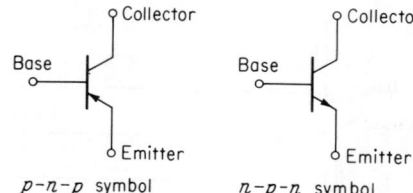

Fig. 1.4-1. Transistor symbols.   p-n-p symbol   n-p-n symbol

(R_O) or all the AND-gate resistors (R_A). The value of each of the resistances in the circuit must be individually calculated [2,9]. Moreover, there is a limit on the number of gates that can be connected in series because it becomes impossible to determine reliably whether E_H or E_L is present at the output. For a large switching system it is usually impractical to calculate individual resistances. The customary practice is to connect only two gates in series and then to amplify the outputs before connecting to any other gates [10]. An amplifier for this purpose could be constructed by connecting in series two of the triode inverter circuits described in Sec. 1.2.

1.4 TRANSISTORS

A transistor† is a germanium or silicon crystal into which impurities have been introduced to form three regions called the *emitter*, the *base*, and the *collector* [1, chap. 18;2;11]. An external connection is made to each of these regions. There are two basic types of transistor—the *p-n-p* and the *n-p-n*, which differ in the impurities used to form the three regions. The two types of transistor are "dual" devices since the general performance of the *p-n-p* is the same as that of the *n-p-n* with the directions of all currents and voltages reversed. The symbols for these transistors are shown in Fig. 1.4-1.

Transistor Characteristics

A typical set of characteristics for an *n-p-n* transistor connected with its base used as the common terminal as in Fig. 1.4-2 (called the *common-base connection*) is shown in Fig. 1.4-3. The corresponding circuit and

† Strictly, a junction transistor.

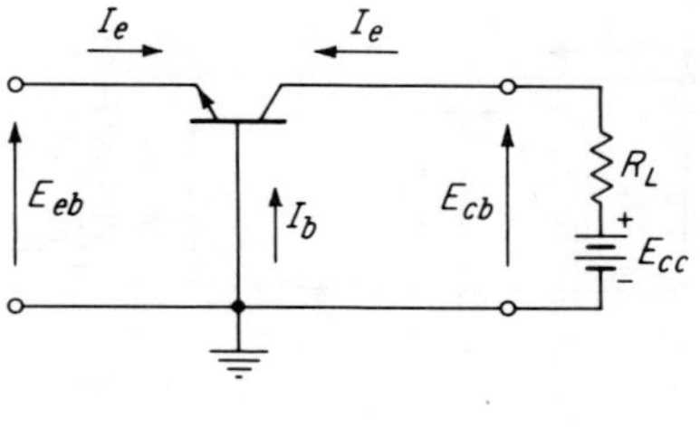

Fig. 1.4-2. *An n-p-n transistor in a common-base connection.*

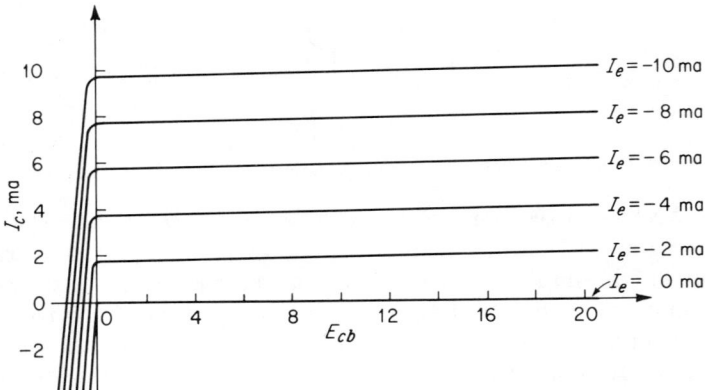

Fig. 1.4-3. *Typical common-base characteristics for an n-p-n transistor.*

characteristics for a *p-n-p* transistor are shown in Figs. 1.4-4 and 1.4-5. Note that they are the same as Figs. 1.4-2 and 1.4-3 except for polarity reversals. The curves of Fig. 1.4-3 show that, when the collector-to-base voltage E_{cb} is positive, the collector current I_c is essentially independent of E_{cb} and proportional to I_e, the emitter current. For an *n-p-n* transistor, when E_{cb} is positive, the collector junction is said to be *reverse-biased* and when E_{cb} is negative the collector junction is said to be *forward-biased*. Similarly, the emitter junction is said to be reverse-biased when E_{eb} is positive and forward-biased when E_{eb} is negative. When the collector junction is reverse-biased (E_{cb} positive) and the emitter junction forward-biased (E_{eb} negative), the transistor is operating in its *active region*, with I_c given by

$$I_c = I_{c0} - \alpha I_e \qquad (1.4\text{-}1)$$

$$\alpha = \left(\frac{\Delta I_c}{\Delta I_e}\right)_{E_{cb}=\text{const}} \qquad (1.4\text{-}2)$$

The symbol I_{c0} represents the reverse-saturation collector current† and is very small, of the order of $2\mu a$ for the type of transistor represented by

† Due to thermally generated minority carriers in the base and collector.

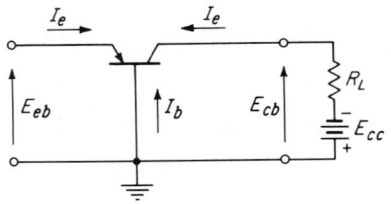

Fig. 1.4-4. *A p-n-p transistor in a common-base connection.*

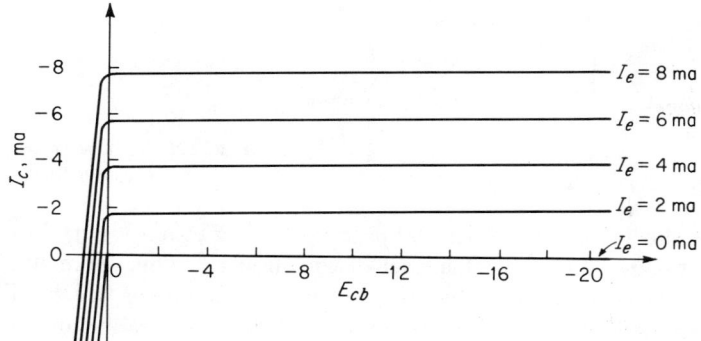

Fig. 1.4-5. *Typical common-base characteristics for a p-n-p transistor.*

the curves of Fig. 1.4-3. The value of α usually lies between 0.95 and 0.99. The minus sign appears in Eq. (1.4-1) because I_e as defined in Fig. 1.4-2 is negative in the active region, while I_c is positive.

When both the emitter and collector junctions are reverse-biased (E_{cb} and E_{eb} positive), only very small currents flow and the transistor is said to be *cut off*.

If both the collector and emitter junctions are forward-biased (E_{cb} and E_{eb} negative), the transistor is in *saturation* and the collector current is approximately independent of changes in the emitter current. The boundary between the active region and the saturation region occurs when E_{cb} is equal to zero.

Basic Transistor Switch

It is also possible to operate a transistor in a *common-emitter* or *common-collector connection* as illustrated in Figs. 1.4-6 and 1.4-7. The connection which is used in the transistor switching circuits to be discussed here is the common-emitter configuration since this is the only one of the three connections which provides both current and voltage gain. The common-base circuit was presented first because the discussion of transistor action is simpler for the common base than for the other connections. In switching circuits the transistor is operated like a switch, with the

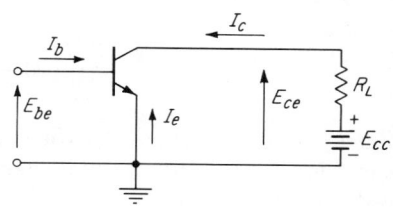

Fig. 1.4-6. *Common-emitter connection.*

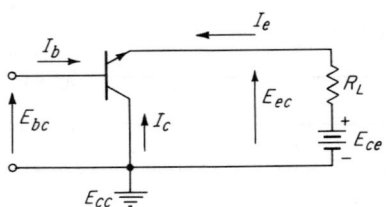

Fig. 1.4-7. *Common-collector connection.*

collector-emitter terminal pair presenting either a very low or a very high resistance to the remainder of the circuit. This operation is obtained by restricting the base current to two values, one of which will cut the transistor off and the other of which will drive the transistor into saturation or just to the edge of saturation. In order to determine the appropriate base currents, it is necessary to determine the relationship between the collector and base currents. By conservation of electrical charge, the three transistor currents must satisfy

$$I_c + I_e + I_b = 0 \tag{1.4-3}$$

When this equation is combined with Eq. (1.4-1) and I_e is eliminated, the following equation results:

$$I_c = \frac{I_{c0}}{1 - \alpha} + \frac{\alpha}{1 - \alpha} I_b \tag{1.4-4}$$

If the symbol β is used for $\alpha/(1 - \alpha)$, this equation can be rewritten as

$$I_c = \frac{I_{c0}}{1 - \alpha} + \beta I_b \tag{1.4-5}$$

For values of α of 0.95 to 0.99 the corresponding values of β vary from 19 to 99. Because of the presence of the $1 - \alpha$ factor in the denominator of β, small variations in α can cause large variations in β. When I_b is made equal to zero, I_c becomes equal to $I_{c0}/(1 - \alpha)$. If α is equal to 0.98 and I_{c0} equals 2 μa, I_c is then equal to 100 μa. Larger values of I_{c0} will produce correspondingly higher values of collector current for zero base current. If this "leakage" current is not tolerable, the usual practice is to make I_b equal to $-I_{c0}$. Then

$$I_c = \frac{I_{c0}}{1 - \alpha} + \frac{\alpha}{1 - \alpha}(-I_{c0}) = I_{c0} \tag{1.4-6}$$

Since I_{c0} is very small, this leakage current is usually tolerable. Thus, the transistor "switch" is opened by making I_b equal to 0 or $-I_{c0}$. In order to close the transistor switch, it is necessary to reduce the collector-emitter voltage to zero. If E_{ce} equals zero, all the voltage E_{cc} must appear across R_L and therefore the collector current must equal E_{cc}/R_L. When $I_c = E_{cc}/R_L$, Eq. (1.4-4) becomes

$$I_c = \frac{E_{cc}}{R_L} = \frac{I_{c0}}{1-\alpha} + \frac{\alpha}{1-\alpha} I_b \tag{1.4-7}$$

or

$$I_b = \frac{E_{cc}}{R_L} \frac{1-\alpha}{\alpha} - \alpha I_{c0} \tag{1.4-8}$$

Usually I_{c0} is small compared with $\frac{E_{cc}}{R_L} \frac{1-\alpha}{\alpha}$ so that

$$I_b \approx \frac{E_{cc}}{R_L} \frac{1-\alpha}{\alpha} = \frac{E_{cc}}{R_L} \frac{1}{\beta} \tag{1.4-9}$$

The collector current cannot increase above E_{cc}/R_L so that for values of I_b greater than E_{cc}/R_L the collector current remains constant and the transistor is driven further into saturation. That saturation actually occurs ($E_{cb} < 0$) can be seen from the characteristics in Fig. 1.4-3. For a fixed I_c, if I_e becomes larger than I_c/α, E_{cb} must be negative. The condition where the collector current is determined by the external load and is independent of I_b is called *bottoming*.

In summary, the transistor is operated in the common-emitter connection as a switch by restricting the base current as follows:

$I_b \leq 0$ Switch open

$I_b \geq \dfrac{E_{cc}}{\beta R_L}$ Switch closed

Values of I_b between 0 and $E_{cc}/\beta R_L$ are not allowed.

Transistor Amplifier

As was pointed out in Sec. 1.3, when diode gates are used it is necessary to provide some form of amplification to make up the current and voltage losses of the gates. This can be done by means of vacuum tubes, as mentioned previously. It is also possible to use transistor amplifiers for this purpose. A circuit for such an amplifier is shown in Fig. 1.4-8. This circuit is just a tandem connection of two of the common-emitter transistor switches just discussed. Two transistors are necessary to eliminate phase inversion and a *p-n-p* and an *n-p-n* transistor are used in order to have the output-signal levels the same as the input-signal level [10]. A complete system of diode logic is customarily constructed by connecting an amplifier to the output of each OR gate. Thus the usual pattern of interconnection consists of AND gates driving OR gates, which in turn drive amplifiers, which connect to other AND gates.

Fig. 1.4-8. Transistor amplifier circuit.

Transistor-diode Logic

It is also possible to connect a single-transistor inverting amplifier to the output of each diode gate [11]. Since only one transistor is used in this amplifier, the total number of transistors in the system is not increased over the number in a system in which a noninverting two-transistor amplifier is provided only for every other gate (OR gate). Moreover, when an inverting amplifier is connected to each diode gate, it is no longer necessary to provide both AND and OR gates—one type of gate is sufficient. This property will be discussed in detail in Chap. 3. The usual practice is to construct standard packages containing a diode gate and a transistor amplifier and then to build a system by interconnecting these packages. This technique has the advantage of requiring only one standard package—flip-flops are formed from two of the basic gate circuits—thereby permitting simple maintenance and supply procedures. A basic transistor-diode gate constructed of an OR gate and amplifier, which will be called a *diode* OR-NOT *gate*, is shown in Fig. 1.4-9. The output of this circuit, e_o, would be connected to the inputs of one or more similar circuits, and each of the inputs of this circuit would be driven by the outputs of similar circuits. When any of the input voltages are equal to E_H, the transistor will be driven at least to the edge of the saturation region; thus the output voltage will be approximately zero, which is thus E_L. If all the input voltages equal zero (E_L), the transistor will be cut off and e_o will equal

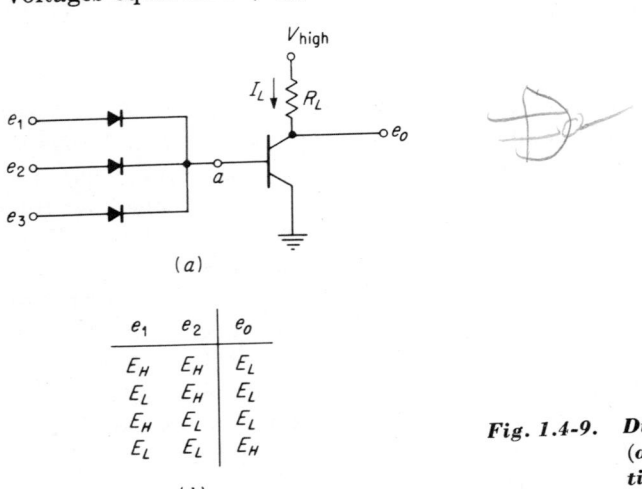

e_1	e_2	e_o
E_H	E_H	E_L
E_L	E_H	E_L
E_H	E_L	E_L
E_L	E_L	E_H

(b)

Fig. 1.4-9. Diode OR-NOT gate. (a) Circuit; (b) operation table.

approximately $V_{high} - I_L R_L$, where I_L is the sum of $I_{c0}/1 - \alpha$ and the input currents to the gates driven by e_o. The actual value of e_o is not important as long as it remains positive; the critical requirement is that enough current be supplied to the transistors of the *driven* gates to ensure that these transistors are saturated or at the edge of the saturation region.

It is also possible to combine a diode AND gate and a transistor inverter as in Fig. 1.4-10. This circuit will be called a *diode* AND-NOT *gate*. The operation of the circuit without the network inside the dotted lines will be considered first. When each of the transistors driving e_1, e_2, and e_3 is cut off, no current will flow through the input diodes and current will flow from V_{high} through R_L and through the base-emitter junction. This will saturate the transistor so that e_o will be approximately zero. When any of the transistors driving e_1, e_2, and e_3 are saturated, current will flow from V_{high} through R_L and the corresponding input diodes. Thus the potential of point a will become $V_{ce} + v_f$, where V_{ce} is the collector-to-emitter voltage of a saturated transistor and v_f is the voltage drop across a forward-biased diode. This potential at point a may be sufficient to allow the transistor to turn on; so it is necessary to add the network containing R_a, R_b, and C. The resistors R_a and R_b are simply a voltage-dividing network to force point a to a slightly negative voltage when any of the input diodes are grounded, and the capacitor is present merely to speed

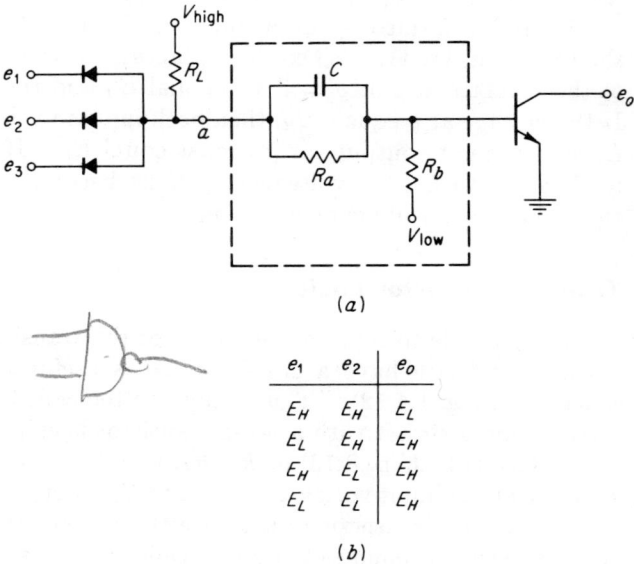

e_1	e_2	e_o
E_H	E_H	E_L
E_L	E_H	E_H
E_H	E_L	E_H
E_L	E_L	E_H

(b)

Fig. 1.4-10. Diode AND-NOT gate. (a) Circuit; (b) operation table.

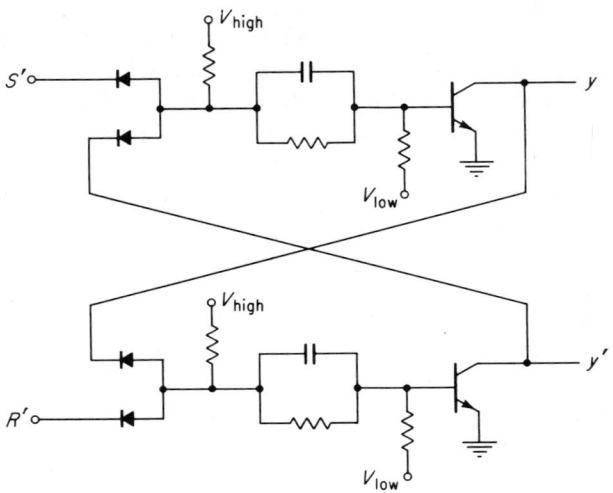

Fig. 1.4-11. *Flip-flop formed of two diode AND-NOT gates.*

up the circuit operation. A similar network is not required in the circuit of Fig. 1.4-9 because the voltage at point a is $V_{ce} - v_f$ rather than $V_{ce} + v_f$ when all the driving transistors are saturated.

A flip-flop formed by interconnecting two diode AND-NOT gates is shown in Fig. 1.4-11. When the voltages at S' and R' both equal E_H, one of the voltages at y or y' will also equal E_H and the other will equal E_L. If the voltage at y equals E_H, then both inputs to the bottom gate are at E_H and the gate output (at y') must equal E_L. If the voltage at R' is made equal to E_L, the voltage at y' must become equal to E_H, which in turn causes the voltage at y to equal E_L.

Transistor-resistor Logic

It is possible to replace the diodes of the transistor-diode circuits by resistors and still have a circuit which is useful for combining binary signals (see Fig. 1.4-12). This circuit is often called a NOR *gate*, and the corresponding circuit with a *p-n-p* transistor is called a NAND *gate* [14]. The values [12, chap. 8;13] of R_1, R_L, and V_{high} are chosen so that if at least one of the input voltages is equal to E_H, enough current is supplied to the base of the transistor to saturate the transistor and reduce e_o to E_L (approximately ground potential). Only if all the input voltages are at E_L will the transistor be cut off and e_o equal E_H. The network consisting of $-V_1$ and R_2 is present to reduce the saturation leakage current by

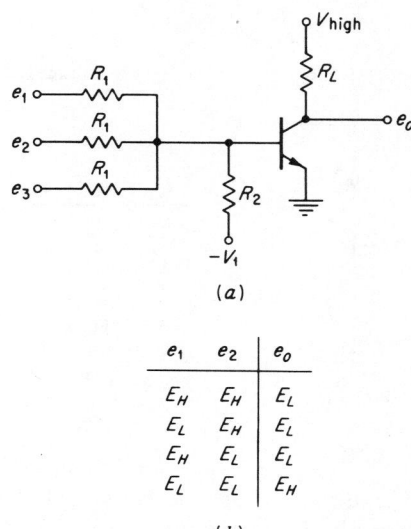

Fig. 1.4-12. Resistor OR-NOT gate. (a) Circuit; (b) operation table.

reversing I_b. Compared with the transistor-diode circuit, the transistor-resistor circuit is less costly and slower in operation.

Direct-coupled Transistor Logic (DCTL)

In the circuits discussed above, the transistors are used to provide amplification and quantization of the signal levels, while the combining of signals is done by means of auxiliary elements such as diodes or resistors.

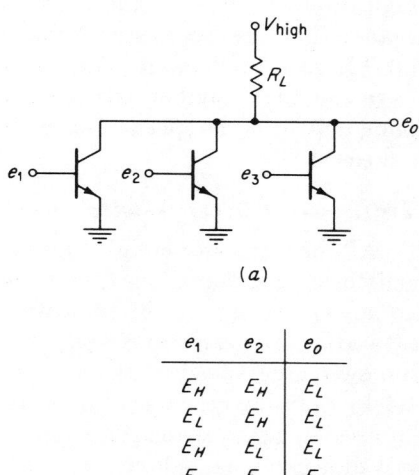

Fig. 1.4-13. DCTL parallel gate. (a) Circuit; (b) circuit operation table.

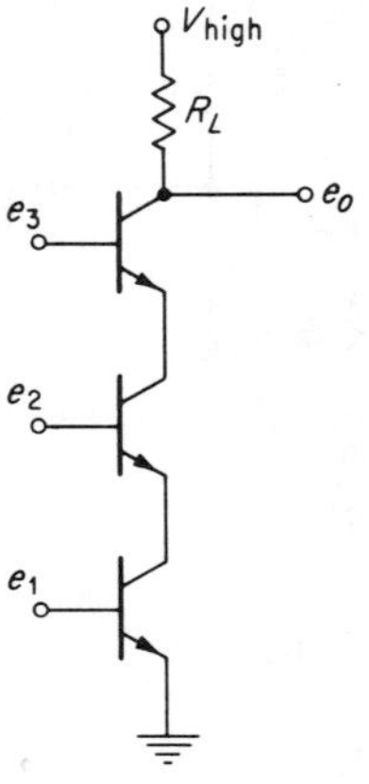

Switching Devices

e_1	e_2	e_o
E_H	E_H	E_L
E_L	E_H	E_H
E_H	E_L	E_H
E_L	E_L	E_H

(b)

Fig. 1.4-14. DCTL series gate. (a) Circuit; (b) circuit operation table.

Circuits can also be constructed in which transistors are used both to combine signals and to provide amplification, and the only additional elements used are power-supply resistors [12, chap. 9;15]. These circuits are shown in Figs. 1.4-13 and 1.4-14. In the DCTL parallel gate (Fig. 1.4-13), if any one of the input voltages is at the high level, the corresponding transistor will be saturated and therefore e_o will be approximately at ground potential. Only if all the transistors are cut off—all input voltages at the low level—will the output voltage remain at the high level. For the DCTL series gate (Fig. 1.4-14) the output will be at the low level (approximately 0 volts) only if all the transistors are saturated owing to high input voltages. These DCTL gates are more costly in transistors than either the transistor-diode or transistor-resistor gates since in a DCTL gate each input requires one transistor, but the DCTL circuits have the advantage of extreme simplicity. By the addition of *RC* coupling networks the speed of operation of the DCTL circuits can be greatly increased.

Transistor Current-switching Gates

All the transistor circuits described previously are similar—a common-emitter connection is used, and the transistors are operated so that they are always either cut off or saturated. When a transistor is operated in saturation or near saturation, i.e., with a low collector-emitter voltage, the switching speed of the transistor is decreased. The circuit of Fig. 1.4-15 permits very fast operation by preventing the transistors from approaching saturation [12, chap. 10;16,17]. The operation of this circuit will first be explained with transistor T_3 omitted. The input voltage e_1 will be equal to either $+0.6$ volt or -0.6 volt. When e_1 is equal to -0.6 volt, transistor T_1 has its emitter-base junction forward-biased and tran-

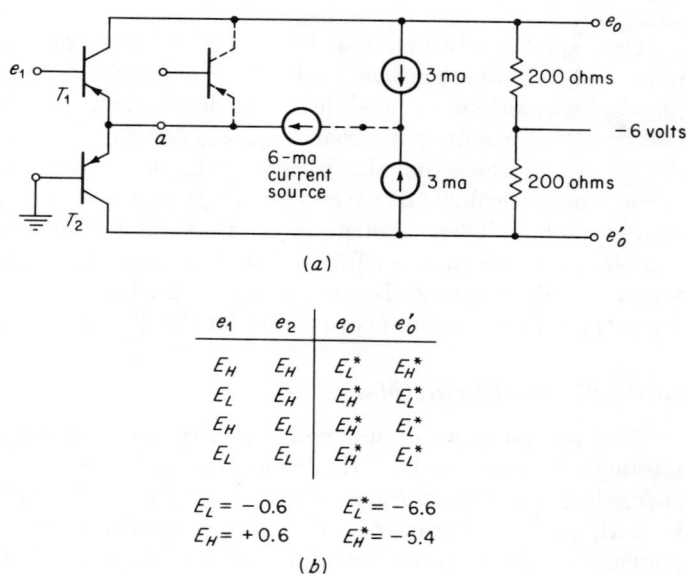

Fig. 1.4-15. *Transistor current-steering gate.* (a) *Circuit;* (b) *circuit operation table.*

sistor T_2 is cut off.† Therefore the entire 6 ma from the current source flows through T_1. In this situation 3 ma flows through the 200-ohm resistors in series, causing e_o to rise to -5.4 volts and e'_o to drop to -6.6 volts. By similar reasoning, when e_1 equals $+0.6$ volt, transistor T_1 is cut off and the entire 6 ma flows through transistor T_2, causing e_o to become -6.6 volts and e'_o to become -5.4 volts. This circuit can be used for combining signals by adding transistors (such as T_3) in parallel with T_1. When any of the input voltages is at the low level (-0.6 volt), current flows through the top path. Only when all the input voltages are equal to E_H ($+0.6$ volt) does the current flow through transistor T_2.

It is not possible to connect the output of one of these gates directly to a similar gate since the input-voltage levels are $+0.6$ and -0.6 volts, while the output-voltage levels are -5.4 and -6.6 volts. This difficulty is overcome by constructing a similar gate with input voltages of -5.4 and -6.6 volts and output voltages of $+0.6$ and -0.6 volts. This other gate uses *n-p-n* transistors and has the direction of current flow reversed from that shown in Fig. 1.4-15. The output levels of one of these gates can also be shifted back to the input levels by means of a passive network.

† The forward-biased emitter voltage for the type of transistor used is approximately 0.3 volt so that the potential of point a is -0.3 volt, which causes transistor T_2 to be cut off.

1.5 CRYOTRONS

One approach which has been used in the continuing search for improved computer components has been *to investigate various fundamental physical phenomena* for possible device development. The principal success resulting from this approach was the development of the *cryotron*—a device which relies on the phenomenon of superconductivity, which various metals exhibit at extremely low temperatures [18,19]. While no computers have been constructed of cryotrons at the present time, there is a sizable development effort devoted to developing practical cryotron systems. The high speeds and small sizes possible with thin-film cryotrons make the cryotron a strong contender for the digital systems of the future.

Basic Physical Principles

The phenomenon of superconductivity was discovered by H. Kammerlingh Onnes in 1911. He found that *the resistance of mercury drops suddenly to zero when its temperature is 4.12°K* or lower. Other metals such as lead, tin, aluminum, etc., also possess zero resistance when the temperature is below a transition temperature, which is a characteristic of the individual metal. That the resistance is actually zero rather than just some exceedingly small value was demonstrated by an experiment carried out at M.I.T. by Prof. S. C. Collins. In this experiment a superconducting lead ring carried an induced current for 2 years without any observable change in the magnitude of the current.

A metal whose temperature is changed from above to below its transition temperature thus exhibits one of the properties which is highly desirable in a switching device: an abrupt transition between two well-defined states. However, temperature control of a superconducting metal

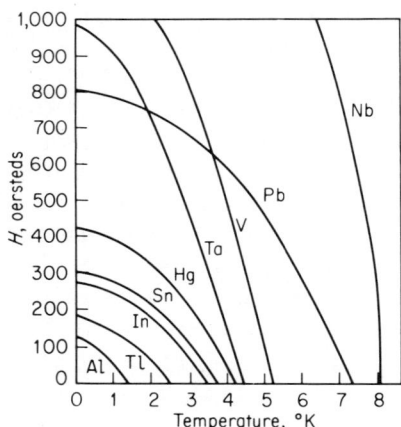

Fig. 1.5-1. Relation between magnetic field and transition temperature for several common superconductors.

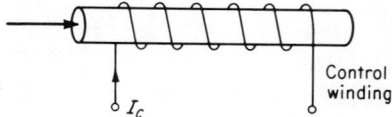

Fig. 1.5-2. Basic wire-wound cryotron.

would clearly result in a slow, unsatisfactory system. The use of superconductivity as the basis for a digital device is possible only because the transition temperature of a metal varies when a magnetic field is present. It is possible to control the transition from zero to finite resistance by controlling a magnetic field, with the temperature remaining constant. Figure 1.5-1 shows the relation between the magnetic field and the transition temperature for several common superconductors.

The Wire-wound Cryotron

The first cryotron, proposed by Dudley Buck in 1956 [18], consisted of a straight piece of tantalum wire about 1 in. long around which was wrapped a single-layer winding of niobium wire (Fig. 1.5-2). Niobium and tantalum wires were chosen because, at 4.2°K, the boiling point of helium, niobium is superconducting even for large magnetic fields, while tantalum is superconducting for zero magnetic fields but can be made resistive by the application of a small magnetic field, between 50 and 100 oersteds. The tantalum is thus used as the controlled switch, and the niobium is used as zero-resistance "hookup" wire. The operation of a basic cryotron is quite straightforward: when the current in the control winding exceeds a value which is sufficient to produce the critical magnetic field in the cryotron, the cryotron resistance changes from zero to a finite value (approximately 0.0075 ohm for Buck's circuits). A digital circuit can be constructed of cryotrons by providing parallel paths through which the supply current can flow and then using cryotrons to force the current into a particular path determined by the control-winding currents.

A Flip-flop Using Wire-wound Cryotrons

The basic cryotron flip-flop is shown in Fig. 1.5-3. In part *a* of this figure, only the two cryotrons necessary to produce two stable circuit conditions are shown. These two stable states correspond directly to the two parallel paths between points 1 and 2 of this circuit. If the supply current I_o flows through the upper cryotron, it also flows through the winding of the lower cryotron, making this cryotron resistive. This means that the path through the top cryotron will be a zero-resistance path, while the parallel path through the lower cryotron will be resistive. Thus all the supply current will continue to flow through the upper cryotron. A similar stable situation occurs when all the current flows through the lower cryotron.

Switching Devices

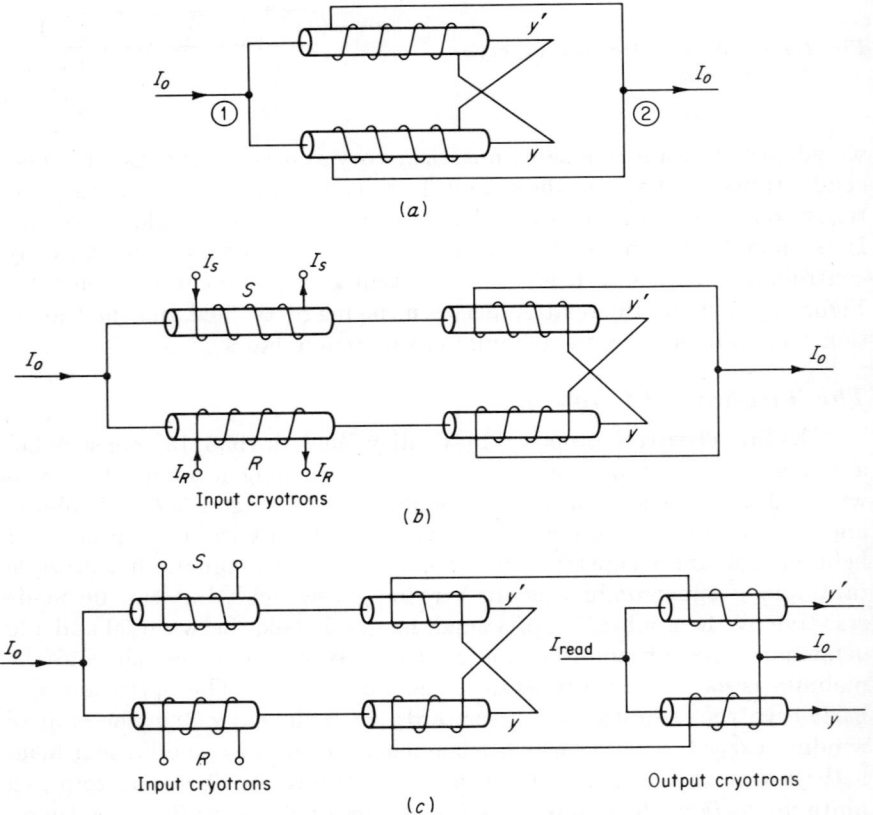

Fig. 1.5-3. *Wire-wound-cryotron flip-flop.* (a) *Basic bistable element;* (b) *bistable element with input cryotrons;* (c) *flip-flop with input and output cryotrons.*

In order to place the flip-flop in a specified state, two additional cryotrons are required as shown in Fig. 1.5-3b. The flip-flop will be said to be SET when the current is flowing through the lower (y) cryotron and RESET when the current is flowing through the upper (y') cryotron. Normally there will be no current in the S and R cryotrons. If the flip-flop is SET (current flowing through the y cryotron, upper cryotron resistive) and current I_S flows so as to make the S cryotron resistive, then no change will take place since there will be two resistive cryotrons (S and y') in the upper path and no resistive cryotrons in the lower path. On the other hand, if the flip-flop is SET and current I_R flows, making R resistive, there will be one resistive cryotron in each path. This will cause the current to divide equally between the two paths. If the magni-

1.5 cryotrons

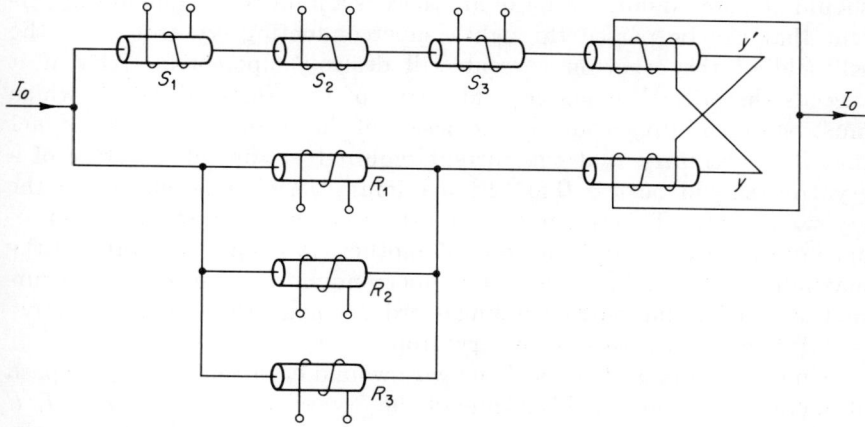

Fig. 1.5-4. Cryotron flip-flop with input logic.

tude of the supply current is less than twice the threshold current for a single cryotron, both the y and y' cryotrons will then become superconducting. This will leave only the R cryotron resistive, which will cause all the supply current to flow in the upper path, making the y cryotron resistive and leaving the flip-flop in the stable RESET condition after the I_R current is removed. A similar analysis applies when the flip-flop is initially in the RESET condition. In summary, current I_R causes the flip-flop to go to the RESET condition and current I_S causes the flip-flop to go to the SET condition.

Reading out the state of the flip-flop is accomplished by means of two additional cryotrons, as shown in Fig. 1.5-3c. If the flip-flop is set, current will flow through the lower path and around the upper output cryotron, making it resistive. The read current I_{read} will thus flow through the superconductive lower output cryotron and appear on lead y. If the flip-flop is reset, current will flow around the lower output cryotron and cause I_{read} to appear on the y' lead.

By using series or parallel connections of the input cryotrons as in Fig. 1.5-4 it is possible to have combinations of several signals participate in the control of the flip-flop. Current flowing around any of the set cryotrons (S_1, S_2, S_3) will cause the flip-flop to be in the set state; thus, the series connection behaves like an OR gate since an action is caused by current around S_1 or S_2 or S_3. The flip-flop can be caused to go to the reset state by having current flow around *all* the reset cryotrons, i.e., around R_1 *and* R_2 *and* R_3. The parallel connection thus performs like an AND gate. Of course, more complicated networks of wire-wound cryotrons are possible, and several typical circuits can be found in the literature.

Before leaving the subject of wire-wound cryotrons, several things should be pointed out. First of all, there is a limit to the amount of current that can be passed through a superconducting cryotron since the self-field of the cryotron current will destroy superconductivity if it exceeds the critical magnetic field. One of the electrical details which must be taken into account in the design of the cryotron itself is the fact that it is necessary that the current required to control the state of a cryotron should be less than the maximum permissible current in the cryotron itself. This requirement must be satisfied so that the output of one cryotron can control the state of another cryotron. The ratio of the maximum permissible current in a superconducting cryotron to the current required in the control winding in order to make the cryotron resistive is called the *current gain* of the cryotron.

One of the major defects of the wire-wound cryotron is its slow speed of operation. The switching time of the cryotron is limited by its L/R time constant. The switching time for the wire-wound-cryotron circuits which have been constructed is of the order of several hundred microseconds.

The Crossed-film Cryotron

In order to decrease the L/R time constant of the cryotron and thereby increase the speed of operation, a cryotron constructed of thin superconducting films has been developed [20,21,22]. The construction of a typical thin-film cryotron is shown in Fig. 1.5-5. This cryotron is constructed of thin (3,000 A) films of tin and lead, with another thin (3,000 A) film of silicon monoxide used as insulation. These films are deposited on a glass substrate as illustrated in the figure. The cryotron is operated with the lead always remaining superconducting and the tin being switched in and out of superconductivity by the field caused by the current

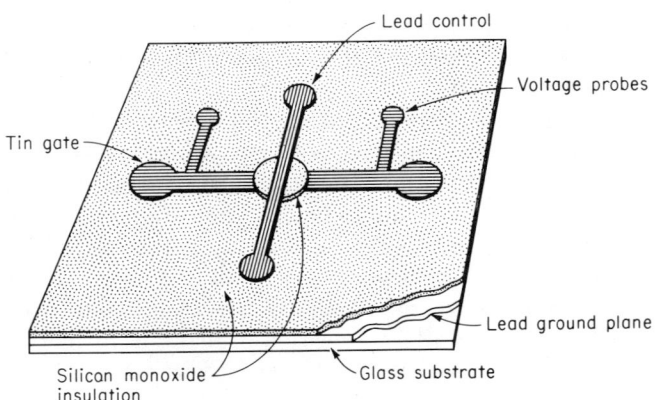

Fig. 1.5-5. *Typical thin-film cryotron.*

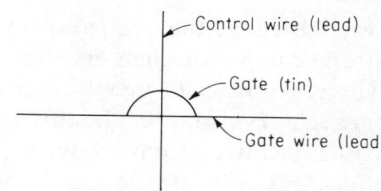

Fig. 1.5-6. *Symbol for crossed-film cryotron.*

in the lead control film. The operating temperature is around 3.6°K. The crossed-film cryotron is approximately 1,000 times faster than the wire-wound cryotron and, in addition, can be fabricated easily and with high densities of gates per unit area by vacuum deposition through suitable metal masks.

Crossed-film-cryotron Circuits

The symbol which will be used for the crossed-film cryotron is shown in Fig. 1.5-6. It is assumed that, when current flows in the vertical (control) wire, the gate is made resistive so that the horizontal (gate) wire contains a resistance. When no current flows in the vertical wire, both wires are superconducting.

A crossed-film-cryotron flip-flop is shown in Fig. 1.5-7. The basic bistable element consists of the heavy lines, and the read-in and read-out circuitry is represented by the lighter lines. The supply current I_o flows

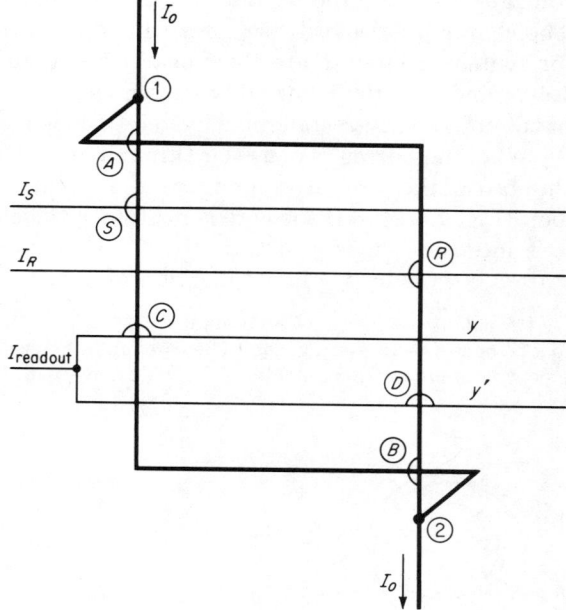

Fig. 1.5-7. *A crossed-film-cryotron flip-flop.*

through one of the two parallel paths between points 1 and 2. If it flows straight down through cryotron A, then it will flow through the control wire of cryotron B, thereby introducing a resistance into the parallel path through cryotron B. In this case the flip-flop will be said to be in the RESET state. Conversely, if I_o flows through the control wire of cryotron A, it will introduce a resistance into the path through cryotron A and ensure that all the current will flow through the zero-resistance path through cryotron B. The flip-flop is then in the SET state. Current I_S through cryotron S introduces a resistance into the path through cryotron A, causing the supply current to flow through the path through cryotron B in a fashion similar to that described for the wire-wound-cryotron flip-flop. Either cryotron C or cryotron D will be resistive, causing all the read-out current to appear on either lead y or lead y'. Parallel or series connections of the input cryotrons can be used in a fashion similar to that described for the wire-wound cryotron to allow combinations of several signals to control the flip-flop setting.

1.6 MAGNETIC CORES

The magnetic cores which are used in switching circuits are toroidal cores constructed either of ceramic ferrite material or of ultrathin ferromagnetic-alloy tape wound on a nonferromagnetic spool (Fig. 1.6-1).†
The characteristics of these cores which make them particularly suitable for switching circuits are their nearly rectangular hysteresis loops (Fig. 1.6-2) and also their low eddy-current losses. Generally the tape cores have more nearly rectangular hysteresis loops than the ferrite cores.

When no current is passing through any of the windings of the core, the state of the core corresponds to either point s or point r of the hysteresis loop (Fig. 1.6-2). If the core is in state s, the core is said to be *set*, and if it is in state r, it is said to be *reset*. It is possible to determine which state a core is in by the application of a sufficiently large current pulse

† A common alloy is 4-79 Permalloy (79% Ni, 17% Fe, 4% Mo), which is rolled to a ribbon of $\frac{1}{4}$- to $\frac{1}{8}$-mil gage. The core might typically consist of 20 wraps of $\frac{1}{4}$-mil tape wound on a ceramic bobbin $\frac{1}{8}$ in. wide and $\frac{3}{16}$ in. in diameter.

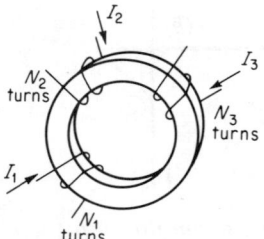

Fig. 1.6-1. A magnetic core.

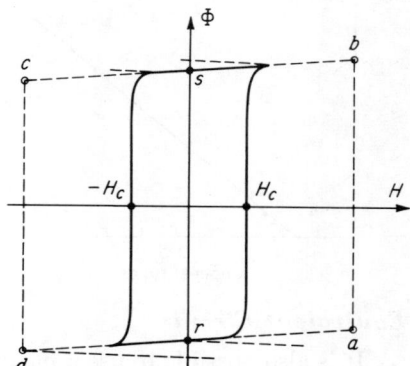

Fig. 1.6-2. *A nearly rectangular hysteresis loop.*

(greater than H_c/N, where N is the number of turns in the winding). If a current corresponding to a positive H, a *positive current*, is applied and the core is in state s, the operating point will move to point b. Since this motion involves only a small change of flux, the voltage induced in other windings on the core will be small and will have a waveform like that of Fig. 1.6-3a. This voltage is called a *shuttle voltage*. If a positive current is applied and the core is in state r, the operating point will move to point a and then to point b, following the dotted lines and returning to point s, when the pulse terminates. The hysteresis loop is not followed, since it is not possible to change the flux instantaneously. In this case a large change of flux occurs, inducing a large voltage in other windings on the core. This voltage, called the *switching voltage*, has a waveform like that of Fig. 1.6-3b. By similar means it is possible to place a core in a predetermined state. A core can thus be used to store a particular setting in a fashion similar to a flip-flop. However, there is a fundamental difference between a core and a flip-flop: it is not possible to determine the state of a core without applying a resetting pulse† to it, thereby always leaving it reset. The setting of a flip-flop can be determined without affecting the state.

† This could equally well be a setting pulse.

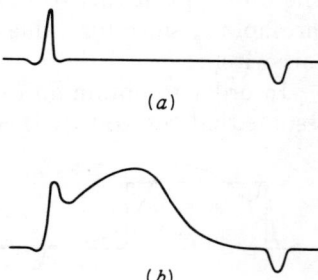

Fig. 1.6-3. *Core voltage waveform.* (a) *Shuttle voltage*; (b) *switching voltage.*

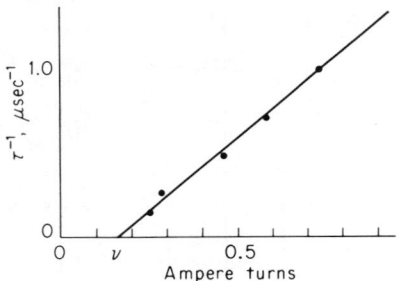

Fig. 1.6-4. *General form of graph relating τ^{-1} and N_1I_1.*

Equivalent Circuit

It is also possible to use a magnetic core for combining binary signals. This use of cores depends on the fact that there is a threshold of magnetic field strength or, equivalently, of current such that the core will not change state, or *switch*, for smaller values of current and will always change state for larger values of current (which are in the proper direction to cause a change of state). Before discussing the combination of binary signals, a mathematical model for the magnetic core will be derived.

In characterizing the operation of a magnetic core, the three quantities of primary interest are:

τ, the time it takes to reverse a large fraction, about 90 percent, of the core flux

E, the magnitude of the electromotive force (emf) induced in the core windings when the core is switched

ν, the magnetic-field-strength threshold

The relation among these quantities can be determined by measuring the switching time τ for various values of I_1, the current in winding 1. A graph is then plotted of $1/\tau$ versus N_1I_1, the magnetic field strength due to I_1. This graph has the general form shown in Fig. 1.6-4. In the region of practical interest the points lie approximately on a straight line. Thus the relation between τ^{-1} and N_1I_1 can be expressed approximately as

$$\tau^{-1} = C(N_1I_1 - \nu) \tag{1.6-1}$$

The intercept of this straight line with the N_1I_1 axis corresponds to the threshold ν, since for values of N_1I_1 less than this intercept the core will not switch.

In order to obtain an expression for the voltage e, it is customary to assume that the voltage is constant during the time the core is switching.

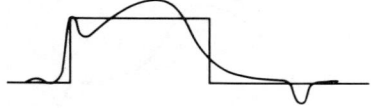

Fig. 1.6-5. *Square-pulse approximation to switching-voltage waveform.*

1.6 magnetic cores

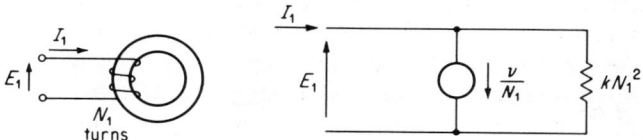

Fig. 1.6-6. *Linear equivalent circuit for a magnetic core during the time of switching.*

This means that the actual switching-voltage waveform is being approximated by a square pulse (Fig. 1.6-5). While this is a crude and very approximate assumption, it has the advantage of leading to a simple expression [Eq. (1.6-4)] which is very useful. In designing core circuits sufficient margins in the induced voltages must be allowed because of the inexactness of Eq. (1.6-4) and the variation in cores. Margins must also be allowed in the values of τ and ν owing to the variation among cores.

The voltage produced by a changing magnetic field is given by $E = N \, (d\Phi/dt)$. For the magnetic core, with the assumption of constant output voltage, this expression becomes

$$E = N \frac{\Phi}{\tau} \tag{1.6-2}$$

where N is the number of turns in the winding and Φ is the total amount of magnetic-flux change used in measuring τ. By using Eq. (1.6-1) to eliminate τ, this expression for E becomes

$$E = N\Phi C(N_1 I_1 - \nu) \tag{1.6-3}$$

which can be expressed as

$$E = kN(N_1 I_1 - \nu) \tag{1.6-4}$$

The voltage induced in the winding in which the input current flows is equal to

$$E_1 = kN_1^2 I_1 - kN_1 \nu = kN_1^2 \left(I_1 - \frac{\nu}{N_1}\right) \tag{1.6-5}$$

This corresponds to an equivalent circuit for the core consisting of a resistance equal to kN_1^2 in parallel with a current sink equal to ν/N_1 as in Fig. 1.6-6. This equivalent circuit is valid only while the core is switching. After the switching is completed the equivalent circuit becomes a short circuit.

Combining Input Signals

All the circuits discussed in Secs. 1.1 to 1.5 had the characteristic that d-c signals could be applied to their inputs and d-c signals would then be available at the outputs. It is not possible to construct circuits with magnetic cores which have this property since a magnetic core generates a voltage in its windings only when it is changing state. Thus, magnetic-core switching circuits must operate either with periodic (sine-wave, square-wave, etc.) or pulsed signals. While some attention has been given to core circuits using periodic signals, the use of the cores in these circuits has been mainly for amplification, the combination of signals being carried out by other devices such as diodes [23].

When pulse signals are used, it is customary to operate the cores in a cyclic manner. Initially the cores are in the reset state. During the input phase each input winding may or may not receive a current pulse. Depending on which windings are pulsed, the cores may or may not be switched to the set state. All cores are reset during the output phase, returning them to their initial state and causing the cores which were set during the input phase to generate pulses in their output windings. One cycle is then completed, and another input phase follows.

If a core has n input windings and the circuit is operated so that the magnitude of the input pulse applied to winding j is greater than ν/N_j, the core will switch whenever at least one winding receives an input. Thus an output pulse will be generated during the output phase, if winding 1 or winding 2 or $\cdots$ or winding n was pulsed during the input phase. Another possible input arrangement is to have the magnitude of the input pulse to winding j slightly larger than ν/mN_j. Then the applied magnetic field strength will not exceed ν unless at least m out of the n input windings are pulsed. Thus an output pulse will occur during the output phase only when m or more cores are pulsed during the input phase. If m is equal to n, winding 1 and winding 2 and $\cdots$ and winding n must receive input pulses for an output pulse to be generated. This mode of operation with $m > 1$ is less desirable than the connection with any input pulse causing the core to be set ($m = 1$). When $m > 1$, the input pulses must coincide and the current I_j in winding j must satisfy the relation $\nu/mN_j < I_j < \nu/(m-1)N_j$ so that currents in only $m-1$ of the windings will not set the core. When $m = 1$, the only restriction that must be satisfied is $I_1 > \nu/N_i$.

Mirror Symbols

When the diagram for a circuit involving several cores has to be drawn, it is inconvenient to represent the cores by circles and coils as in Fig. 1.6-1. A neater and clearer diagram results when the so-called *mirror symbols* are used (Fig. 1.6-7). In this figure, the heavy vertical line represents the core, and each horizontal line represents a winding. The sense of the winding is indicated by the 45° line at the intersection of the horizontal and vertical lines. If current I_1 flows as shown by the arrow it will apply a magnetomotive force (mmf) of N_1I_1 to the core in a direction to set the core. This is determined by "reflecting" I_1 upward,

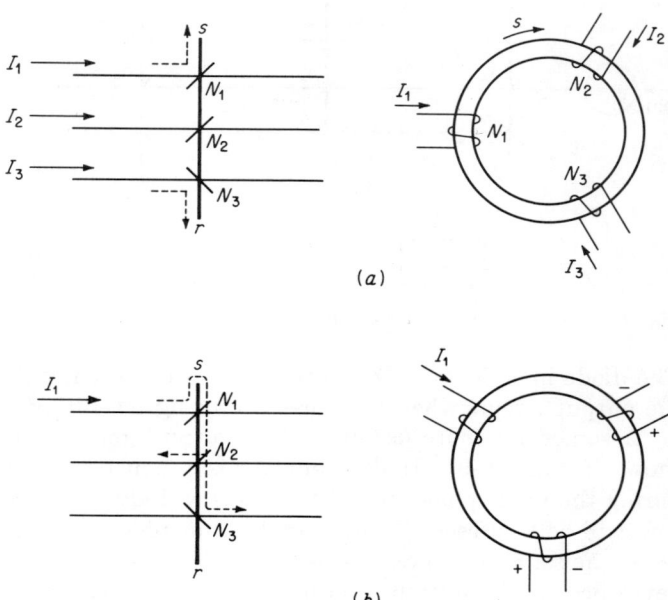

Fig. 1.6-7. Mirror symbols for magnetic cores. (a) Current-flux relations; (b) induced voltages.

with the 45° bar at the intersection of the horizontal and vertical lines. Similarly, a current I_3 flowing in the direction of the arrow will tend to reset the core. When a current causes the core to switch, the voltage induced in the windings is found by reflecting the input current around the core as shown for I_1 in Fig. 1.6-7b. This input current would produce a voltage in N_2 tending to drive current to the left and a voltage in N_3 tending to drive current to the right.

Output Connections

There are many techniques used for interconnecting magnetic cores. The simplest type of output connection is the so-called T type shown in Fig. 1.6-8. During the input phase, the signals present in the input windings will determine which cores are switched to the set state. During the output phase, all the cores will be reset by the advance pulse. Any cores which were set (during the input phase) will be switched by the advance pulse and will develop voltages across their output windings N_o (dotted line in Fig. 1.6-8). These output voltages will cause a current to flow through the output impedance Z, which may be the input winding to another core, the input to a flip-flop, etc. Output current will flow through Z if one or more of the cores was set during the input phase.

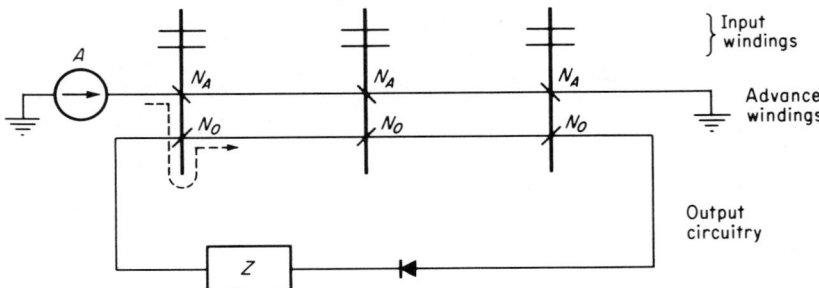

Fig. 1.6-8. T-type output connection.

The diode in series with Z is present to prevent current from flowing in the output circuit when the cores are being *set* during the input phase.

A somewhat more useful type of output connection is the AB type shown in Fig. 1.6-9. In developing this circuit it has been assumed that during the input phase one of the cores will always be set and the other core will be left reset. During the output phase, the advance pulse will cause the set core to be switched to the reset state, and a voltage will be developed in the output winding of the switched core (dotted lines in Fig. 1.6-9). There are two parallel paths for the advance current to follow between points B and G. The voltage of the output winding of the switched core appears as an emf opposing the advance current in one of these branches. Consequently the advance current will flow through the output impedance ($Z_{x'}$, or Z_x), which is *not* in series with the output

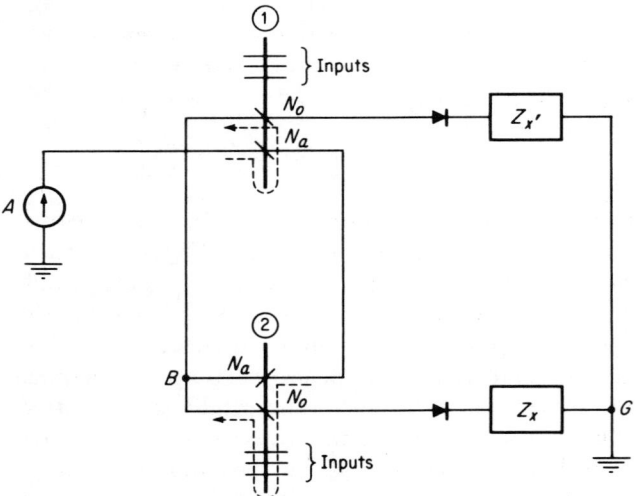

Fig. 1.6-9. AB-type output connection.

1.7 symbols for electronic gates

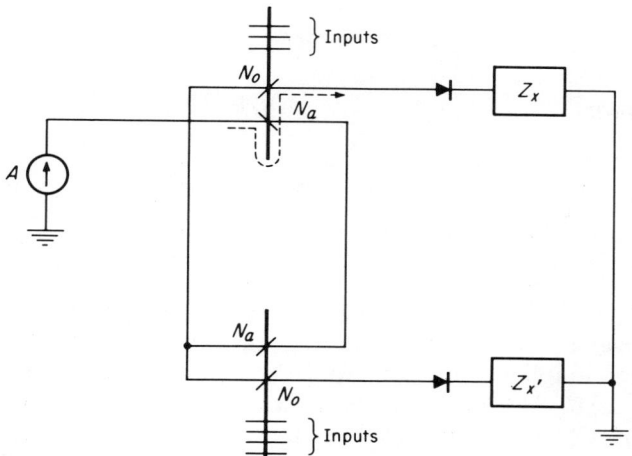

Fig. 1.6.10. AF-type output connection.

winding of the switched core. In general, there will be many parallel cores, all cores except one will have a back emf present during the output phase, and the advance current will flow only through the core which does *not* have the back emf.

A similar connection, the AF type, is shown in Fig. 1.6-10. This connection also has several parallel paths for the advance current, but differs from the AB type in that the output-winding voltage is in a direction which aids rather than opposes the advance current. In this connection, only one core is set during each input phase. The advance current flows through the impedance in series with the core switched (reset) during the output phase.

1.7 SYMBOLS FOR ELECTRONIC GATES

The remaining chapters of this book will be concerned with the problem of how to interconnect the basic circuits or gates described in this chapter in order to achieve a specified performance. In forming these interconnections of gates the detailed electrical characteristics will not be of interest. Rather the relationship between the combinations of high and low signals appearing at the network outputs will be of prime concern. Since the detailed structure of the gates is not of interest, it is appropriate to have a symbolic representation for each gate that describes only the relation between the binary input and output signals. The symbols which will be used in the remaining chapters are described here.

Usually it will not be necessary to use symbols for any amplifiers which

Switching Devices

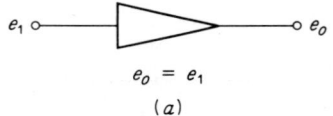

Fig. 1.7-1. Amplifier symbols. (a) Amplifier symbol; (b) inverter symbol.

may be included in a network, since an amplifier presents the same binary signal at its output as is present at the input. In those few cases in which an amplifier must be explicitly shown, the symbol of Fig. 1.7-1a will be used. Whenever an inverting amplifier, or INVERTER, such as shown in Fig. 1.2-4 is included in a network, it will be necessary to use an explicit symbol to represent it. The output of an INVERTER is always at the opposite binary level from the input. The INVERTER symbol is shown in Fig. 1.7-1b, where the inversion is indicated by the circle at the output terminal.

Figure 1.7-2 shows the two symbols which will be used to represent the AND gate and OR gate of Fig. 1.3-2. The AND gate symbol represents a circuit in which there will be a high voltage E_H at the output terminal only when there is a high voltage E_H present at *all* the input terminals. The OR-gate symbol represents a circuit in which the presence of a high voltage E_H at any one (or more) of the input terminals will cause a high voltage E_H to be present at the output terminal.

Symbols are now required for the OR-NOT gate (Figs. 1.2-5, 1.4-9, 1.4-12, 1.4-13) and the AND-NOT gate (Figs. 1.2-7, 1.4-10, 1.4-14). One possibility would be to form symbols for these gates by combining the

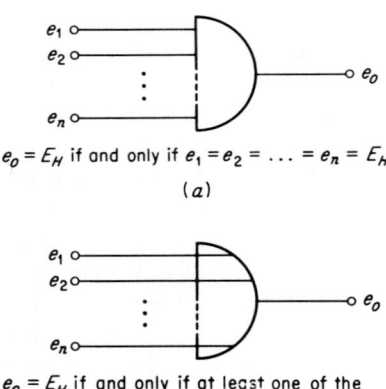

Fig. 1.7-2. Symbols for AND gate (a) and OR gate (b).

1.7 symbols for electronic gates

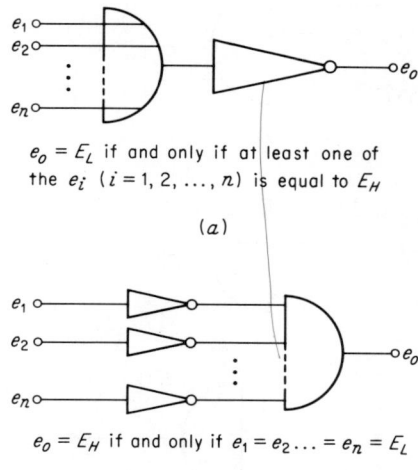

Fig. 1.7-3. Two possible representations for OR-NOT gates. (a) Inverter at output; (b) inverters at inputs.

(a) $e_o = E_L$ if and only if at least one of the e_i $(i = 1, 2, \ldots, n)$ is equal to E_H

(b) $e_o = E_H$ if and only if $e_1 = e_2 \ldots = e_n = E_L$

INVERTER symbol with the AND-gate or OR-gate symbol. Since the OR-NOT gate has a *low* output voltage (E_L) only when at least one input has a high voltage (E_H) and the OR gate has a *high* output voltage when at least one input has a high voltage, it is appropriate to represent the OR-NOT gate by an OR gate followed by an INVERTER as in Fig. 1.7-3a. An alternative is to represent the OR-NOT gate by an AND gate with an INVERTER connected to each input (Fig. 1.7-3b). An OR-NOT gate has a high output voltage only when there is a *low* voltage at all the input terminals, and an AND gate has a high output voltage only when there is a *high* voltage at all the input terminals. While these symbols follow logically from the previous symbols, they are too cumbersome for extensive use. More useful symbols are obtained by removing the INVERTER and attaching the circles directly to the gate symbol as in Fig. 1.7-4. Two symbols are shown in this figure even though only one is really

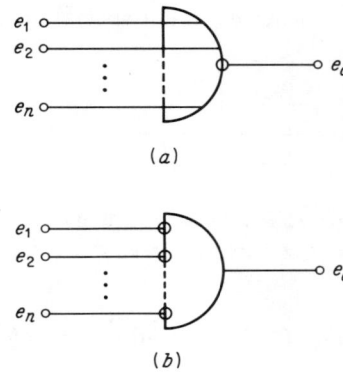

Fig. 1.7-4. OR-NOT gate symbols. (a) Output inversion; (b) input inversion.

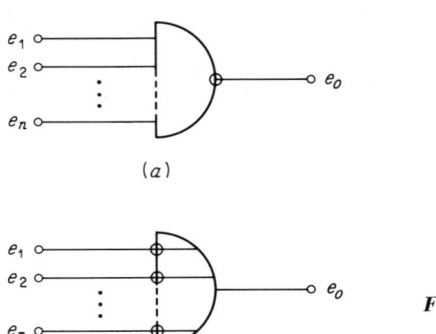

Fig. 1.7-5. **AND-NOT gate symbols.** (a) Output inversion; (b) input inversion.

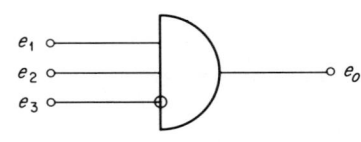

$e_o = 1$ if and only if $e_1 = e_2 = E_H$ and $e_3 = E_L$

Fig. 1.7-6. **Generalized gate symbol.**

required. Both these symbols will be used in the following chapters for reasons which will be discussed in Chap. 3 in connection with analyzing networks of OR-NOT gates. The same general principles apply to AND-NOT gates, and a similar line of reasoning results in the symbols shown in Fig. 1.7-5 for AND-NOT gates.

It is possible to generalize these symbols to gates in which some but not all of the inputs have inversion present. This is illustrated in Fig. 1.7-6. Occasionally a need will arise for a symbol for a gate which cannot be represented by such a generalization. For example, it is possible to have a gate for which the output is high if an odd number of the inputs are high. Special symbols, in this case the symbol $\oplus$, will be adopted for such specialized operations, and the gate symbols will be the AND-gate symbol with the special operation symbol placed inside it as in Fig. 1.7-7a. Another common special gate is the gate for which the output is high if

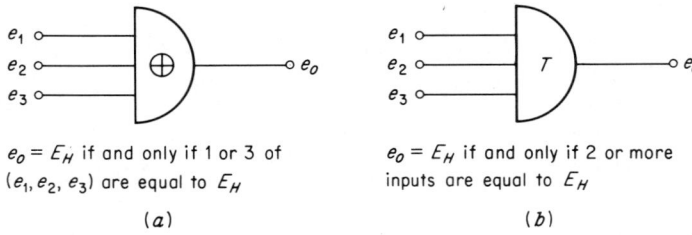

$e_o = E_H$ if and only if 1 or 3 of (e_1, e_2, e_3) are equal to E_H

(a)

$e_o = E_H$ if and only if 2 or more inputs are equal to E_H

(b)

Fig. 1.7-7. **Special gate symbols.** (a) $\oplus$ gate; (b) threshold gate.

and only if T or more inputs are high. The symbol for this gate will be the AND-gate symbol with the number T placed inside it, as in Fig. 1.7-7b.

Problems

1. Design a circuit using keys to control a light from two locations. Specify the type of switch that would be used by means of the symbols given in the text.
2. What happens in the circuit of Fig. P1-2?

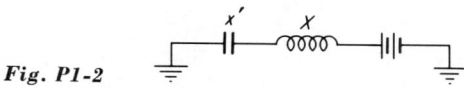

Fig. P1-2

3. For the circuit of Fig. P1-3 in which $e_i = E_H$ or E_L:
    (a) Give a word statement for the conditions when e_o will be equal to E_H.
    (b) Give a word statement for the conditions when e_o will be equal to E_L.

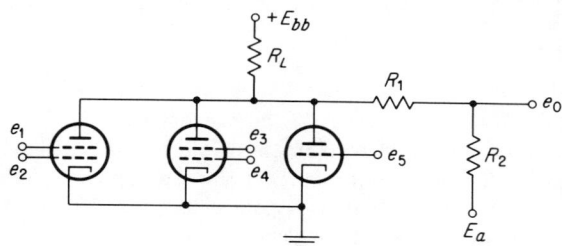

Fig. P1-3

4. In Sec. 1.2 it was demonstrated that triodes and pentodes sharing a common plate resistor could be used for combining binary signals.
    (a) Which of the transistor-diode and transistor-resistor circuits of Sec. 1.4 can be combined by means of a common-collector resistor? Explain.
    (b) Which of the connections discussed in (a) are useful? Explain.
5. (a) Derive an equivalent circuit for e_2 when the core of Fig. P1-5 is switching (ν and k are core parameters).
    (b) When is k measured at N_1 equal to k measured at N_2?

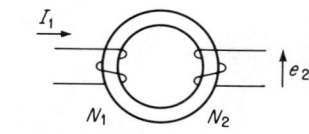

Fig. P1-5

6. The circuit of Fig. P1-6 is to be analyzed. The generator produces a train of alternating positive and negative rectangular current pulses whose amplitude is sufficient to switch the core back and forth completely.
    When R_L is infinite, we assume that the following relations hold during the switching time:

$$E_1 = N_1 \frac{\phi}{\tau}$$

$$\tau^{-1} = \frac{k}{\phi}(N_1 I_1 - \nu)$$

Given that for the configuration above

$N_1 = 5$ turns
$N_o = 10$ turns
$R_L = 20$ ohms
$\phi = 5 \times 10^{-1}$ volt-sec/turn
$k = 0.2$ ohm/turn^2
$\nu = 0.25$ amp-turns

find the equivalent linear circuit of the core as seen by the current-pulse generator during switching time.

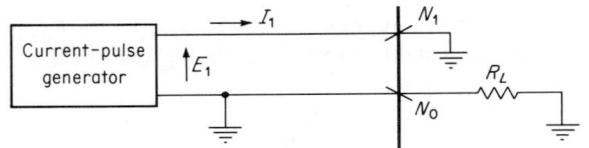

Fig. P1-6

7. Given an AND-NOT gate constructed as shown in Fig. P1-7. The gate output saturates (at essentially 0 volts) if nine inputs are grounded and -3 volts is impressed on the tenth input. What is the maximum number of inputs of similar gates which can be driven by the output of one such gate if stray loads are neglected?

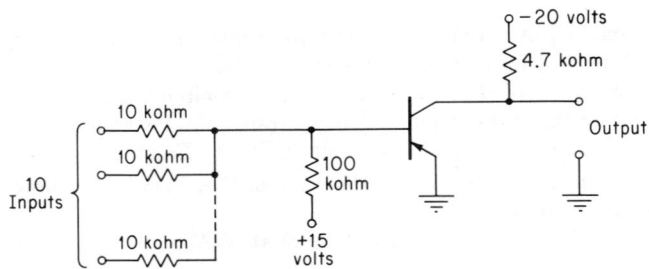

Fig. P1-7

REFERENCES

1. Millman, J., and H. Taub: "Pulse and Digital Circuits," McGraw-Hill Book Company, New York, 1956.
2. Richards, R. K.: "Digital Computer Components and Circuits," D. Van Nostrand Company, Inc., Princeton, N.J., 1957.
3. Keister, W., A. E. Ritchie, and S. H. Washburn: "The Design of Switching Circuits," D. Van Nostrand Company, Inc., Princeton, N.J., 1951.
4. Caldwell, S. H.: "Switching Circuits and Logical Design," John Wiley & Sons, Inc., New York, 1958.

5. Burks, A. W.: Electronic Computing Circuits of the ENIAC, *Proc. IRE*, vol. 36, no. 1, pp. 121–161, January, 1948.
6. Eccles, W. H., and F. W. Jordan: A Trigger Relay, *Radio Rev.*, vol. 1, no. 3, pp. 143–148, December, 1919.
7. Staff of the Harvard Computation Laboratory: "Synthesis of Electronic Computing and Control Circuits," Harvard University Press, Cambridge, Mass., 1951.
8. Engineering Research Associates: "High Speed Computing Devices," McGraw-Hill Book Company, New York, 1950.
9. Yokelson, B. J., and W. Ulrich: Engineering Multistage Diode Logic Circuits. *AIEE Trans.*, pt. 1, vol. 74, no. 20, pp. 466-475, September, 1955.
10. Yokelson, B. J., W. B. Cagle, and M. D. Underwood: Semiconductor Circuit Design Philosophy for the Central Control of an Electronic Switching System, *Bell System Tech. J.*, vol. 37, no. 5, pp. 1125–1160, September, 1958.
11. Cagle, W. B., and W. H. Chen: Designing Low-level High-speed Semiconductor Logic Circuits, 1957 *IRE Wescon Conv. Record*, pt. 2, pp. 3–9.
12. Pressman, A. I.: "Design of Transistorized Circuits for Digital Computers," John F. Rider, Publisher, Inc., New York, 1959.
13. Rowe, W. D.: The Transistor NOR Circuit, 1957 *IRE Wescon Conv. Record*, pt. 4, pp. 231–245.
    ——— and G. H. Royer: Transistor NOR Circuit Design, *AIEE Trans.*, pt. 1, vol. 76, no. 31, pp. 263–267, July, 1957.
14. Maley, G. A., and J. Earle: "The Logic Design of Transistor Digital Computers," Prentice-Hall, Inc., Englewood Cliffs, N.J., 1963.
15. Beter, R. H., W. E. Bradley, R. B. Brown, and M. Rubinoff: Surface Barrier Transistor Switching Circuits, *IRE Conv. Record*, pt. 4, pp. 139–145, 1955.
16. Yourke, H. S.: Millimicrosecond Transistor Current Switching Circuits, *IRE Trans. on Circuit Theory*, vol. CT-4, no. 3, pp. 236–240, September, 1957.
17. Henle, R. A., and J. L. Walsh: Application of Transistors to Computers, *Proc. IRE*, vol. 46, no. 6, pp. 1240–1254, June, 1958.
18. Buck, D. A.: The Cryotron—A Superconductive Computer Component, *Proc. IRE*, vol. 44, no. 4, pp. 482–493, April, 1956.
19. Bremer, J. W.: "Superconductive Devices," McGraw-Hill Book Company, New York, 1962.
20. Newhouse, V. L., and J. W. Bremer: High-speed Superconductive Switching Element Suitable for Two-dimension Fabrication, *J. Appl. Phys.*, vol. 30, p. 1458, September, 1959.
21. Newhouse, V. L., J. W. Bremer, and H. H. Edwards: An Improved Film Cryotron and Its Application to Digital Computers, *Proc. IRE*, vol. 48, no. 8, pp. 1395–1404, August, 1960.
22. Smallman, C. R., A. E. Slade, and N. L. Cohen: Thin-film Cryotrons, *Proc. IRE*, vol. 48, no. 9, pp. 1562-1582, September, 1960.
23. Ramey, R. A.: The Single-core Magnetic Amplifier as a Computer Element, *AIEE Trans., Communication and Electronics*, pt. I, vol. 71, pp. 442–446, January, 1953.

2 NUMBER SYSTEMS AND CODES

Arithmetic operations using decimal numbers are quite common. However, in logical design it is necessary to perform manipulations in the so-called binary system of numbers because of the on-off nature of the physical devices used. The present chapter is intended to acquaint the reader with the fundamental concepts involved in dealing with number systems other than decimal. In particular, the binary system is covered in considerable detail.

2.1 POSITIONAL NOTATION [1,2,3]

An ordinary decimal number can be regarded as a polynomial in powers of 10. For example, 423.12 can be regarded as $4 \times 10^2 + 2 \times 10^1 + 3 \times 10^0 + 1 \times 10^{-1} + 2 \times 10^{-2}$. Decimal numbers like this are said to be expressed in a number system with *base*, or *radix*, 10 because there are 10 basic digits (0,1,2, . . . ,9) from which the number system is formulated. In a similar fashion we can express any number N in a system using any base b. We shall write such a number as $(N)_b$. Whenever $(N)_b$ is written, the convention of always expressing b in base 10 will be followed. Thus $(N)_b = (p_n p_{n-1} \ldots p_1 p_0 . p_{-1} p_{-2} \ldots p_{-m})_b$, where b is an integer greater than 1 and $0 \leq p_i \leq b - 1$. The value of a number represented in this fashion, which is called *positional notation*, is given by

$$(N)_b = p_n b^n + p_{n-1} b^{n-1} + \cdots + p_0 b^0 + p_{-1} b^{-1} + p_{-2} b^{-2} + \cdots + p_{-m} b^{-m} \quad (2.1\text{-}1)$$

$$(N)_b = \sum_{i=-m}^{n} p_i b^i \quad (2.1\text{-}2)$$

For decimal numbers, the symbol "." is called the *decimal point;* for more general base-b numbers, it is called the *radix point*. That

2.1 positional notation

Table 2.1-1. Integers in Various Bases

	2	3	4	5	Base b ...	10	11	12	...	15
	0000	000	00	00		00	00	00		00
	0001	001	01	01		01	01	01		01
	0010	002	02	02		02	02	02		02
	0011	010	03	03		03	03	03		03
	0100	011	10	04		04	04	04		04
	0101	012	11	10		05	05	05		05
	0110	020	12	11		06	06	06		06
$(N)_b$	0111	021	13	12		07	07	07		07
	1000	022	20	13		08	08	08		08
	1001	100	21	14		09	09	09		09
	1010	101	22	20		10	0A	0A		0A
	1011	102	23	21		11	10	0B		0B
	1100	110	30	22		12	11	10		0C
	1101	111	31	23		13	12	11		0D
	1110	112	32	24		14	13	12		0E
	1111	120	33	30		15	14	13		10

portion of the number to the right of the radix point $(p_{-1}p_{-2} \cdots p_{-m})$ is called the *fractional part*, and the portion to the left of the radix point $(p_n p_{n-1} \cdots p_0)$ is called the *integral part*.

The integers from 0 to 15 are given in Table 2.1-1 for several bases. Since there are no coefficient values for the range 10 to b-1 when $b > 10$, the letters $A, B, C, \ldots$ are used in the table.

Conversion of Base

In order to make use of nondecimal number systems, it is necessary to be able to convert a number expressed in one base into the correct representation of the number in another base. One way of doing this makes direct use of the polynomial expression (2.1-1). For example, consider the base-2 number $(1011.101)_2$. The corresponding polynomial expression is

$$1 \times 2^3 + 0 \times 2^2 + 1 \times 2^1 + 1 \times 2^0 + 1 \times 2^{-1} + 0 \times 2^{-2} + 1 \times 2^{-3}$$

or

$$8 + \qquad 2 + 1 + \tfrac{1}{2} + \qquad\qquad \tfrac{1}{8}$$

or

$$11\tfrac{5}{8} = 11.625$$

This technique of directly evaluating the polynomial expression for a number is a general method for converting from an arbitrary base b_1 to another arbitrary base b_2. For convenience it will be called the *polynomial method*. It consists in:

1. Expressing the number $(N)_{b_1}$ as a polynomial, with base-b_2 numbers used in the polynomial
2. Evaluating the polynomial, base-b_2 arithmetic being used

This polynomial method is most often used by human beings when a number is to be converted to base 10, since it is then possible to use decimal arithmetic.

This method for converting numbers from one base to another is the first example of one of the major goals of this book: the development of algorithms. In general terms, an algorithm is a list of instructions specifying a sequence of operations which will give the answer to any problem of a given type [4,5]. The important characteristics of an algorithm are (1) that it is fully specified and does not rely on any skill or intuition on the part of the person applying it and (2) that it always works, i.e., that a correct answer is always obtained.

It is not always convenient to use base-b_2 arithmetic in converting from base b_1 to base b_2. An algorithm for carrying out this conversion by using base-b_1 arithmetic will be discussed next. This discussion is specifically for the situation in which $b_1 = 10$, but it can easily be extended to the more general case. This will be called the *iterative method*, since it involves iterated multiplication or division.

In converting $(N)_{10}$ to $(N)_b$ the portions to the left and to the right of the decimal point are converted separately. First, consider the portion to the left. The general conversion procedure is to divide $(N)_{10}$ by b, giving $(N)_{10}/b$ and a remainder. The remainder, call it p_0, is the least significant (rightmost) digit of $(N)_b$. The next least significant digit p_1 is the remainder of $(N)_{10}/b$ divided by b, and succeeding digits are obtained by continuing this process. A convenient form for carrying out this conversion is illustrated in the following example.

Example 2.1-1

(a) $(23)_{10} = (10111)_2$

2	23	(Remainder)
2	11	1
2	5	1
2	2	1
2	1	0
	0	1

2.1 positional notation

(b) $(23)_{10} = (27)_8$

8	23	(Remainder)
8	2	7
	0	2

(c) $(410)_{10} = (3120)_5$

5	410	(Remainder)
5	82	0
5	16	2
5	3	1
	0	3

Now consider the portion of the number to the right of the decimal point, i.e., the fractional part. The procedure for converting this is to multiply $(N)_{10}$ (fractional) by b. If the resulting product is less than 1, then the most significant (leftmost) digit of the fractional part is 0. If the resulting product is greater than 1, the most significant digit of the fractional part is the integral part of the product. The next most significant digit is formed by multiplying the fractional part of this product by b and taking the integral part. The remaining digits are formed by repeating this process. The process may or may not terminate. A convenient form for carrying out this conversion is illustrated below.

Example 2.1-2

(a) $(0.625)_{10} = (0.5)_8$ $0.625 \times 8 = 5.000 \mid 0.5$

(b) $(0.23)_{10} = (0.001110 \cdots)_2$

$0.23 \times 2 = 0.46 \mid 0.0$
$0.46 \times 2 = 0.92 \mid 0.00$
$0.92 \times 2 = 1.84 \mid 0.001$
$0.84 \times 2 = 1.68 \mid 0.0011$
$0.68 \times 2 = 1.36 \mid 0.00111$
$0.36 \times 2 = 0.72 \mid 0.001110$

(c) $(27.68)_{10} = (11011.101011 \cdots)_2 = (33.53 \cdots)_8$

2	27	
2	13	1
2	6	1
2	3	0
2	1	1
	0	1

$0.68 \times 2 = 1.36 \mid 0.1$
$0.36 \times 2 = 0.72 \mid 0.10$
$0.72 \times 2 = 1.44 \mid 0.101$
$0.44 \times 2 = 0.88 \mid 0.1010$
$0.88 \times 2 = 1.76 \mid 0.10101$
$0.76 \times 2 = 1.52 \mid 0.101011 \cdots$

8	27	
8	3	3
	0	3

$0.68 \times 8 = 5.44 \mid 0.5$
$0.44 \times 8 = 3.52 \mid 0.53$

Number Systems and Codes

This latter example illustrates the simple relationship between the base-2 (binary) system and the base-8 (octal) system. The binary digits, called *bits*, are taken three at a time in each direction from the binary point and are expressed as decimal digits to give the corresponding octal number. For example, 101 in binary is equivalent to 5 in decimal; so the octal number in part *c* above has a 5 for the most significant digit of the fractional part. The conversion between octal and binary is so simple that the octal expression is sometimes used as a convenient shorthand for the corresponding binary number.

When a fraction is converted from one base to another, the conversion may not terminate, since it may not be possible to represent the fraction exactly in the new base with a finite number of digits. For example, consider the conversion of $(0.1)_3$ to a base-10 fraction. The result is clearly $(0.33 \cdots 3)_{10}$, which can be written as $(0.\overline{3})_{10}$ to indicate that the 3's are repeated indefinitely. It is always possible to represent the result of a conversion of base in this notation, since the nonterminating fraction must consist of a group of digits which are repeated indefinitely. For example, $(0.2)_{11} = 2 \times 11^{-1} = (0.1818 \cdots)_{10} = (0.\overline{18})_{10}$.

It should be pointed out that by combining the two conversion methods it is possible to convert between any two arbitrary bases by using only arithmetic of a third base. For example, to convert $(16)_7$ to base 3, first convert to base 10,

$$(16)_7 = 1 \times 7^1 + 6 \times 7^0 = 7 + 6 = (13)_{10}$$

Then convert $(13)_{10}$ to base 3,

3	13	(Remainder)	
3	4	1	$(16)_7 = (13)_{10} = (111)_3$
3	1	1	
3	0	1	

2.2 BINARY ARITHMETIC

Many modern digital computers employ the binary (base-2) number system to represent numbers, and carry out the arithmetic operations using binary arithmetic. While a detailed treatment of computer arithmetic [6,7] is not within the scope of this book, it will be useful to have the elementary techniques of binary arithmetic available. In performing decimal arithmetic it is necessary to memorize the tables giving the results of the elementary arithmetic operations for pairs of decimal digits. Similarly, for binary arithmetic the tables for the elementary operations for the binary digits are necessary.

2.2 binary arithmetic

Binary Addition

The binary addition table is as follows:

	Carry
$0 + 0 = 0$	0
$0 + 1 = 1$	0
$1 + 0 = 1$	0
$1 + 1 = 0$	1

Addition is performed by writing the numbers to be added in a column with the binary points aligned. The individual columns of binary digits, or *bits*, are added in the usual order according to the above addition table. Note that in adding a column of bits, there is a 1 carry for each pair of 1's in that column. These 1 carries are then treated as bits to be added in the next column to the left. A general rule for addition of a column of numbers (using any base) is to add the column decimally and divide by the base. The remainder is entered as the sum for that column, and the quotient is carried to be added in the next column.

Example 2.2-1

Base 2

$$
\begin{array}{rl}
10011\ 11 = & \text{Carries} \\
1001.011 = & (9.375)_{10} \\
+\ 1101.101 = & (13.625)_{10} \\
\hline
10111.000 = & (23)_{10} = \text{Sum}
\end{array}
$$

Binary Subtraction

The binary subtraction table is as follows:

	Borrow
$0 - 0 = 0$	0
$0 - 1 = 1$	1
$1 - 0 = 1$	0
$1 - 1 = 0$	0

Subtraction is performed by writing the minuend over the subtrahend with the binary points aligned and carrying out the subtraction according to the above table. If a borrow occurs and the next leftmost digit of the minuend is a 1, it is changed to a 0 and the process of subtraction is then continued from right to left.

	Base 2	Base 10
Borrow	1	
	0	
Minuend	1̶0	2
Subtrahend	−01	−1
Difference	01	1

If a borrow occurs and the next leftmost digit of the minuend is a 0, then this 0 is changed to a 1, as is each successive minuend digit to the left which is equal to 0. The first minuend digit to the left which is equal to 1 is changed to 0, and then the subtraction process is resumed.

	Base 2	Base 10
Borrow	1	
	011	
Minuend	1̶1̶0̶0̶0̶	24
Subtrahend	−10001	−17
Difference	00111	7
Borrow	1 1	
	01011	
Minuend	1̶0̶1̶0̶0̶0̶	40
Subtrahend	−011001	−25
Difference	001111	15

Complements

It is possible to avoid this subtraction process by using a complement representation for negative numbers. This will be discussed specifically for binary *fractions*, although it is easy to extend the complement techniques to integers and mixed numbers. The 2's complement (^2B) of a binary fraction B is defined as follows:

$$^2B = (2 - B)_{10} = (10 - B)_2$$

Thus $^2(0.1101) = 10.0000 - 0.1101 = 1.0011$. A particularly simple means of carrying out the subtraction indicated in the expression for $^2(0.1101)$ is obtained by noting that $10.0000 = 1.1111 + 0.0001$. Thus $10.0000 - 0.1101 = (1.1111 - 0.1101) + 0.0001$. The subtraction $1.1111 - 0.1101$ is particularly easy, since all that is necessary is to reverse each of the digits of 0.1101 to obtain 1.0010. Finally the addition

2.2 binary arithmetic

of 0.0001 is also relatively simple and yields 1.0011. In general the process of forming ^2B involves reversing the digits of B and then adding $.00 \cdots 01$.

The usefulness of the 2's complement stems from the fact that it is possible to obtain the difference $A - B$ by adding ^2B to A. Thus $A + {}^2B = (A + 10 - B)_2 = (10 + (A - B))_2$. If $(A - B) > 0$, then $(10 + A - B)_2$ will be 10 plus the positive fraction $(A - B)$. It is thus possible to obtain $A - B$ by dropping the leftmost 1 in $A + {}^2B$. For example,

$$\begin{array}{rl} A = & 0.1110 \\ -B = & 0.1101 \\ \hline & .0001 \end{array} \qquad \begin{array}{rl} A = & 0.1110 \\ +{}^2B = & 1.0011 \\ \hline & 10.0001 \end{array}$$

If $(A - B) < 0$, then $A + {}^2B = (10 - |A - B|)_2$, which is just equal to $^2(A - B)$, the 2's-complement representation of $A - B$. For example,

$$\begin{array}{rl} A = & 0.1101 \\ -B = & -0.1110 \\ \hline & -0.0001 \end{array} \qquad \begin{array}{rl} A = & 0.1101 \\ +{}^2B = & 1.0010 \\ \hline & 1.1111 \end{array} \qquad {}^2(0.0001) = 1.1111$$

The 1's complement is also very commonly used. This is defined as
$$^1B = (10 - 0.000 \cdots 1 - B)_2$$
where the location of the 1 in $0.000 \cdots 1$ corresponds to the least significant digit of B. Since $(10 - .000 \cdots 1)_2$ is equal to $01.111 \cdots 1$, it is possible to form ^1B by reversing the digits of B and adding a 1 before the radix point. Thus $^1(.1101) = 1.0010$.

If $A + {}^1B$ is formed, the result is $(A - B + 10 - .000 \cdots 1)_2$. If $(A - B) > 0$, this can be converted to $A - B$ by removing the $(10)_2$ and adding a 1 to the least significant digit of $A + {}^1B$. This is called an *end-around carry*. For example:

$$\begin{array}{rl} A = & 0.1110 \\ -B = & 0.1101 \\ \hline & .0001 \end{array} \qquad \begin{array}{rl} A = & 0.1110 \\ +{}^1B = & +\ 1.0010 \\ \hline A + {}^1B = & 10.0000 \\ +\ & .0001 \hookleftarrow \\ \hline & .0001 \end{array} \qquad \text{End-around carry}$$

so that $\qquad A - B = \qquad .0001$

If $(A - B) < 0$, then $A + {}^1B$ will be the 1's complement of $|A - B|$. For example,

$$\begin{array}{rl} A = & 0.1101 \\ -B = & -0.1110 \\ \hline & -0.0001 \end{array} \qquad \begin{array}{rl} A = & 0.1101 \\ {}^1B = & 1.0001 \\ \hline A + {}^1B = & 1.1110 \end{array} \qquad {}^1(.0001) = 1.1110$$

Number Systems and Codes

The radix complement of a base-b fraction F is defined as

$$^{b}F = (10 - F)_b$$

and the diminished radix complement is defined as

$$^{b-1}F = (10 - F - .000 \cdots 1)_b$$

Similar procedures hold for the formation of the complements and their use for subtraction.

When integers or mixed numbers are involved in the subtractions, the definitions of the complements must be generalized to

$$^{b}N = (100 \cdots 0. - N)_b$$

and

$$^{b-1}N = (100 \cdots 0. - N - .00 \cdots 1)_b$$

where $100 \cdots 0$ contains two more digits than any integer to be encountered in the subtractions. For example if $(N)_2 = 11.01$, then

$$
\begin{aligned}
^{2}(N)_2 &= 1000.00 - 11.01 \\
&= 111.11 - 11.01 + .01 \\
&= 100.10 + .01 \\
&= 100.11
\end{aligned}
$$

$$
\begin{array}{rl}
M = & 11.10 \\
-N = & -11.01 \\ \hline
& 0.01
\end{array}
\qquad
\begin{array}{rl}
M = & 11.10 \\
^{2}N = & 100.11 \\ \hline
& 1000.01
\end{array}
$$

$\uparrow$ Discard

Shifting

In carrying out multiplication or division there are intermediate steps which require that numbers be shifted to the right or the left. Shifting a base-b number k places to the right has the effect of multiplying the number by b^{-k}, and shifting k places to the left is equivalent to multiplication by b^{+k}. Thus, if

$$(N)_b = \sum_{i=-m}^{n} p_i b^i = (p_n p_{n-1} \cdots p_1 p_0 . p_{-1} p_{-2} \cdots p_{-m})_b$$

shifting $(N)_b$ k places to the right yields

$$(p_n p_{n-1} \cdots p_1 p_0 p_{-1} \cdots p_{-k} . p_{-k-1} \cdots p_{-m})_b = \sum_{i=-m}^{n} p_i b^{i+k}$$

and

$$\sum_{i=-m}^{n} p_i b^{i+k} = b^k \sum_{i=-m}^{n} p_i b^i = b^k (N)_b$$

A similar manipulation shows the corresponding situation for left shifts. Shifting the binary point k places (k positive for right shifts and negative

2.2 binary arithmetic

for left shifts) in a binary number multiplies the value of the number by 2^k. For example,

$(110.101)_2 = (6.625)_{10}$

$(1.10101)_2 = 2^{-2}(6.625)_{10} = \left(\dfrac{6.625}{4}\right)_{10} = (1.65625)_{10}$

$(11010.1)_2 = 2^{+2}(6.625)_{10} = (4 \times 6.625)_{10} = (26.5)_{10}$

Binary Multiplication

The binary multiplication table is as follows:

$0 \times 0 = 0$
$0 \times 1 = 0$
$1 \times 0 = 0$
$1 \times 1 = 1$

The process of binary multiplication is illustrated by the following example:

110.10	Multiplicand
10.1	Multiplier
11010	Partial product
00000	Partial product
11010	Partial product
10000.010	

For every digit of the multiplier which is equal to 1, a partial product is formed consisting of the multiplicand shifted so that its least significant digit is aligned with the 1 of the multiplier. An all-zero partial product is formed for each 0 multiplier digit. Of course, the all-zero partial products can be omitted. The final product is formed by summing all the partial products. The binary point is placed in the product by using the same rule as for decimal multiplication: the number of digits to the right of the binary point of the product is equal to the sum of the numbers of digits to the right of the binary points of the multiplier and the multiplicand.

The commonest technique for handling the multiplication of negative numbers is to use the process just described to multiply the magnitudes of the numbers. The sign of the product is determined separately, and the product is made negative if either the multiplier or the multiplicand but not both are negative. It is possible to carry out multiplication directly with negative numbers represented in complement form [6, pp. 161–165; 7, secs. 3.2 and 3.3], but this is a specialized technique and will not be discussed here.

Binary Division

Division is the most complex of the four basic arithmetic operations. Decimal long division as taught in grade school is a trial-and-error process. For example, in dividing 362 by 46 one must first recognize that 46 is larger than 36 and then must guess how many times 46 will go into 362. If an initial guess of 8 is made and the multiplication $8 \times 46 = 368$ is carried out, the result is seen to be larger than 362 so that the 8 must be replaced by a 7. This process of trial and error is simpler for binary division because there are fewer possibilities in the binary case.

In order to implement binary division in a digital computer a division algorithm must be specified. Two different algorithms, called *restoring* and *nonrestoring division*, are used.

Restoring division is carried out as follows: In the first step the divisor is subtracted from the dividend with their leftmost digits aligned. If the result is positive, a 1 is entered as the quotient digit corresponding to the rightmost digit of the dividend from which a digit of the divisor was subtracted. The next rightmost digit of the dividend is appended to the result, which then becomes the next partial dividend. The divisor is then shifted one place to the right so that its least significant digit is aligned with the rightmost digit of the partial dividend, and the process just described is repeated.

If the result of subtracting the divisor from the dividend is negative, a 0 is entered in the quotient and the divisor is added back to the negative result so as to *restore* the original dividend. The divisor is then shifted one place to the right, and subtraction is carried out again. The process of restoring division is illustrated in the following example:

```
Divisor = 1 1 1 1      Dividend = 1 1 0 0
                                   q₀ q₋₁ q₋₂ q₋₃ q₋₄ q₋₅
                                   0  .1  1   0   0   1

                    1 1 1 1/1 1 0 0 .0 0 0 0 0
Subtract                    1 1 1 1
Negative result  q₀ = 0   −0 0 1 1
Restore                   +1 1 1 1
                           1 1 0 0 0
Subtract                     1 1 1 1
Positive result  q₋₁ = 1     1 0 0 1 0
Subtract                       1 1 1 1
Positive result  q₋₂ = 1       0 0 0 1 1 0
Subtract                           1 1 1 1
Negative result  q₋₃ = 0         −1 0 0 1
Restore                          +1 1 1 1
                                  0 1 1 0 0
Subtract                            1 1 1 1
Negative result  q₋₄ = 0           −0 0 1 1
Restore                            +1 1 1 1
                                    1 1 0 0 0
Subtract                              1 1 1 1
Positive result  q₋₅ = 1              1 0 0 1  (Remainder)
```

2.3 binary codes

In nonrestoring division, the step of adding the divisor to a negative partial dividend is omitted, and instead the *shifted* divisor is added to the negative partial dividend. This step of adding the shifted divisor replaces the two steps of adding the divisor and then subtracting the shifted divisor. This can be justified as follows: If X represents the negative partial dividend and Y the divisor, then $\frac{1}{2}Y$ represents the divisor shifted one place to the right. Adding the divisor and then subtracting the shifted divisor yields $X + Y - \frac{1}{2}Y = X + \frac{1}{2}Y$, while adding the shifted divisor yields the same result, $X + \frac{1}{2}Y$. The steps which occur in using nonrestoring division to divide 1100 by 1111 are shown in the following example:

```
Divisor = 1 1 1 1    Dividend = 1 1 0 0
                                        q₀  q₋₁ q₋₂ q₋₃ q₋₄ q₋₅
                                        0   .1   1   0   0   1
                              1 1 1 1/1 1 0 0 .0 0 0 0 0
Subtract                              1 1 1 1
Negative result    q₀ = 0            −0 0 1 1 0
Shift and add                        + 1 1 1 1
Positive result    q₋₁ = 1           +1 0 0 1 0
Shift and subtract                   −    1 1 1 1
Positive result    q₋₂ = 1           +0  0 1 1 0
Shift and subtract                   −    1 1 1 1
Negative result    q₋₃ = 0           −1   0 0 1 0
Shift and add                        +    1 1 1 1
Negative result    q₋₄ = 0           −0    0 1 1 0
Shift and add                        +     1 1 1 1
Positive result    q₋₅ = 1           +1    0 0 1   (Remainder)
```

2.3 BINARY CODES

The binary number system has many advantages and is widely used in digital systems. However, there are times when binary numbers are not appropriate. Since we think much more readily in terms of decimal numbers than binary numbers, facilities are usually provided so that data can be entered into the system in decimal form, the conversion to binary being performed automatically inside the system. In fact many computers have been designed which work entirely with decimal numbers. For this to be possible, a scheme for representing each of the 10 decimal digits as a sequence of binary digits must be used.

Binary-coded-decimal Numbers

In order to represent 10 decimal digits, it is necessary to use at least 4 binary digits, since there are 2^4, or 16, different combinations of 4 binary

Number Systems and Codes

Table 2.3-1. Some Common 4-bit Decimal Codes

Decimal digits	8 4 2 1 b_3 b_2 b_1 b_0	8 4−2−1	2 4 2 1	Excess-3
0	0 0 0 0	0 0 0 0	0 0 0 0	0 0 1 1
1	0 0 0 1	0 1 1 1	0 0 0 1	0 1 0 0
2	0 0 1 0	0 1 1 0	0 0 1 0	0 1 0 1
3	0 0 1 1	0 1 0 1	0 0 1 1	0 1 1 0
4	0 1 0 0	0 1 0 0	0 1 0 0	0 1 1 1
5	0 1 0 1	1 0 1 1	1 0 1 1	1 0 0 0
6	0 1 1 0	1 0 1 0	1 1 0 0	1 0 0 1
7	0 1 1 1	1 0 0 1	1 1 0 1	1 0 1 0
8	1 0 0 0	1 0 0 0	1 1 1 0	1 0 1 1
9	1 0 0 1	1 1 1 1	1 1 1 1	1 1 0 0

digits but only 2^3, or 8, different combinations of 3 binary digits. If 4 binary digits, or *bits*, are used and only one combination of bits is used to represent each decimal digit, there will be six unused or invalid code words. In general any arbitrary assignment of combinations of bits to digits can be used so that there are 16!/6! or approximately 2.9×10^{10} possible codes. Only a few of these codes have ever been used in any system, since the arithmetic operations are very difficult in all but a few of these codes. Several of the more common 4-bit decimal codes are shown in Table 2.3-1.

The 8,4,2,1 code is obtained by taking the first 10 binary numbers and assigning them to the corresponding decimal digits. This code is an example of a *weighted code*, since the decimal digits can be determined from the binary digits by forming the sum $d = 8b_3 + 4b_2 + 2b_1 + b_0$. The coefficients 8, 4, 2, 1 are known as the *code weights*. The number 462 would be represented as 0100 0110 0010 in the 8,4,2,1 code. It has been shown [8] that there are only 17 different sets of weights possible for a positively weighted code: (3,3,3,1), (4,2,2,1), (4,3,1,1), (5,2,1,1), (4,3,2,1), (4,4,2,1), (5,2,2,1), (5,3,1,1), (5,3,2,1), (5,4,2,1), (6,2,2,1), (6,3,1,1), (6,3,2,1), (6,4,2,1), (7,3,2,1), (7,4,2,1), (8,4,2,1).

It is also possible to have a weighted code in which some of the weights are negative, as in the 8,4,−2,−1 code shown in Table 2.3-1. This code has the useful property of being *self-complementing*: if a code word is formed by taking the complement of each bit, then this new code word represents the 9's complement of the digit to which the original code word corresponds. For example, 0101 represents 3 in the 8,4,−2,−1 code, and 1010 represents 6 in this code. In general, if b_i' denotes the complement of

2.4 geometric representation of binary numbers

Table 2.3-2. *Some Decimal Codes Using More than 4 Bits*

Decimal digits	2-out-of-5	Biquinary 5043210
0	00011	0100001
1	00101	0100010
2	00110	0100100
3	01001	0101000
4	01010	0110000
5	01100	1000001
6	10001	1000010
7	10010	1000100
8	10100	1001000
9	11000	1010000

b_i, then a code is self-complementing if, for any code word $b_3b_2b_1b_0$ representing a digit d_i, the code word $b'_3b'_2b'_1b'_0$ represents $9 - d_i$. The 2,4,2,1 code of Table 2.3-1 is an example of a self-complementing code having all positive weights, and the excess-3 code is an example of a code which is self-complementing but not weighted. The excess-3 code is obtained from the 8,4,2,1 code by adding (using binary arithmetic) 0011 (or 3) to each 8,4,2,1 code word to obtain the corresponding excess-3 code word.

Although 4 bits are sufficient for representing the decimal digits, it is sometimes expedient to use more than 4 bits in order to achieve arithmetic simplicity or ease in error detection. The 2-out-of-5 code shown in Table 2.3-2 has the property that each code word has exactly two 1's. A single error which complements 1 of the bits will always produce an invalid code word and is therefore easily detected. This is an unweighted code. The biquinary code shown in Table 2.3-2 is a weighted code in which 2 of the bits specify whether the digit is in the range 0 to 4 or the range 5 to 9 and the other 5 bits identify where in the range the digit occurs. This code was used in the IBM 650 computer.

2.4 GEOMETRIC REPRESENTATION OF BINARY NUMBERS

An n-bit binary number may be represented by what is called a *point in n-space*. To see just what is meant by this, consider the set of 1-bit binary numbers, that is, 0 and 1. This set can be represented by two points in 1-space, i.e., by two points on a line. Such a representation is

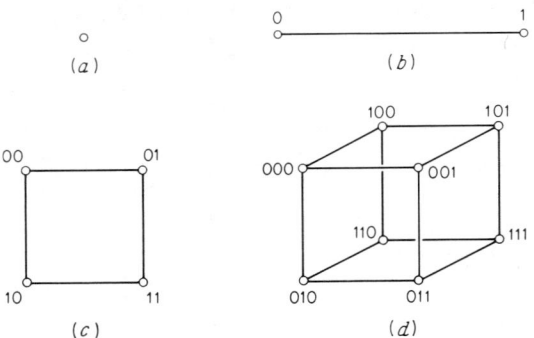

Fig. 2.4-1. *n-Cubes for n = 0, 1, 2, 3.* (a) *0-Cube;* (b) *1-cube;* (c) *2-cube;* (d) *3-cube.*

called a 1-*cube* and is shown in Fig. 2.4-1b. (A 0-*cube* is a single point in 0-space.)

Now consider the set of 2-bit binary numbers, that is, 00, 01, 10, 11 (or, decimally, 0,1,2,3). This set can be represented by four points (also called *vertices*, or *nodes*) in 2-space. This representation is called a 2-*cube* and is shown in Fig. 2.4-1c. Note that this figure can be obtained by projecting the 1-cube (i.e., the horizontal line with two points) downward and by prefixing a 0 to the 0 and 1 on the original 1 cube and a 1 to 0 and 1 on the projected 1-cube. A similar projection procedure can be followed in obtaining any next-higher-dimensional figure. For example, the representation for the set of 3-bit binary numbers is obtained by projecting the 2-cube representation of Fig. 2.4-1c. A 0 is prefixed to the bits on the original 2-cube, and a 1 is prefixed to the bits on the projection of the 2-cube. Thus, the 3-bit representation, or 3-*cube*, is shown in Fig. 2.4-1d.

A more formal statement for the projection method of defining an n-cube is as follows:

1. A 0-cube is a single point with no designation.
2. An n-cube is formed by projecting an $(n-1)$-cube. A 0 is prefixed to the designations of the points of the original $(n-1)$-cube, and a 1 is prefixed to the designations of the points of the projected $(n-1)$-cubes.

There are 2^n points in an n-cube. A *p-subcube* of an n-cube $(p \leq n)$ is defined as a collection of any 2^p points which have exactly $(n-p)$ corresponding bits all the same. For example, the points 100, 101, 000, and 001 in the 3-cube (Fig. 2.4-1d) form a 2-subcube, since there are $2^2 = 4$ total points and $3 - 2 = 1$ of the bits (the second) is the same for all four

2.4 geometric representation of binary numbers

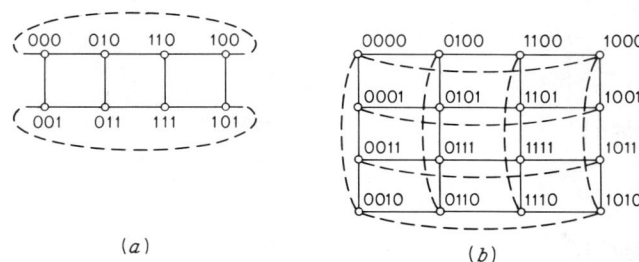

Fig. 2.4-2. *Alternative representations of 3-cube (a) and 4-cube (b).*

points. In general, there are $\frac{n!2^{n-p}}{(n-p)!p!}$ different p-subcubes in an n-cube, since there are $C^n_{n-p} = \frac{n!}{(n-p)!p!}$ (number of ways of selecting n things taken $n-p$ at a time) ways in which $n-p$ of the bits may be the same and there are 2^{n-p} combinations which these bits may take on. For example, there are $\frac{3!2^2}{2!1!} = 12$ 1-subcubes (line segments) in a 3-cube, and there are $\frac{3!2^1}{1!2!} = 6$ 2-subcubes ("squares") in a 3-cube.

Besides the form shown in Fig. 2.4-1, there are two other methods of drawing an n-cube which are frequently used. The first of these is shown in Fig. 2.4-2 for the 3- and 4-cubes. It is seen that these still agree with the projection scheme and are merely a particular way of drawing the cubes. The lines which are dotted are usually omitted for convenience in drawing.

If in the representation of Fig. 2.4-2 we replace each dot by a square area, we have what is known as an *n-cube map*. This representation is shown for the 3- and 4-cubes in Fig. 2.4-3. Maps will be of considerable use to us later. Notice that the appropriate entry for each cell of the maps of Fig. 2.4-3 can be determined from the corresponding row and column labels.

It is sometimes convenient to represent the points of an n-cube by the decimal equivalents of their binary designations. For example, Fig. 2.4-4 shows the 3- and 4-cube maps represented in this way. It is of interest to note that, if a point has the decimal equivalent N_i in an n-cube, in an $(n+1)$-cube this point and its projection (as defined) become N_i and $N_i + 2^n$.

Distance

A concept which will be of later use is that of the distance between two points on an n-cube. Briefly, the *distance* between two points on an n-

	00	01	11	10
0	000	010	110	100
1	001	011	111	101

	00	01	11	10
00	0000	0100	1100	1000
01	0001	0101	1101	1001
11	0011	0111	1111	1011
10	0010	0110	1110	1010

(a)   (b)

Fig. 2.4-3. *n*-Cube maps for $n = 3$ (a) and $n = 4$ (b).

cube is simply the number of coordinates (bit positions) in which the binary representations of the two points differ. For example, 10110 and 01101 differ in all but the third coordinate (from left or right). Since the points differ in four coordinates, the distance between them is 4. A more formal definition is as follows:

First, define the *mod 2 sum* of two bits, $a \oplus b$, by

$$0 \oplus 0 = 0$$
$$1 \oplus 0 = 1$$
$$0 \oplus 1 = 1$$
$$1 \oplus 1 = 0$$

That is, the sum is 0 if the 2 bits are alike, and it is 1 if the 2 bits are different. Now consider the binary representations of two points, $P_i = (a_{n-1} a_{n-2} \cdots a_0)$ and $P_j = (b_{n-1} b_{n-2} \cdots b_0)$, on the *n*-cube. The mod 2 sum of these two points is defined as

$$P_k = P_i \oplus P_j = (a_{n-1} \oplus b_{n-1}, a_{n-2} \oplus b_{n-2}, \ldots, a_0 \oplus b_0)$$

This sum P_k is the binary representation of another point on the *n*-cube. If we define $|P_i|$ to represent the number of 1's in the binary representation P_i, then the distance (or *metric*) between two points is defined as

$$D(P_i, P_j) = |P_i \oplus P_j|$$

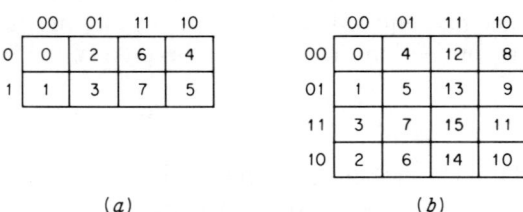

Fig. 2.4-4. Decimal labels in *n*-cube maps. (a) 3-cube map; (b) 4-cube map.

2.4 geometric representation of binary numbers

The distance function satisfies the following three properties:

$D(P_i, P_j) = 0$ if and only if $P_i = P_j$
$D(P_i, P_j) = D(P_j, P_i) > 0$ if $P_i \neq P_j$
$D(P_i, P_j) + D(P_j, P_k) \geq D(P_i, P_k)$ Triangle inequality

To return to the more intuitive approach, since two adjacent points (connected by a single line segment) on an n-cube form a 1-subcube, they differ in exactly one coordinate and thus are distance 1 apart. We see then that, to any two points which are distance D apart, there corresponds a *path* of D connected line segments on the n-cube joining the two points. Furthermore, there will be more than one path of length D connecting the two points (for $D > 1$ and $n < 2$), but there will be no path shorter than length D connecting the two points. A given shortest path connecting the two points, thus, cannot intersect itself, and $D + 1$ nodes (including the end points) will occur on the path.

Unit-distance Codes

In terms of the geometric picture, a code is simply the association of the decimal integers (0,1,2, . . .) with the points on an n-cube. There are two types of codes which are best described in terms of their geometric properties. These are so-called *unit-distance codes* and *error-detecting* and *-correcting codes*.

A unit-distance code is simply the association of the decimal integers (0,1,2, . . .) with the points on a connected path in the n-cube such that the distance is 1 between the point corresponding to any integer i and the point corresponding to integer $i + 1$ (see Fig. 2.4-5). That is, if P_i is

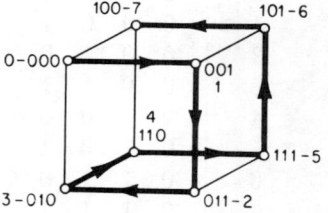

Fig. 2.4-5. Path on a 3-cube corresponding to a unit-distance code.

Table 2.4-1. *Unit-distance Code of Fig. 2.4-5*

0	000
1	001
2	011
3	010
4	110
5	111
6	101
7	100

the binary-code word for decimal integer i, then we must have

$$D(P_i, P_{i+1}) = 1 \quad i = 0, 1, 2, \ldots$$

Unit-distance codes are used in devices for converting analog or continuous signals such as voltages or shaft rotations into binary numbers which represent the magnitude of the signal. Such a device is called an *analog-digital converter* [9]. In any such device there must be boundaries between successive digits, and it is always possible for there to be some misalignment among the different bit positions at such a boundary. For example, if the seventh position is represented by 0111 and the eighth position by 1000, misalignment could cause signals corresponding to 1111 to be generated at the boundary between 7 and 8. If binary numbers were used for such a device, large errors could thus occur. By using a unit-distance code in which adjacent positions differ only in 1 bit, the error due to misalignment can be eliminated.

The highest integer to be encoded may or may not be required to be distance 1 from the code word for 0. If it is distance 1, then the path is closed. Of particular interest is the case of a closed nonintersecting path which goes through all 2^n points of the n-cube. In graph theory such a

Table 2.4-2. The Gray Code

Decimal	Binary				Gray			
	b_3	b_2	b_1	b_0	g_3	g_2	g_1	g_0
0	0	0	0	0	0	0	0	0
1	0	0	0	1	0	0	0	1
2	0	0	1	0	0	0	1	1
3	0	0	1	1	0	0	1	0
4	0	1	0	0	0	1	1	0
5	0	1	0	1	0	1	1	1
6	0	1	1	0	0	1	0	1
7	0	1	1	1	0	1	0	0
8	1	0	0	0	1	1	0	0
9	1	0	0	1	1	1	0	1
10	1	0	1	0	1	1	1	1
11	1	0	1	1	1	1	1	0
12	1	1	0	0	1	0	1	0
13	1	1	0	1	1	0	1	1
14	1	1	1	0	1	0	0	1
15	1	1	1	1	1	0	0	0

2.4 geometric representation of binary numbers

path is known as a (closed) *Hamilton line*. Any unit-distance code associated with such a path is sometimes called a *Gray code*, although this term is usually reserved for a particular one of these codes. To avoid confusing terminology, we shall refer to a unit-distance code which corresponds to a closed Hamilton line as a *closed n code*. This is a unit-distance code containing 2^n code words in which the code word for the largest integer ($2^n - 1$) is distance 1 from the code word for the least integer (0). An *open n code* is similar except that the code words for the least and largest integer, respectively, are not distance 1 apart.

The most useful unit distance code is the Gray code, which is shown in Table 2.4-2. The attractive feature of this code is the simplicity of the algorithm for translating from the binary number system into the Gray code. This algorithm is described by the expression

$$g_i = b_i \oplus b_{i+1}$$

Thus the Gray code word corresponding to 1100 in binary is formed as follows:

$$g_0 = b_0 \oplus b_1 = 0 \oplus 0 = 0$$
$$g_1 = b_1 \oplus b_2 = 0 \oplus 1 = 1$$
$$g_2 = b_2 \oplus b_3 = 1 \oplus 1 = 0$$
$$g_3 = b_3 \oplus b_4 = b_3 = 1 \quad b_4 \text{ understood to be 0}$$

Symmetries of the *n*-Cube

A *symmetry of the n-cube* is defined to be any one-to-one translation of the binary point representations on the n cube which leaves all pairwise distances the same. If we consider the set of binary numbers, we see that there are only two basic translation schemes which leave pairwise distances the same. (1) The bits of one coordinate may be interchanged with the bits of another coordinate in all code words. (2) The bits of one coordinate may be complemented (i.e., change 1's to 0's and 0's to 1's) in all code words. Since there are $n!$ permutations of n coordinates, there are $n!$ translation schemes possible using (1), and since there are 2^n ways in which coordinates may be complemented, there are 2^n translation schemes possible using (2). Thus, in all there are $2^n(n!)$ symmetries of the *n*-cube. This means that for any *n*-bit code there are $2^n(n!) - 1$ rather trivial modifications of the original code (in fact, some of these may result in the original code) which can be obtained by interchanging and complementing coordinates. The pairwise distances are the same in all these codes.

It is sometimes desired to enumerate the different types of a class of codes. Two codes are said to be of the same *type* if a symmetry of the *n*-cube translates one code into the other (i.e., by interchanging and

Table 2.4-3. Nine Different Types of Unit-distance 4-bit Code

Type	Coordinate-change sequence
1 (Gray)	0 1 0 2 0 1 0 3 0 1 0 2 0 1 0 3
2	1 0 1 3 1 0 1 2 0 1 0 3 0 1 0 2
3	1 0 1 3 0 1 0 2 1 0 1 3 0 1 0 2
4	1 0 1 3 2 3 1 0 1 3 1 0 2 0 1 3
5	1 0 1 3 2 0 1 3 1 0 1 3 2 0 1 3
6	1 0 1 3 2 3 1 3 2 0 1 2 1 3 1 2
7	1 0 1 3 2 0 2 1 0 2 0 3 0 1 0 2
8	1 0 1 3 2 1 2 0 1 2 1 3 0 1 0 2
9	1 0 1 3 2 3 1 0 3 0 2 0 1 2 3 2

complementing coordinates). As an example, we might ask: What are the types of closed n codes? It turns out that for $n < 4$ there is just one type, and this is the type of the conventional Gray code. For $n = 4$, there are nine types. Rather than specify a particular code of each type, we can list these types by specifying the sequence of coordinate changes for a closed path of that type. On the assumption that the coordinates are numbered (3210), the nine types are shown in Table 2.4-3.

Error-detecting and -correcting Codes

Another useful class of codes consists of those points of the n-cube with the property that the distance between any two points is 2. For example, the points 001, 010, 100, and 111 have this property. It is easily seen that an error in a single bit of any of these code words (points) gives a point which is not in the code. A code with this property is said to be *single-error-detecting*. Note, however, that given a binary representation due to a single error it is not possible to tell what the original code word was. But suppose that the code is such that the distance between any two points is 3. Then a single error in a bit will result in a point which is distance 1 from the original point but which is at least distance 2 from any other code point. Thus, seeing an incorrect point due to a single error, we can always deduce what the original code point was. A code with this property is said to be *single-error-correcting*. In general a code for which the minimum distance between any pair of code words is $2e + 1$ is capable of correcting e errors. Error-correcting codes have been covered extensively in the literature [10] and will not be discussed further here.

Problems

1. Convert:
   (a) $(523.1)_{10}$ to base 8
   (b) $(523.1)_{10}$ to base 2
   (c) $(101.11)_2$ to base 8
   (d) $(101.11)_2$ to base 10
   (e) $(1100.11)_2$ to base 7
   (f) $(101.11)_2$ to base 4
   (g) $(321.40)_6$ to base 7
   (h) $(\frac{25}{3})_{10}$ to base 2

2. In base 10 the highest number which can be obtained by multiplying together two single digits is $9 \times 9 = 81$, which can be expressed with two digits. What is the maximum number of digits required to express the product of two single digits in an arbitrary base-b system?

3. Given that $(79)_{10} = (142)_b$, determine the value of b.

4. Given that $(301)_b = (I^2)_b$, where I is an integer in base b and I^2 is its square, determine the value of b.

5. Let

$$N^* = (n_4 n_3 n_2 n_1 n_0)^* = 2 \cdot 3 \cdot 4 \cdot 5 \cdot n_4 + 3 \cdot 4 \cdot 5 \cdot n_3 + 4 \cdot 5 \cdot n_2 + 5 \cdot n_1 + n_0$$

$$= 120 n_4 + 60 n_3 + 20 n_2 + 5 n_1 + n_0$$

where

$$0 \le n_0 \le 4 \quad 0 \le n_1 \le 3 \quad 0 \le n_2 \le 2 \quad 0 \le n_3 \le 1$$
$$0 \le n_4 \le 1$$

with all the n_i positive integers.
   (a) Convert $(11111)^*$ to base 10.
   (b) Convert $(11234)^*$ to base 10.
   (c) Convert $(97)_{10}$ to its equivalent $(n_4 n_3 n_2 n_1 n_0)^*$.
   (d) Which decimal numbers can be expressed in the form $(n_4 n_3 n_2 n_1 n_0)^*$?

6. In order to write a number in base 16 the following symbols will be used for the numbers from 10 to 15:

   | 10 | t | 12 | w | 14 | u |
   | 11 | e | 13 | h | 15 | f |

   (a) Convert $(4tu)_{16}$ to base 10.
   (b) Convert $(2tfu)_{16}$ to base 2 directly (without first converting to base 10).

7. Convert $(1222)_3$ to base $5 (N)_5$, using only binary arithmetic:
   (a) Convert $(1222)_3$ to $(N)_2$.
   (b) Convert $(N)_2$ to $(N)_5$.

Number Systems and Codes

8. Perform the following binary-arithmetic operations:

    $11.10 + 10.11 + 111.00 + 110.11 + 001.01 = ?$
    $111.00 - 011.11 = ?$
    $011.11 - 111.00 = ?$
    $111.001 \times 1001.1 = ?$
    $101011.1 \div 1101.11 = ?$

9. Form the radix complement and the diminished radix complement for each of the following numbers:
    (a) $(.10111)_2$
    (b) $(.110011)_2$
    (c) $(0.5231)_{10}$
    (d) $(0.32499)_{10}$
    (e) $(0.3214)_6$
    (f) $(0.32456)_7$

10. (a) Write out the following weighted decimal codes:
    (i) $7,4,2,-1$
    (ii) $8,4,-2,-1$
    (iii) $4,4,1,-2$
    (iv) $7,5,3,-6$
    (v) $8,7,-4,-2$

    (b) Which codes of part a are self-complementing?
    (c) If a weighted binary-coded-decimal code is self-complementing, what necessary condition is placed on the sum of the weights?
    (d) Is the condition of part c sufficient to guarantee the self-complementing property? Give an example to justify your answer.

11. Write out the following weighted decimal codes: $(7,3,1,-2)$, $(8,4,-3,-2)$, $(6,2,2,1)$. Which of these, if any, are self-complementing?

12. Sketch a 4-cube, and label the points. List the points in the p-subcubes for $p = 2, 3$.

13. Compute all the pairwise distances for the points in a 3-cube. Arrange these in a matrix form where the rows and columns are numbered 0, 1, . . . , 7, corresponding to the points of the 3-cube. The 0-, 1-, and 2-cube pairwise distances are given by submatrices of this matrix. By observing the relationship between these matrices, what is a scheme for going from the n-cube pairwise-distance matrix to the $(n + 1)$-cube pairwise-distance matrix?

14. What is a scheme for going from the Gray code to the ordinary binary code using the mod 2 operation only?

15. For the Gray code, a weighting scheme exists in which the weights associated with the bits are constant except for sign. The signs alternate with the occurrence of 1's, left to right. What is the weighting scheme?

16. List the symmetries of the 2-cube.

17. Write out a typical type-6 closed-unit-distance 4 code (Table 2.4-3).

18. Write out two open unit-distance 4 codes of different type (i.e., one is not a symmetry of the other).

19. Write out a set of six code words which have the single-error-detecting and single-error-correcting property.
20. A closed error-detecting unit-distance code is defined as follows: There are k ($k < 2^n$) ordered binary n-bit code words with the property that changing a single bit in any word will change the original word into either its predecessor or its successor in the list (the first word is considered the successor for the last word) or into some other n-bit word *not* in the code. Changing a single bit cannot transform a code word into any code word other than its predecessor or successor.
    (a) You are to determine a 4-bit closed error-detecting unit-distance code having eight code words ($k = 8$). List the code words for such a code.
    (b) List the code words for such a code with $k = 6$, $n = 3$. Is there more than one symmetry type of code for these specifications? Why?

REFERENCES

1. Chrystal, G.: "Algebra; an Elementary Text-book," pt. I, Dover Publications, Inc., New York, 1961.
2. Stein, M. L., and W. D. Munro: "Computer Programming: A Mixed Language Approach," Academic Press Inc., New York, 1964.
3. Ware, W. H.: "Digital Computer Technology and Design," vol. I, Mathematical Topics, Principles of Operation, and Programming, John Wiley & Sons, Inc., New York, 1963.
4. Trakhtenbrot, B. A.: "Algorithms and Automatic Computing Machines," Topics in Mathematics, D. C. Heath and Company, Boston, 1963.
5. Davis, Martin: "Computability and Unsolvability," McGraw-Hill Book Company, New York, 1958.
6. Richards, R. K.: "Arithmetic Operations in Digital Computers," D. Van Nostrand Company, Inc., Princeton, N.J., 1955.
7. Flores, I.: "The Logic of Computer Arithmetic," Prentice-Hall, Inc., Englewood Cliffs, N.J., 1963.
8. White, Garland S.: Coded Decimal Number Systems for Digital Computers, *Proc. IRE*, vol. 41, no. 10, pp. 1450–1452, October, 1953.
9. Susskind, Alfred K. (ed.): "Notes on Analog-Digital Conversion Techniques," The Technology Press of the Massachusetts Institute of Technology, Cambridge, Mass., 1957.
10. Peterson, W. W.: "Error-correcting Codes," The M.I.T. Press, Cambridge, Mass., John Wiley & Sons, Inc., New York, 1961.

3 SWITCHING ALGEBRA

It has already been pointed out that the distinguishing feature of the circuits to be discussed here is the use of two-valued, or binary, signals. There will be some deviation of the signals from their nominal values, but within certain limits this variation will not affect the performance of the circuit. If the variations exceed these limits, the circuit will not behave properly and steps must be taken to confine the signals to the proper ranges. When the statement is made that the signals are two-valued, what is really meant is that the value of each signal is within one of two (nonoverlapping) continuous ranges. Since the operation of the circuit does not depend on exactly which value within a given range the signal takes on, a particular value is chosen to represent the range and the signal is said to be equal to this value.

The exact numerical value of the signal is not important. It is possible to have two circuits perform the same function and have completely different values for their signals. In order to avoid any possible confusion which might arise because of this situation and to simplify the design procedures, it is customary to carry out the *logical* design without specifying the actual values of the signals. Once the logical design has been completed, actual values must be assigned to the signals in the course of designing the detailed electrical circuit. For the purposes of the logical design, *arbitrary* symbols are chosen to represent the two values to which the signals are to be restricted. An algebra† using these symbols is then developed as the basis for formal design techniques. The development of such an algebra is the subject of this chapter.

---

† This algebra will here be called *switching algebra*. It is identical with a Boolean algebra and was originally applied to switching circuits [1] by reinterpreting Boolean algebra in terms of switching circuits rather than by developing a switching algebra directly, as will be done here.

3.1 POSTULATES

The two symbols most commonly chosen to represent the two values taken on by binary signals are 0 and 1. It should be emphasized that there is no numerical significance attached to these symbols. For an electronic circuit which has its signals equal to either 0 or -15 volts it is perfectly appropriate to assign the symbol 1 to 0 volts and the symbol 0 to -15 volts. On the other hand, there would be nothing wrong in assigning the symbol 0 to 0 volts and 1 to -15 volts. Some other set of symbols such as H and L or $+$ and $-$ could be used instead of 0 and 1, but there is a strong tradition behind the use of 0 and 1.

It is next necessary to associate with each binary signal a switching variable which will be equal to either 0 or 1 depending on the actual value of the binary signal. Each relay will have associated with it a variable whose value will depend on whether or not the relay is operated. This assignment can be made in two possible ways, as shown in the following table, in which the variable X is associated with the relay. It is customary to use the variable associated with the relay also as a name, such as X relay, Y relay, etc. Either scheme for assigning values to X is valid, and the two schemes are equally useful. The early writing on this subject [2,3] used the hindrance concept, but the transmission concept is now universally used and will be the assignment employed here.

X relay	Hindrance concept	Transmission concept
Unoperated	$X = 1$	$X = 0$
Operated	$X = 0$	$X = 1$

For electronic circuits a switching variable is associated with each voltage (or current) signal. Frequently the higher voltage is assigned the symbol 1, but there are times when the other assignment is more useful.

The first postulate of the switching algebra can now be presented. This is merely a formal statement of the fact that switching variables are always equal to either 0 or 1. In the statements of postulates and theorems which follow, the symbols $X, Y, Z, X_1, X_2, \ldots, X_n$ will be used to represent switching variables.

(P1) $X = 0$ if $X \neq 1$ (P1') $X = 1$ if $X \neq 0$

Fig. 3.1-1. *Series control of a relay.*

It is necessary to associate a switching variable with each relay contact as well as with the relay itself. This is done by using the same variable for both the relay and its normally open (make) contacts. When the relay is unoperated, the variable representing a make contact is equal to 0 and there is an open circuit between the contact's terminals. When the relay is operated, the make-contact variable equals 1 and the contact's terminals present a closed circuit. Thus in reference to contacts, *a 0 represents an open circuit and a 1 represents a closed circuit*. This is consistent with using a 0 to represent an unoperated relay and a 1 to represent an operated relay because of the method ordinarily used to control the operation of a relay. This method, called *series control*, is shown in Fig. 3.1-1. When the network shown in the box presents a closed circuit between points a and b, ground is connected to point b and current flows through the relay, causing it to operate. Thus a closed circuit in the network corresponds to the relay being operated. Similarly an open circuit in the control network corresponds to the relay being unoperated.

The switching variable associated with a contact is called the *transmission* of the contact. A switching variable, called the *network transmission*, is also commonly associated with a two-terminal network of contacts. A transmission of 1 corresponds to a closed circuit, and a transmission of 0 corresponds to an open circuit.

A switching variable must also be associated with the normally closed, or break, contacts. Since a break contact is closed when a make contact on the same relay is open, and vice versa, the transmission of the break contact is 1 when the make-contact transmission is 0, and vice versa. In order to take this relation into account, the symbol x' is used for the transmission of a break contact on relay X, with the understanding that $x' = 0$ when $x = 1$ and $x' = 1$ when $x = 0$. Postulate 2 is a formal statement of this convention,

(P2)   If $X = 0$, then $X' = 1$ (P2')   If $X = 1$, then $X' = 0$

The two symbols X and X' are not two different variables, since they involve only X. In order to distinguish them the term literal is used, where a *literal* is defined as a variable with or without an associated prime and X and X' are different literals. The literal X' is called the complement† of X, 0 is called the complement of 1, and 1 is called the complement of 0.

---

† Some authors use the symbols $\bar{x}$ or $\sim x$, rather than x', to indicate the complement of x.

3.1 postulates

The logical operation of the inverter can now be described in terms of switching algebra. If x_1 represents the input signal and x_0 the output signal, then $x_0 = x_1'$, since x_0 is high when x_1 is low, and vice versa.

In order to represent the action of the other switching devices described in Chap. 1 it is necessary to define additional algebraic operations. A two-input AND gate will be considered first. If the two inputs are represented by x_1 and x_2 and the output by x_0, the logical performance of the circuit is represented by the accompanying table, in which 1 corresponds to a high voltage and 0 corresponds to a low voltage. This table

	x_1	
x_2	0	1
0	0	0
1	0	1

$x_0 = x_1 x_2$

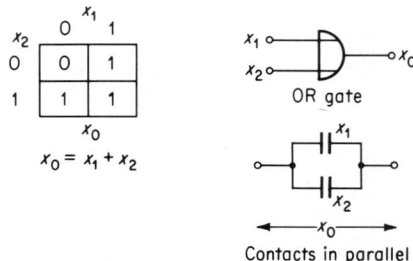

AND gate

Contacts in series

is also correct for the transmission x_0 of two contacts x_1 and x_2 in series. The operation represented is identical with ordinary multiplication and will be defined as multiplication in the switching algebra developed here. Thus, the equation for an AND gate, or two contacts in series, is $x_0 = x_1 x_2$.

The table for an OR gate, or two contacts in parallel, is shown herewith.

	x_1	
x_2	0	1
0	0	1
1	1	1

$x_0 = x_1 + x_2$

OR gate

Contacts in parallel

This table will be taken as the definition of switching-algebra addition.† It is identical with ordinary addition except for the case $1 + 1 = 1$. The remaining postulates are merely restatements of the definitions of multiplication and addition.

(P3) $0 \cdot 0 = 0$ (P3') $1 + 1 = 1$
(P4) $1 \cdot 1 = 1$ (P4') $0 + 0 = 0$
(P5) $1 \cdot 0 = 0 \cdot 1 = 0$ (P5') $0 + 1 = 1 + 0 = 1$

† This operation is also called *logical addition*, and some writers use the symbol $x_1 \vee x_2$ rather than $x_1 + x_2$.

Switching Algebra

All the postulates have been stated in pairs. These pairs have the property that when (0 and 1) and (+ and ·) are interchanged, one member of the pair is changed into the other. This is an example of the general *principle of duality*, which is true for all switching-algebra theorems since it is true for all the postulates [4, pp. 8, 28; 5, p. 7]. This algebraic duality arises from the fact that either 1 or 0 can be assigned to a high voltage and either 1 or 0 can be assigned to an open circuit. If 0 is chosen to represent a high voltage, the expression for an AND gate becomes $x_0 = x_1 + x_2$. Similarly, the transmission of two contacts in series is equal to $x_1 + x_2$ if 0 is chosen to represent a closed circuit.

3.2 ANALYSIS OF SWITCHING CIRCUITS

Analysis of a circuit consists in examining the circuit and somehow determining what the behavior of the circuit will be for all possible inputs. The switching algebra as developed in Sec. 3.1 is sufficient for the analysis of those switching circuits which are known as *combinational switching circuits*. A *combinational switching circuit* is defined as one for which the outputs depend only on the present inputs to the circuit. The other type of circuit, called a *sequential switching circuit*, is one for which the outputs depend not only on the present inputs but also on the past history of inputs. Networks of relay contacts are combinational circuits. An example of a sequential circuit is one which contains a relay which can "lock up," i.e., remain operated under control of one of its own contacts after the input which caused it to operate has been removed.

The analysis of a combinational circuit consists in writing an algebraic function for each output. These are functions of the input variables from which the condition of each output can be determined for each combination of input conditions. For a contact network these output functions are the transmissions between each pair of external terminals. Only two-terminal networks will be considered for the present. In an electronic network, an output function specifies those input conditions for which the voltage of an output node will be at the high level.

Contact Networks

It has already been shown that the transmission of two contacts in series is x_1x_2 and the transmission of two contacts in parallel is $x_1 + x_2$. Similarly it is true that the transmission of two networks, T_1 and T_2, in series is T_1T_2 and the transmission of two networks in parallel is $T_1 + T_2$. By making use of this fact, it is possible to write down the transmission of any planar nonbridge (series-parallel) network directly. Any series-parallel network can be split into either two networks in series (Fig. 3.2-1a) or two networks in parallel (Fig. 3.2-1b). If the transmissions of

3.2 analysis of switching circuits

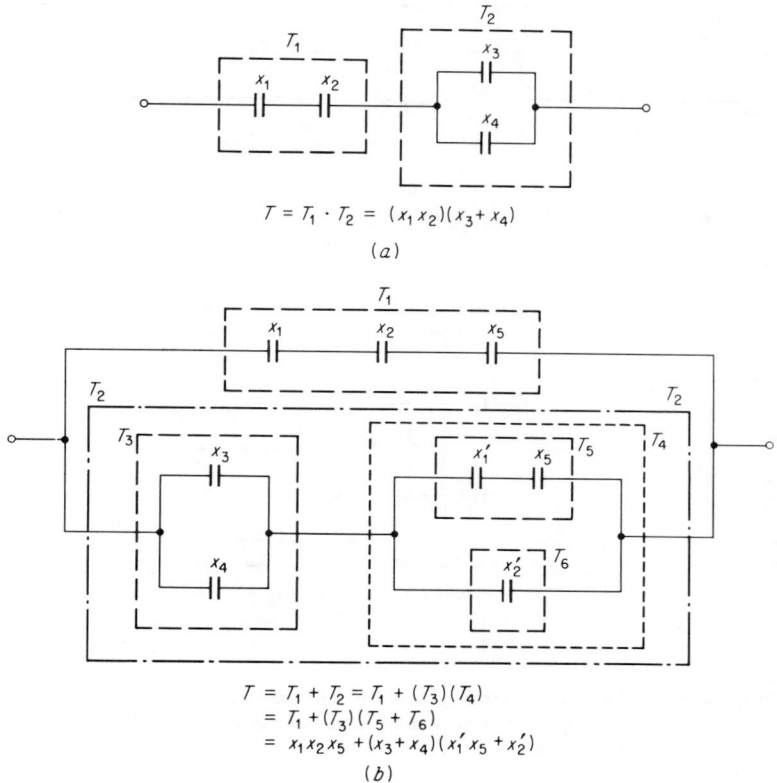

Fig. 3.2-1. *Analysis of series-parallel contact networks.* (a) *Two networks in series;* (b) *two networks in parallel.*

the two subnetworks are T_1 and T_2, the transmission of the original network is $T_1 T_2$ or $T_1 + T_2$. Each of the subnetworks can again be split into two subnetworks, etc. This process of splitting into subnetworks is continued until each subnetwork consists only of contacts in series or contacts in parallel. The transmission for each subnetwork can next be written directly, and then the transmission of the original network can be obtained by combining the transmissions of the subnetworks. It is now necessary to continue splitting networks until there are only two contacts in each subnetwork, for it can be shown very easily that the transmission of n contacts in parallel is $x_1 + x_2 + x_3 + \cdots + x_n$ and the transmission of n contacts in series is $x_1 x_2 x_3 \cdots x_n$. This entire procedure is illustrated in Fig. 3.2-1.

The procedure just described for analyzing contact networks is ambiguous in the sense that there is no specification of the order in which the transmission of two parallel networks is to be written, for example,

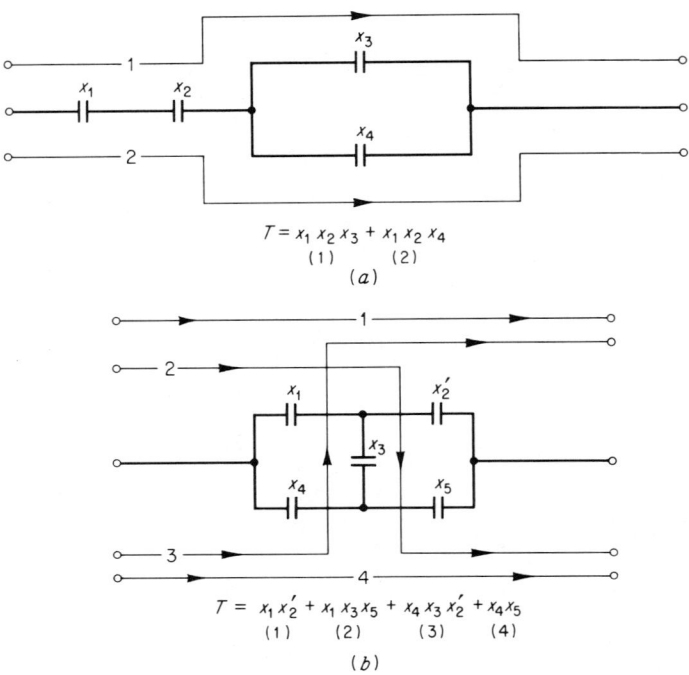

Fig. 3.2-2. *Determination of overall transmissions by path tracing.* (a) *Series-parallel network;* (b) *bridge network.*

$T_1 + T_2$ or $T_2 + T_1$. This is tolerable only because of the following pair of theorems, which show the order of such terms to be unimportant:

(T6) $X + Y = Y + X$ (T6') $XY = YX$ (Commutative law)

This theorem states that the two operations of addition and multiplication as defined here are both commutative. The numbering of the theorems follows that given in Sec. 3.4, in which the method of proof is discussed and the theorems are listed. For the present it is sufficient to note that, since X and Y each can have only two possible values (0 and 1), these theorems can be verified by trying each of the four possible combinations of values for X and Y.

This method is not applicable to non-series-parallel networks. Another method, which works for all networks, consists in tracing all paths from input to output, writing for each path a transmission equal to the product of the labels of all contacts occurring in the path, and then forming the overall transmission as the sum of the individual path transmissions. This is illustrated in Fig. 3.2-2. The overall transmission

3.2 analysis of switching circuits

obtained by this method will be a sum of products of literals. The two external terminals are connected together only when all contacts of at least one of the paths through the network are closed. The overall transmission can equal 1 only when at least one of the product terms is equal to 1, that is, when all literals of at least one product term are equal to 1.

An alternative method of describing the network performance is to list each set of literals which corresponds to a path between the two external terminals of the network. Each such set of literals is called a P set, and the P sets for Fig. 3.2-2b are

$x_1 x_2'$
$x_1 x_3 x_5$
$x_4 x_3 x_2'$
$x_4 x_5$

Definition. A set of literals is a P set of a network if and only if (1) there is a closed path between the external terminals of the network whenever all the contacts corresponding to literals of the set are closed (all literals are equal to 1); and (2) when any literal is removed condition 1 no longer holds.

The second part of this definition is included to rule out paths which are not proper paths in the sense that they include other paths which also connect the external terminals and include fewer contacts.

It is also possible to describe the network by listing the ways in which the network terminals can be cut apart. This is done by tracing "cut paths," paths through contacts which separate the external terminals, as in Fig. 3.2-2b. Each set of literals occurring in such a cut path is called an S set; those for Fig. 3.2-2b are

$x_1 x_4$
$x_1 x_3 x_5$
$x_2' x_3 x_4$
$x_2' x_5$

Definition. A set of literals is an S set of a network if and only if (1) there is no closed path between the external terminals of the network whenever all the contacts corresponding to literals of the S set are open; and (2) when any literal is removed from the S set condition 1 no longer holds.

An expression for the network transmission can be written down directly from the S sets. Corresponding to each S set a sum of literals in the S set is formed. The network transmission is equal to the product of these sums.

AND Gates and OR Gates

Networks made up of AND and OR gates can be analyzed directly. There will always be some gates which have only inputs which are also circuit inputs. The first step is to label the outputs of these gates with the proper function—step 1 in Fig. 3.2-3. There will now be some gates all of whose inputs are either circuit inputs or leads labeled in step 1. The outputs of these gates are now labeled, treating the functions obtained in step 1 in the same manner as individual circuit inputs. This is step 2 in Fig. 3.2-3. This process is continued until finally the output lead is labeled. This label on the output lead is the output function.

Gate networks can also be described by means of P sets and S sets. A P set is a set of literals such that, whenever all the inputs corresponding to these literals equal 1, the network output will equal 1. For Fig. 3.2-3, the P sets are

$$
\begin{array}{ll}
(a) & (b) \\
x_4 x_2 x_1 & x_1 x_2 x_5 \\
x_3 x_2 x_1 & x_3 x_2' \\
 & x_3 x_1' x_5 \\
 & x_4 x_2' \\
 & x_4 x_1' x_5
\end{array}
$$

These can be determined by inspection of the circuit diagram. Algebraic techniques will be discussed in Chap. 7. An S set is a set of literals such

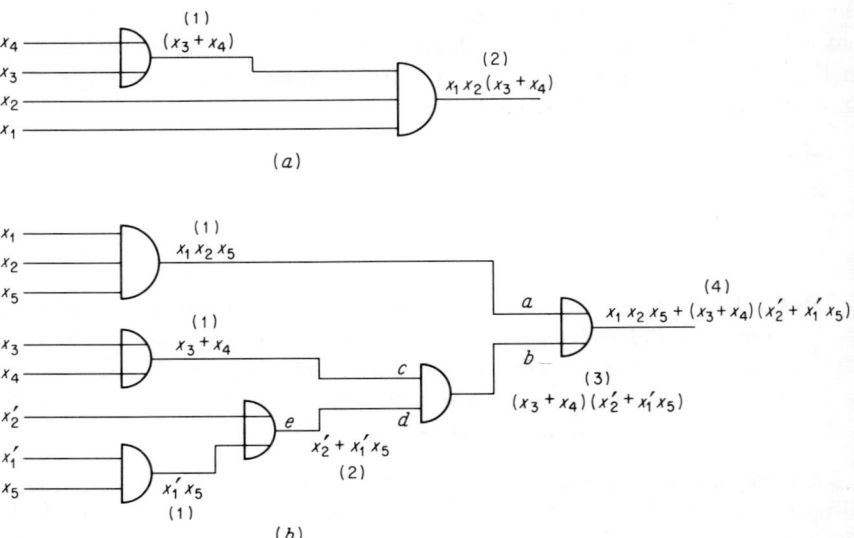

Fig. 3.2-3. *Analysis of gate networks.*

3.2 analysis of switching circuits

that the network output will be zero whenever all the inputs corresponding to these literals equal zero. For Fig. 3.2-2a the S sets are

x_1
x_2
$x_3 x_4$

Table of Combinations

The procedure for using these output functions to determine the circuit performance is to substitute a set of values for the input variables and then use the postulates to simplify the resulting expression. A value will be obtained for the output variable which specifies the output for the particular input combination chosen. For example, if the input combination $x_1 = 0$ ($x_1' = 1$), $x_2 = 1$ ($x_2' = 0$), $x_3 = 0$, $x_4 = 1$, $x_5 = 1$ is chosen, the transmission of Fig. 3.2-1b becomes

$$T = x_1 x_2 x_5 + (x_3 + x_4)(x_2' + x_1' x_5)$$
$$= 0 \cdot 1 \cdot 1 + (0 + 1)(0 + 1 \cdot 1) = 0 + (1)(1) = 0 + 1 = 1$$

By carrying out this procedure for all possible input combinations it is possible to form a table which lists the output for each input combination.

Table 3.2-1. *Table of Combinations for* $T = x_1 x_2 (x_3 + x_4)$

	x_1	x_2	x_3	x_4	T
0	0	0	0	0	0
1	0	0	0	1	0
2	0	0	1	0	0
3	0	0	1	1	0
4	0	1	0	0	0
5	0	1	0	1	0
6	0	1	1	0	0
7	0	1	1	1	0
8	1	0	0	0	0
9	1	0	0	1	0
10	1	0	1	0	0
11	1	0	1	1	0
12	1	1	0	0	0
13	1	1	0	1	1
14	1	1	1	0	1
15	1	1	1	1	1

Such a table describes completely the circuit performance and is called a *table of combinations*.† Table 3.2-1 is the table of combinations for the circuits of Figs. 3.2-1a and 3.2-3a.

3.3 SYNTHESIS

In designing a combinational circuit it is necessary to carry out in reverse the procedure just described. The desired circuit performance is specified by means of a table of combinations. From this table an algebraic function is formed, and finally the circuit is derived from the function. A concise means of specifying the table of combinations, called a *decimal specification*, is to list the numbers of the rows for which the output is to equal 1. For Table 3.2-1 this specification is

$$T(x_1,x_2,x_3,x_4) = \Sigma(13,14,15)$$

where the Σ signifies that the rows for which the function equals 1 are being listed. It is also possible to list the rows for which the function equals 0, such a list being preceded by the symbol Π to indicate that it is the zero rows which are listed. This specification for the preceding table is $T(x_1,x_2,x_3,x_4) = \Pi(0,1,2,3,4,5,6,7,8,9,10,11,12)$. In order to avoid any ambiguity in these specifications, it is necessary to adopt some rule for numbering the rows of the table of combinations. The usual procedure is to regard each row of the table as a binary number and then use the decimal equivalent of this binary number as the row number. The output-column entries are not included in forming the binary row numbers. There is nothing special about decimal numbers other than the fact that they are the most familiar—any other number base such as octal could be used. The reason for using a number system other than binary is simply that binary numbers take too much space to write down.

It has been pointed out that the table of combinations is a complete specification for a combinational circuit. The first step in designing a circuit is to formulate such a table. There are no general formal techniques for doing this. When a sequential circuit is being designed, it is customary to reduce the sequential-design problem to (several) combinational problems, and formal techniques exist for doing this. However, when a combinational circuit is being designed, no formal techniques are available and it is necessary to rely on common sense. This is not too surprising, since any formal technique must start with a formal statement of the problem, and this is precisely what the table of combinations is. As an example of how this is done, the table of combinations for a circuit to check binary-coded-decimal digits is shown below for the 8, 4, 2, 1

† In logic the table is called a *truth table*, and some writers use this term when discussing switching circuits.

Table 3.3-1. Table of Combinations for Circuit to Check Binary-coded-decimal Digits

	b_8	b_4	b_2	b_1	T	
0	0	0	0	0	0	
1	0	0	0	1	0	
2	0	0	1	0	0	
3	0	0	1	1	0	
4	0	1	0	0	0	
5	0	1	0	1	0	
6	0	1	1	0	0	
7	0	1	1	1	0	
8	1	0	0	0	0	
9	1	0	0	1	0	
10	1	0	1	0	1	
11	1	0	1	1	1	Invalid
12	1	1	0	0	1	code
13	1	1	0	1	1	words
14	1	1	1	0	1	
15	1	1	1	1	1	

code. This circuit is to deliver an output whenever a digit having an invalid combination of bits is received.

In forming a table of combinations there very often are rows for which it is unimportant whether the function equals 0 or 1. The usual reason for this situation is that the combination of inputs corresponding to these rows can never occur (when the circuit is functioning properly). As an example of this consider a circuit to translate from the 8, 4, 2, 1 BCD code to a Gray (cyclic binary) code. When the circuit is working correctly, the input combinations represented by rows 10 through 15 of the table of combinations cannot occur. Therefore, the output need not be specified for these rows. The symbol d will be used to indicate the output condition for such a situation† (see Table 3.3-2). It is possible to include the d rows in the decimal specification of a function by listing them after the symbol d. Thus the decimal specification for g_1 of Table 3.3-2 would be

$$g_1(b_8,b_4,b_2,b_1) = \Sigma(1,2,5,6,9) + d(10,11,12,13,14,15)$$
$$= \Pi(0,3,4,7,8) + d(10,11,12,13,14,15)$$

† In the literature, the symbol ϕ is also used, and the output conditions so denoted are called *don't-care conditions*.

Table 3.3-2. Table of Combinations for Circuit to Translate from BCD 8,4,2,1 Code to Gray Code

	BCD-code inputs b_8 b_4 b_2 b_1	Gray-code outputs g_4 g_3 g_2 g_1
0	0 0 0 0	0 0 0 0
1	0 0 0 1	0 0 0 1
2	0 0 1 0	0 0 1 1
3	0 0 1 1	0 0 1 0
4	0 1 0 0	0 1 1 0
5	0 1 0 1	0 1 1 1
6	0 1 1 0	0 1 0 1
7	0 1 1 1	0 1 0 0
8	1 0 0 0	1 1 0 0
9	1 0 0 1	1 1 0 1
10	1 0 1 0	d d d d
11	1 0 1 1	d d d d
12	1 1 0 0	d d d d
13	1 1 0 1	d d d d
14	1 1 1 0	d d d d
15	1 1 1 1	d d d d

The handling of d rows will be discussed further in Chap. 4. For the present it will be assumed that the output specified for each row of the table of combinations is either 0 or 1.

Canonical Expressions

After forming the table of combinations, the next step in designing a circuit is to write an algebraic expression for the output function. The simplest output functions to write are those which equal 1 for only one row of the table of combinations or those which equal 0 for only one row. It is possible to associate with each row two functions—one which equals 1 only for the row and one which equals 0 only for the row (Table 3.3-3). These functions are called *fundamental products* and *fundamental sums*, respectively. Each fundamental product or sum contains all the input variables. The rule for forming the fundamental product for a given row is to prime any variables which equal 0 for the row and leave unprimed any variables which equal 1 for the row. The fundamental product equals the product of the literals so formed. The fundamental sum is formed by a completely reverse or dual procedure. Each variable

Table 3.3-3. Table of Combinations Showing Fundamental Products and Fundamental Sums

	x_1	x_2	x_3	Fundamental products	Fundamental sums
0	0	0	0	$x_1'x_2'x_3'$	$x_1 + x_2 + x_3$
1	0	0	1	$x_1'x_2'x_3$	$x_1 + x_2 + x_3'$
2	0	1	0	$x_1'x_2x_3'$	$x_1 + x_2' + x_3$
3	0	1	1	$x_1'x_2x_3$	$x_1 + x_2' + x_3'$
4	1	0	0	$x_1x_2'x_3'$	$x_1' + x_2 + x_3$
5	1	0	1	$x_1x_2'x_3$	$x_1' + x_2 + x_3'$
6	1	1	0	$x_1x_2x_3'$	$x_1' + x_2' + x_3$
7	1	1	1	$x_1x_2x_3$	$x_1' + x_2' + x_3'$

which equals 0 for the row is left unprimed, and each variable which equals 1 for the row is primed. The fundamental sum is the sum of the literals obtained by this process. The algebraic expression for any table for which the output is equal to 1 (or 0) for only one row can be written down directly by choosing the proper fundamental product (or sum). For example, the output function specified by $f(x_1,x_2,x_3) = \Sigma(6)$ is written algebraically as $f = x_1x_2x_3'$, and the output function $f(x_1,x_2,x_3) = \Pi(6)$ is written as $f = x_1' + x_2' + x_3$. The fundamental product corresponding to row i of the table of combinations will be denoted by p_i, and the fundamental sum corresponding to row i will be denoted by s_i.

The algebraic expression which equals 1 (or 0) for more than one row of the table of combinations can be written directly as a sum of fundamental products or as a product of fundamental sums. A function f which equals 1 for two rows, i and j, of the table of combinations can be expressed as a sum of the two fundamental products p_i and p_j: $f = p_i + p_j$. When the inputs correspond to row i, $p_i = 1$ and $p_j = 0$ so that $f = 1 + 0 = 1$. When the inputs correspond to row j, $p_i = 0$, $p_j = 1$, and $f = 0 + 1 = 1$. When the inputs correspond to any other row, $p_i = 0$, $p_j = 0$, and $f = 0 + 0 = 0$. This shows that the function $f = p_i + p_j$ does equal 1 only for rows i and j. This argument can be extended to output functions which equal 1 for any number of input combinations—they can be represented algebraically as a sum of the corresponding fundamental products (see Table 3.3-4). An algebraic expression which is a sum of fundamental products is called a *canonical sum*. An arbitrary function can also be expressed as a product of fundamental sums. This form is called the *canonical product*. The canonical product for a function which is equal to 0 only for rows i and j of the table of com-

Table 3.3-4. $f(x_1,x_2,x_3) = \Sigma(1,2,3,4)$

(a) Table of combinations

	x_1	x_2	x_3	f
0	0	0	0	0
1	0	0	1	1
2	0	1	0	1
3	0	1	1	1
4	1	0	0	1
5	1	0	1	0
6	1	1	0	0
7	1	1	1	0

(b) Canonical sum
$$f = x_1'x_2'x_3 + x_1'x_2x_3' + x_1'x_2x_3 + x_1x_2'x_3'$$

(c) Canonical product
$$f = (x_1 + x_2 + x_3)(x_1' + x_2 + x_3')$$
$$(x_1' + x_2' + x_3)(x_1' + x_2' + x_3')$$

binations is given by $f = s_i \cdot s_j$. For row i, $s_i = 0$ so that $f = 0$. For row j, $s_j = 0$ so that $f = 0$, and for any other row $s_i = s_j = 1$ so that $f = 1$. In general, the canonical product is equal to the product of all fundamental sums which correspond to input conditions for which the function is to equal 0.

It is possible to write a general expression for the canonical sum by making use of the following theorems:

(T2') $\quad x \cdot 0 = 0$
(T1) $\quad x + 0 = x \quad$ (T1') $\quad x \cdot 1 = x$

If the value of the function $f(x_1,x_2,\ldots,x_n)$ for the ith row of the table of combinations is f_i ($f_i = 0$ or 1), then the canonical sum is given by

$$f(x_1,x_2,\ldots,x_n) = f_0 p_0 + f_1 p_1 + \cdots + f_{2^n-1} p_{2^n-1}$$
$$= \sum_{i=0}^{2^n-1} f_i p_i$$

For the function $f(x_1,x_2) = \Sigma(0,2)$ the values of the f_i are $f_0 = f_2 = 1$, $f_1 = f_3 = 0$ so that

$$\begin{align}f(x_1,x_2) &= 1 \cdot p_0 + 0 \cdot p_1 + 1 \cdot p_2 + 0 \cdot p_3 \\ &= p_0 + 0 + p_2 + 0 \\ &= p_0 + p_2 \\ &= x_1'x_2' + x_1 x_2'\end{align}$$

In a similar fashion a general expression for the canonical product can be obtained by using the theorems

(T2) $x + 1 = 1$
(T1) $x + 0 = x$ (T1') $x \cdot 1 = x$

The resulting expression is

$$f(x_1, x_2, \ldots, x_n) = (f_0 + s_0)(f_1 + s_1) \cdots (f_{2^n-1} + s_{2^n-1})$$
$$= \prod_{i=0}^{2^n-1} (f_i + s_i)$$

For $f(x_1, x_2) = \Sigma(0,2) = \Pi(1,3)$ this becomes

$$f(x_1, x_2) = (1 + s_0)(0 + s_1)(1 + s_2)(0 + s_3)$$
$$= 1 \cdot s_1 \cdot 1 \cdot s_3$$
$$= s_1 \cdot s_3$$
$$= (x_1 + x_2')(x_1' + x_2')$$

Networks

A technique for obtaining an algebraic expression from a table of combinations has just been described. A circuit can be drawn directly from this expression by reversing the procedures used to analyze series-parallel contact networks or gate networks. The circuit for a single fundamental product is just a series connection of contacts or an AND gate

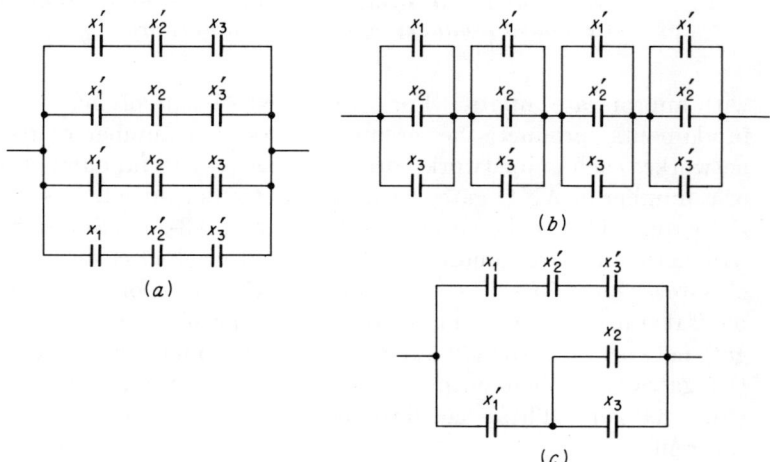

Fig. 3.3-1.   Contact networks for $\Sigma(1,2,3,4)$. (a) Network derived from canonical sum; (b) network derived from canonical product; (c) economical network.

Switching Algebra

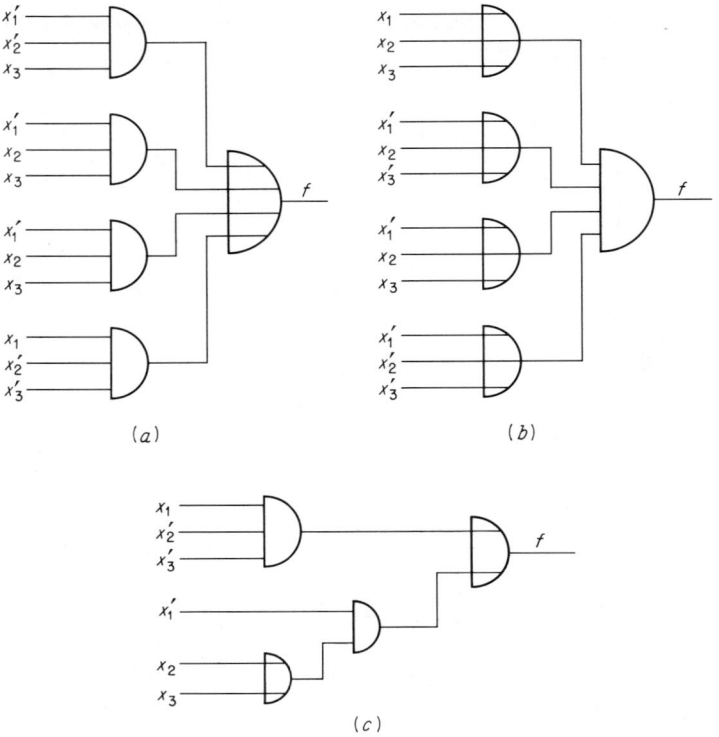

Fig. 3.3-2. *Gate networks for* $\Sigma(1,2,3,4)$. *(a) Network derived from canonical sum; (b) network derived from canonical product; (c) economical network.*

with appropriate inputs. For a canonical sum involving more than one fundamental product, the circuit consists of a number of parallel subnetworks, each subnetwork corresponding to one fundamental product, or a number of AND gates with their outputs connected as inputs to an OR gate. This is shown in Figs. 3.3-1a and 3.3-2a. Similarly, the contact network corresponding to a canonical product consists of a number of subnetworks in series, each subnetwork corresponding to one fundamental sum and consisting of contacts in parallel (see Fig. 3.3-1b). The gate network corresponding to a canonical product consists of a number of OR gates with their outputs connected as the inputs of an AND gate (Fig. 3.3-2b). These conclusions are summarized in the following theorem:

Theorem 3.3-1. For any arbitrary table of combinations, a network whose performance corresponds to the table of combinations can be constructed of:

1. *Relay contacts* if a sufficient number of make and break contacts are available on each relay
2. *AND gates, OR gates*, and *inverters* if there are no restrictions on the number of gates available

In a certain sense our design procedure is now complete. A method has been presented for going from a table of combinations to a circuit diagram. However, the canonical circuits so designed are usually very uneconomical and therefore unsatisfactory. An example of this can be seen by comparing the circuits of Figs. 3.3-1a and b and 3.3-2a and b with those of Figs. 3.3-1c and 3.3-2c. In order to design satisfactory circuits, it is necessary to have procedures for simplifying them so that they correspond to simpler circuits. This is discussed in Chap. 4.

Number of Functions

An alternative approach to practical circuit design would be to make use of a table which listed economical circuits for all switching functions with fewer than some specified number of inputs. All circuits which did not have too many inputs could then be designed merely by looking them up in the table. Theoretically, there is no reason why this could not be done, since for any given number of input variables there are only a finite number of switching functions possible. In fact the number of switching functions of n input variables is exactly 2^{2^n}. There are 2^n rows in an n-input table of combinations, and since the output for each row can be either 0 or 1, there are 2^{2^n} different output functions. Table 3.3-5 lists 2^{2^n} for values of n from 1 to 5. It is clear from this table that it would be impractical to list functions of more than three variables. The situation is not quite so bad as the listed values of 2^{2^n} would seem to indicate, since many of the functions counted differ only by an interchange of input variables. For example, the two functions $f_1 = x_1 + x_2x_3$ and $f_2 = x_2 + x_1x_3$ are different, but one can be converted into the other by interchanging the variables x_1

Table 3.3-5. Numbers of Functions

n	2^n	2^{2^n}	N_n†
1	2	4	3
2	4	16	6
3	8	256	22
4	16	65,536	402
5	32	4,294,967,296	1,228,158

† N_n = number of types of functions of n variables.

and x_2. Interchanging two variables corresponds to relabeling two leads, and therefore the same circuit can be used for both functions f_1 and f_2 by merely changing the labels on two leads. Similarly the functions $f_1 = x_1 + x_2 x_3$ and $f_3 = x_1' + x_2 x_3$ can be interchanged by interchanging x_1 and x_1'. In a relay circuit this corresponds to interchanging make-and-break contacts on the X_1 relay, and in an electronic circuit it again corresponds to relabeling leads. When one function can be changed into another by permuting and/or complementing input variables, the two functions are said to be of the same *type*. Clearly a table of standard circuits need list circuits for only one function of each type. At first glance it might seem that the number of types of n-input functions would be $2^{2^n} \div n!2^n$ since these are $n!$ permutations and 2^n different ways of complementing variables. This is not correct, for permutating and/or complementing variables does not necessarily convert a given function into a different function. For example, $f_1 = x_1 + x_2 x_3$ is left unchanged when the variables x_2 and x_3 are interchanged. The number of types of functions for n input variables have been computed [6] and are listed as N_n in Table 3.3-6. If only types of functions are listed, it is practical to catalogue circuits for functions of four variables. This has actually been done for contact networks [7] and for electronic tube circuits [8]. Since circuits having more than five input variables are often required, it is necessary to develop a general design procedure. This will be the subject of Chap. 4.

3.4 THEOREMS

The method of using switching algebra in designing switching circuits is to formulate the desired circuit performance as an algebraic expression and then to manipulate the expression into a form from which a desirable circuit can be arrived at directly. The manipulations are carried out by means of the theorems which will be presented in this section.

The usual procedure for proving theorems is to prove some theorems by using only the postulates, then to prove further theorems by using both the theorems already proved and the postulates, and so on. When such a method of proof is used, it is customary to present the theorems in an order such that each theorem can be proved by using only the postulates and previous theorems. Since the algebra being developed here involves only the constants 0 and 1, it is possible to prove almost all theorems directly from the postulates by using *perfect induction*. A proof by perfect induction involves substituting all possible combinations of values for the variables occurring in a theorem and then verifying that the theorem gives the correct result for all combinations. For example, the theorem $1 + X = 1$ is proved by first substituting 0 for X, which yields $1 + 0 = 1$

Table 3.4-1. Switching-algebra Theorems Involving One Variable

(T1)	$X + 0 = X$	(T1')	$X \cdot 1 = X$	(Identities)	
(T2)	$X + 1 = 1$	(T2')	$X \cdot 0 = 0$	(Null elements)	
(T3)	$X + X = X$	(T3')	$X \cdot X = X$	(Idempotency)	
(T4)	$(X')' = X$			(Involution)	
(T5)	$X + X' = 1$	(T5')	$X \cdot X' = 0$	(Complements)	

and then substituting 1 for X, obtaining $1 + 1 = 1$. The theorem is true since these results are precisely postulates P5' and P3'. Because perfect induction is available as a method of proof, the theorems will be presented in an order designed to emphasize the structure of the algebra rather than to facilitate formal proofs. The presentation of the theorems will be in pairs, where one member of the pair is obtained from the other by interchanging (0 and 1) and (addition and multiplication). This is done for the same reason that it was done for the postulates. Strictly speaking, X and X' should also be interchanged. This is not done, for the variables appearing in the theorems are generic variables and a theorem such as $X' + 1 = 1$ is no different from the theorem $X + 1 = 1$. Actually it is possible to substitute an entire expression for the variables appearing in a theorem. Thus, the theorem $X + XY = X$ implies that $ab + c + (ab + c)(de + fg) = ab + c$.

Single-variable Theorems

The switching-algebra theorems which involve only a single variable are shown in Table 3.4-1. Three of these theorems, T2, T3, and T3', are especially noteworthy since they are false for ordinary algebra. Theorems T3 and T3' can be extended to $X + X + X + \cdots + X = nX = X$ and $X \cdot X \cdot X \cdots X = X^n = X$. Theorem T5' can be used to simplify the path-tracing method of contact-network analysis. Whenever the same variable occurs both primed and unprimed in a path (Fig. 3.4-1), it is not necessary to include a term corresponding to this path in the algebraic expression for the network transmission.

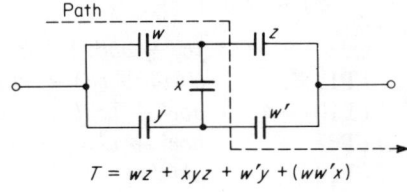

Fig. 3.4-1. Example of a network in which a variable w occurs both primed and unprimed in the same path.

$T = wz + xyz + w'y + (ww'x)$

Two- and Three-variable Theorems

Table 3.4-2 lists the switching-algebra theorems which involve two or three variables. Theorems T7 and T7′ are useful in eliminating terms from algebraic expressions and thereby eliminating elements from the corresponding networks. Theorem T10′ is noteworthy in that it is not true for ordinary algebra even though its dual T10 is.

Table 3.4-2. Switching-algebra Theorems Involving Two or Three Variables

(T6)	$X + Y = Y + X$	(T6′)	$XY = YX$	(Commutative)
(T7)	$X + XY = X$	(T7′)	$X(X + Y) = X$	(Absorption)
(T8)	$(X + Y')Y = XY$	(T8′)	$XY' + Y = X + Y$	
(T9)	$(X + Y) + Z = X + (Y + Z) = X + Y + Z$			(Associative)
	(T9′) $(XY)Z = X(YZ) = XYZ$			
(T10)	$XY + XZ = X(Y + Z)$			(Distributive)
	(T10′) $(X + Y)(X + Z) = X + YZ$			
(T11)	$(X + Y)(X' + Z)(Y + Z) = (X + Y)(X' + Z)$			(Consensus)
	(T11′) $XY + X'Z + YZ = XY + X'Z$			
(T12)	$(X + Y)(X' + Z) = XZ + X'Y$			

In reducing algebraic expressions, Theorems T11 and T11′ are very important and are used frequently, as is illustrated in the example which follows. This example shows how the theorems are used to manipulate a given algebraic expression into some other form. Very frequently the form desired is one which has as few literals occurring as possible. The number of literal occurrences corresponds directly to the number of contacts in a contact network and roughly to the number of diodes in a gate network.

Example 3.4-1. By use of the theorems, the expression $(c' + abd + b'd + a'b)(c + ab + bd)$ is to be shown equal to $b(a + c)(a' + c') + d(b + c)$.

	$(c' + abd + b'd + a'b)(c + ab + bd)$
(T12)	$c'(ab + bd) + c(abd + b'd + a'b)$
(T10)	$abc' + bc'd + abcd + b'cd + a'bc$
(T6)	$abc' + abcd + bc'd + a'bc + b'cd$
(T10)	$ab(c' + cd) + bc'd + a'bc + b'cd$

(T8')	$ab(c' + d) + bc'd + a'bc + b'cd$
(T11)	$ab(c' + d) + bc'd + a'bc + a'bd + b'cd$
(T10)	$abc' + abd + bc'd + a'bc + a'bd + b'cd$
(T6)	$(abd + a'bd) + abc' + bc'd + a'bc + b'cd$
(T10)	$bd(a + a') + abc' + bc'd + a'bc + b'cd$
(T5)	$bd(1) + abc' + bc'd + a'bc + b'cd$
(T1')	$bd + abc' + bc'd + a'bc + b'cd$
(T6)	$bd + bc'd + abc' + a'bc + b'cd$
(T7)	$bd + abc' + a'bc + b'cd$
(T6), (T10)	$d(b + b'c) + abc' + a'bc$
(T8')	$d(b + c) + abc' + a'bc$
(T10)	$d(b + c) + b(ac' + a'c)$
(T12)	$d(b + c) + b(a + c)(a' + c')$

Theorem T12 has no dual (T12'), for the dual would be identical with the original theorem. This theorem shows how a break-make transfer contact can be used to provide the same circuit operation as a continuity transfer contact. In the network corresponding to $XZ + X'Y$, there is a momentary open circuit when $Z = 1$, $Y = 1$ and a break-make transfer contact used for X and X' is changing position. If a continuity transfer is used for X and X' or if the network corresponding to $(X + Y)(X' + Z)$ is used with a break-make contact, there is no momentary open circuit.

n-Variable Theorems

The switching-variable theorems that involve an arbitrary number of variables are shown in Table 3.4-3. Three of these theorems (T13, T13', and T14) cannot be proved by perfect induction. For these theorems, the proofs require the use of finite induction [9, p. 11]. Theorems T13 and T13' are proved by first letting $n = 2$ and using perfect induction to

Table 3.4-3. *Switching-variable Theorems Involving n Variables*

(DeMorgan's theorems)
(T13) $(X_1 + X_2 + \cdots + X_n)' = X'_1 X'_2 \cdots X'_n$
  (T13') $(X_1 X_2 \cdots X_n)' = X'_1 + X'_2 + \cdots + X'_n$
(Generalized DeMorgan's theorem)
(T14) $f(X_1, X_2, \ldots, X_n, +, \cdot)' = f(X'_1, X'_2, \ldots, X'_n, \cdot, +)$
(Expansion theorem)
(T15) $f(X_1, X_2, \ldots, X_n) = X_1 f(1, X_2, \ldots, X_n)$
$\qquad\qquad\qquad\qquad\qquad\qquad + X'_1 f(0, X_2, \ldots, X_n)$
  (T15') $f(X_1, X_2, \ldots, X_n) = [X_1 + f(0, X_2, \ldots, X_n)]$
$\qquad\qquad\qquad\qquad\qquad\qquad [X'_1 + f(1, X_2, \ldots, X_n)]$

Switching Algebra

prove their validity for this special case. It is then assumed that the theorems are true for $n = k$, and this is shown to imply that they must then be true for $n = k + 1$. This completes the proof, the details of which are given in [9, p. 340]. Theorem T14 is proved by using Theorems T13 and T13′ along with the fact that every function can be split into the sum of several functions or the product of several functions (the argument here is similar to that used in connection with Fig. 3.2-1). By successively splitting the function into subfunctions and using T13 and T13′ it is possible to prove T14.

Theorem T14, which is a generalization of T13, forms the basis of a method for constructing complementary networks. Two networks having outputs T_1 and T_2 are said to be *complementary* if $T_1 = T_2'$. The complementary network for any given network can be designed by writing the output T_1 for the first network, then forming T_1' by means of T14, and then designing a network having ouput T_1'. For example, if $T_1 = (x + y)[w(y' + z) + xy]$, then $T_1' = x'y' + (w' + yz')(x' + y')$.

Theorems T15 and T15′ show that, for any arbitrary contact network, a design is always possible which requires only one transfer contact on one of the relays. An example of this is shown in Fig. 3.4-2, where the network is modified by Theorem T15 to require only a single transfer contact on the X relay.

It was pointed out in connection with Figs. 3.3-1 and 3.3-2 that the canonical networks are generally uneconomical. By manipulating the

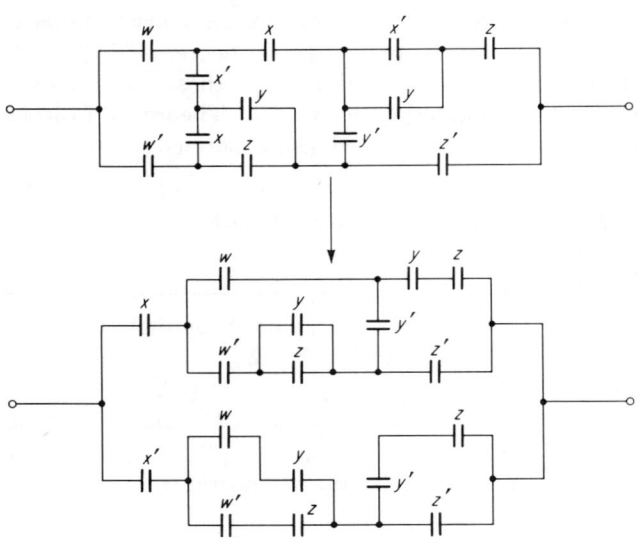

Fig. 3.4-2. *Modification of a network by means of Theorem T15 to require only a single transfer contact on the X relay.*

3.5 general gate networks

canonical sum or product with the aid of the theorems just presented, it is usually possible to obtain algebraic expressions which correspond to more economical networks than the canonical networks. The following example shows how this is done for the networks of Figs. 3.3-1 and 3.3-2. The final expressions correspond to the networks of Figs. 3.3-1c and 3.3-2c.

Example 3.4-2

$$f = X_1'X_2'X_3 + X_1'X_2X_3' \quad\quad\quad + X_1'X_2X_3 + X_1X_2'X_3'$$
$$f = X_1'X_2'X_3 + X_1'X_2X_3 \quad\quad\quad + X_1'X_2X_3' + X_1X_2'X_3'$$
$$f = X_1'X_2'X_3 + X_1'X_2X_3 + X_1'X_2X_3 + X_1'X_2X_3' + X_1X_2'X_3'$$
$$f = X_1'X_3(X_2' + X_2) \quad\quad + X_1'X_2(X_3 + X_3') \quad + X_1X_2'X_3'$$
$$f = X_1'X_3(1) \quad\quad\quad\quad\quad + X_1'X_2(1) \quad\quad\quad\quad + X_1X_2'X_3'$$
$$f = X_1'X_3 \quad\quad\quad\quad\quad\quad + X_1'X_2 \quad\quad\quad\quad\quad + X_1X_2'X_3'$$
$$f = X_1'(X_3 + X_2) \quad\quad\quad\quad\quad\quad\quad\quad\quad\quad + X_1X_2'X_3'$$

Many of the theorems of ordinary algebra are also valid for switching algebra. One which is not is the cancellation law. In ordinary algebra it follows that $X = Z$ if $X + Y = Y + Z$. In switching algebra this is not true. For example, it is generally true that $X + XY = X + 0$, but it is not necessarily true that $XY = 0$. This can be easily verified by writing out the tables of combinations for $f_1(X,Y) = X + XY$, $f_2(X,Y) = X + 0$, and $f_3(X,Y) = XY$. Similar remarks apply to the situation in which $XY = XZ$ does not imply that $Y = Z$.

More precise techniques for simplifying algebraic expressions will be presented in the following chapter.

3.5 GENERAL GATE NETWORKS

The previous discussion of gate networks in this chapter has been concerned solely with networks constructed of AND gates and OR gates. This can be considered only an introduction to the topic of gate networks, for other types of gates are equally important. In this section other types of gate networks will be considered.

Complete Sets

The discussion in connection with Theorem 3.3-1 showed that any arbitrary switching function could be realized by a network of AND gates, OR gates, and inverters. A natural question to ask in this connection is whether all three types of elements are necessary. Inverters are required if the inputs to the network consist of signals representing the input variables but not of signals representing the complements of the input variables. The situation when signals representing the complements are available is called *double-rail logic*, and when the complements are not

Switching Algebra

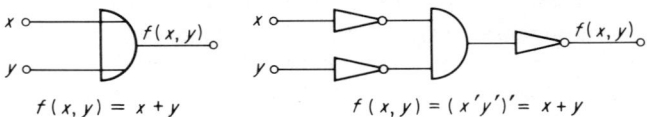

Fig. 3.5-1. *Realization of an OR gate by means of an AND gate and inverters.*

available, the term *single-rail logic* is used. Both techniques are employed, but for the purposes of the present discussion it will be assumed that complements are not directly available (single-rail logic). Any function which can be realized by a network of AND gates and OR gates but no inverters must have a corresponding algebraic expression which does not contain any complemented variables or parentheses. It has been shown [10, 11] that only certain functions, called *frontal functions*, can be so expressed and that the majority of functions require complementation. For example, the function $f(x) = x'$ cannot be realized by a network of AND gates and OR gates only. In fact, a technique has been developed for calculating the minimum number of inverters required for a function [12]. In this development it is shown that a network having three inputs x_1, x_2, and x_3 and three outputs corresponding to the functions $f_1(x_1,x_2,x_3) = x'_1$, $f_2(x_1,x_2,x_3) = x'_2$, and $f_3(x_1,x_2,x_3) = x'_3$ can be realized with only two inverters and a quantity of AND gates and OR gates.

It is clear that inverters are required to realize arbitrary functions, but the possibility of using only AND gates and inverters still exists. That the OR gates are not necessary is easily demonstrated, for it is possible to construct a network having the function of an OR gate and using only AND gates and inverters. This is done by making use of DeMorgan's theorem—$X + Y = (X'Y')'$—as is illustrated in Fig. 3.5-1. Thus any network consisting of AND gates, OR gates, and inverters can be changed into a network containing only AND gates and inverters by using the replacement shown in Fig. 3.5-1 to remove the OR gates. By duality, a similar technique can be used to remove the AND gates instead.

Since it is not possible to use only inverters to realize arbitrary functions, a minimum set of elements has now been determined. Because it is possible to construct a network containing only AND gates and inverters for any arbitrary function, the AND gate and inverter are said to form a *complete gate set*. Similarly, the OR gate and inverter form a complete gate set.

An important characteristic of the OR-NOT gate described in Chap. 1 is that it forms a complete gate set by itself. In Sec. 1.7 it was shown that the OR-NOT gate was equivalent to an OR gate whose output was connected to an inverter. Thus the algebraic expression for the output of an OR-NOT gate with inputs $x_1, x_2, \ldots, x_m$ is $(x_1 + x_2 + \cdots$

3.5 general gate networks

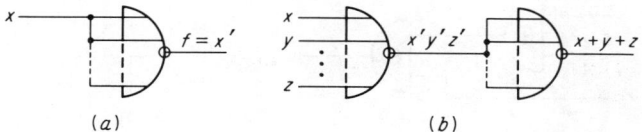

Fig. 3.5-2. *OR-NOT gates connected as an inverter and as an OR gate.* (a) *Inverter;* (b) *OR gate.*

$+ x_m)' = x_1' x_2' \cdots x_m'$. If all inputs of an OR-NOT gate are connected together to the same input x as shown in Fig. 3.5-2a, the output f is given by $f = (x + x + \cdots + x)' = x'$. Clearly the OR-NOT gate has been converted into an inverter. By connecting an OR-NOT gate with all its inputs tied together to the output of another OR-NOT gate it is possible to obtain the same performance as that of an OR gate. This is illustrated in Fig. 3.5-2b. It would thus be possible to replace all OR gates and inverters in a network by OR-NOT gates. Since it is possible to realize any function by means of a network containing only OR gates and inverters, it follows that it is possible to realize any function by means of a network containing only OR-NOT gates.

A dual argument applies to AND-NOT gates. An AND-NOT gate with inputs $x_1, x_2, \ldots, x_m$ has an output equal to $(x_1 x_2 \cdots x_m)' = x_1' + x_2' + \cdots + x_m'$. The AND-NOT gate is a complete set by itself.

Another important type of gate is the *sum modulo two*, or EXCLUSIVE OR gate,† which has a high output only when an odd number of its inputs are high. The table of combinations for a two-input EXCLUSIVE OR gate is given in Table 3.5-1. This table shows that the output of an EXCLUSIVE OR gate with inputs x_1 and x_2 is given by $x_1 \oplus x_2 = x_1' x_2 + x_1 x_2'$. It is easily demonstrated that the EXCLUSIVE OR opera-

† The term EXCLUSIVE OR is used since a two-input gate has a 1 output if one *but not both* of the inputs is equal to 1. The OR gate (sometimes called INCLUSIVE OR) has a 1 output if one or both inputs are equal to 1.

Table 3.5-1. *Table of Combinations for a Two-input EXCLUSIVE OR Gate* $f = x_1' x_2 + x_1 x_2'$

x_1	x_2	f
0	0	0
0	1	1
1	0	1
1	1	0

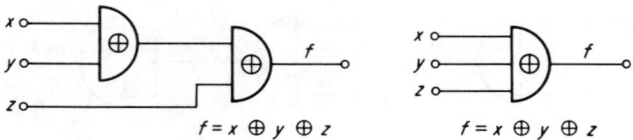

Fig. 3.5-3. Equivalent networks of EXCLUSIVE OR gates.

tion is commutative and associative; that is,

$$x \oplus y = y \oplus x \quad \text{and} \quad (x \oplus y) \oplus z = x \oplus (y \oplus z)$$

If both inputs of an EXCLUSIVE OR gate are connected together to x, the output is given by $x \oplus x = xx' + x'x = 0$. The fact that this operation is associative means that any network composed only of EXCLUSIVE OR gates is equivalent to a single EXCLUSIVE OR gate with all inputs connected directly to it. This is illustrated in Fig. 3.5-3. Because of the facts that $x \oplus x = 0$ and $x \oplus 0 = x$ it is possible to remove any duplicated inputs so that the resulting network contains only one gate for which no input appears more than once. Since this gate, called the *reduced gate*, cannot function as an inverter, it follows that no network containing only EXCLUSIVE OR gates can be equivalent to an inverter. Thus the EXCLUSIVE OR gate is not a complete set by itself.

The EXCLUSIVE OR gate can perform as an inverter if a signal representing a constant 1 is available, since $x \oplus 1 = x1' + x'1 = x'$ (see Fig. 3.5-4). It is still not possible to construct an OR gate by using EXCLUSIVE OR gates and a 1 signal, for the output of the reduced gate will still be of the form $X \oplus Y \oplus 1$, which does not equal 1 when either one or both of X and Y are equal to 1. A complete set can be formed by using both AND gates and EXCLUSIVE OR gates. As demonstrated above, any arbitrary function can be realized by a network containing only AND gates and inverters. By using EXCLUSIVE OR gates as in Fig. 3.5-4 to replace the inverters it is possible to obtain a network containing only AND gates and EXCLUSIVE OR gates.

In Sec. 3.3, canonical expressions involving AND, OR, and inverter operations were derived. It is possible to obtain similar canonical expressions for any complete set. The canonical expression using AND and EXCLUSIVE OR for two-variable functions is

$$f(x,y) = g_0 \oplus g_1 \cdot x \oplus g_2 \cdot y \oplus g_3 \cdot x \cdot y$$

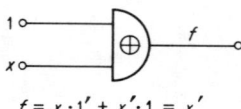

Fig. 3.5-4. Use of a constant 1 signal to form an inverter from an EXCLUSIVE OR gate.

3.5 general gate networks

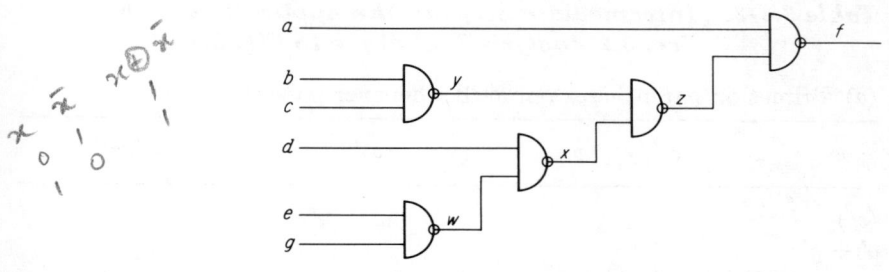

Fig. 3.5-5. *A network of AND-NOT gates.*

where

$$g_0 = f_0 \qquad g_1 = f_0 \oplus f_2 \qquad g_2 = f_0 \oplus f_1 \qquad g_3 = f_0 \oplus f_1 \oplus f_2 \oplus f_3$$

Thus for

$$f(x,y) = \Sigma(0,3) = x'y' + xy$$
$$g_0 = 1 \qquad g_1 = 1 \qquad g_2 = 1 \qquad g_3 = 0$$

so that

$$f(x,y) = 1 \oplus x \oplus y$$

Analysis of AND-NOT Networks and OR-NOT Networks

The analysis technique for gate networks which was presented in Sec. 3.2 is not restricted to AND and OR gate networks but is more general, as was pointed out in Sec. 3.2. This technique can thus be used to analyze networks constructed of AND-NOT gates or OR-NOT gates. Direct applications of the analysis technique to such gates can be quite cumbersome and inefficient. A network of AND-NOT gates is shown in Fig. 3.5-5. The outputs of the internal gates of this network are labeled with the variables w, x, y, and z to facilitate the following discussion of the analysis of the networks. The steps in a straightforward application of the Sec. 3.2 analysis technique to the network of Fig. 3.5-5 are summarized in Table 3.5-2. Part *a* of this table shows the intermediate results when DeMorgan's theorem is used to remove primed parentheses as soon as possible, and part *b* shows the results when no primes are removed from parentheses until the output expression has been obtained. Both approaches have undesirable features. That of part *a* requires unnecessary effort in converting the same expression (*eg*, for instance) back and forth several times, and that of part *b* requires that many levels of primed parentheses be kept, with the consequent danger of error. These drawbacks can be eliminated by use of the modification to be described next [13].

Switching Algebra

Table 3.5-2. Intermediate Steps in the Application of the Sec. 3.2 Analysis Technique to Fig. 3.5-5

(a) Primes on parentheses removed whenever possible

w	$x = (dw)'$	y	$z = (xy)'$	$f = (az)'$

$(eg)'$
$e' + g'$
$\quad\quad [d(e' + g')]'$
$\quad\quad d' + eg$
$\quad\quad\quad\quad (bc)'$
$\quad\quad\quad\quad b' + c'$
$\quad\quad\quad\quad\quad\quad [(d' + eg)(b' + c')]'$
$\quad\quad\quad\quad\quad\quad d(e' + g') + bc$
$\quad\quad\quad\quad\quad\quad\quad\quad (a[d(e' + g') + bc])'$
$\quad\quad\quad\quad\quad\quad\quad\quad a' + [d(e' + g') + bc]'$
$\quad\quad\quad\quad\quad\quad\quad\quad a' + (d' + eg)(b' + c')$

(b) Primes on parentheses not removed until end

w	$x = (dw)'$	y	$z = (xy)'$	$f = (az)'$

$(eg)'$
$\quad\quad [d(eg)']'$
$\quad\quad\quad\quad (bc)'$
$\quad\quad\quad\quad\quad\quad ([d(eg)']'[bc]')'$
$\quad\quad\quad\quad\quad\quad\quad\quad [(a)([d(eg)']'[bc]')'$
$\quad\quad\quad\quad\quad\quad\quad\quad a' + ([d(eg)']'[bc]')$
$\quad\quad\quad\quad\quad\quad\quad\quad a' + [d' + eg][b' + c']$

Figure 3.5-6a shows the network of Fig. 3.5-5 redrawn so as to make use of the gate symbol corresponding to the expression $x' + y'$ as well as that corresponding to the $(xy)'$ form. In Fig. 3.5-6b a fragment of this network is again redrawn, this time with detached inverters used rather than circles on gate inputs or gate outputs. This figure illustrates the fact that there are, in effect, two inverters in series between the output of the eg gate and the input of the x gate. Because of the involution theorem, $(X')' = X$, these inverters have no logical effect and can be ignored. This observation leads to the following conclusion: Whenever a lead has a circle output at one end and a circle input at the other end, *both* circles can be ignored. Thus the network of Fig. 3.5-6a can be transformed to

3.5 general gate networks

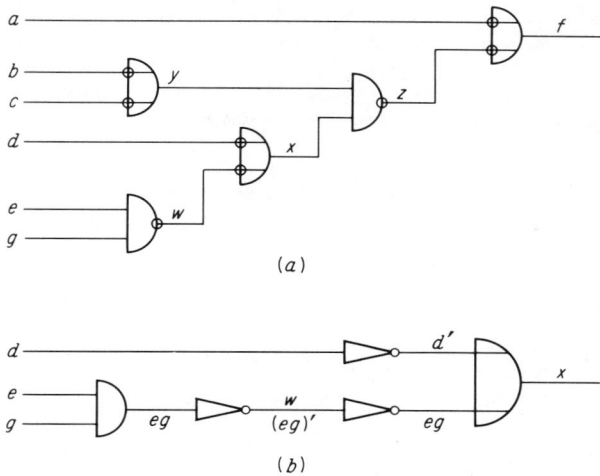

Fig. 3.5-6. Network of Fig. 3.5-5 using two types of gate symbols. (a) Network; (b) fragment.

the network of Fig. 3.5-7 *for the purposes of analysis.* The analysis of Fig. 3.5-7 is shown in Table 3.5-3. It should be emphasized that the technique just presented is merely a trick to avoid some unnecessary work in writing down an expression for the output of an AND-NOT gate network. An exactly analogous technique can be used for networks of OR-NOT gates.

Whenever an AND-NOT or OR-NOT gate network has the property that each gate output is connected to only one gate input, it is possible to transform the network diagram so that no internal inversion symbols remain [14]. If the network does not have this property, some inversion symbols may still remain. This is illustrated in Fig. 3.5-8.

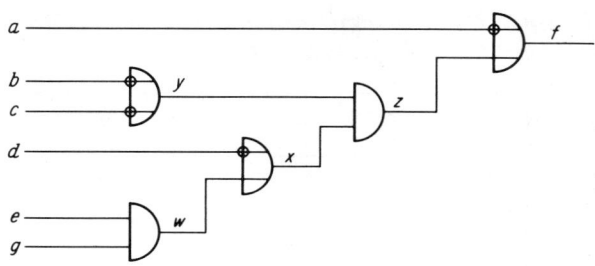

Fig. 3.5-7. Network of Fig. 3.5-5 redrawn so as to eliminate internal inversion symbols.

95

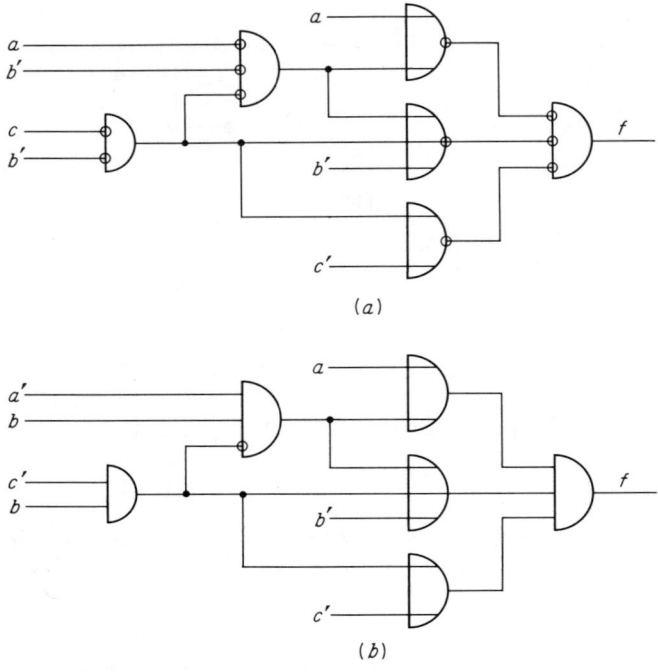

Fig. 3.5-8. *OR-NOT gate network for which it is not possible to remove all inversion symbols. (a) Before removal of inversion circles; (b) after removal of inversion circles.*

The technique just described for analyzing networks of OR-NOT or AND-NOT gates is also useful in the design of such networks. If a network has been designed using OR gates and AND gates, it can be converted into an OR-NOT or an AND-NOT gate network by inserting inversion circles in the appropriate places. This is illustrated in Fig. 3.5-9. Notice

Table 3.5-3. *Intermediate Steps in the Analysis of the Network of Fig. 3.5-7*

W	$X = d' + W$	Y	$Z = XY$	$f = a' + z$
eg				
	$d' + eg$			
		$b' + c'$		
			$(d' + eg)(b' + c')$	
				$a' + (d' + eg)(b' + c')$

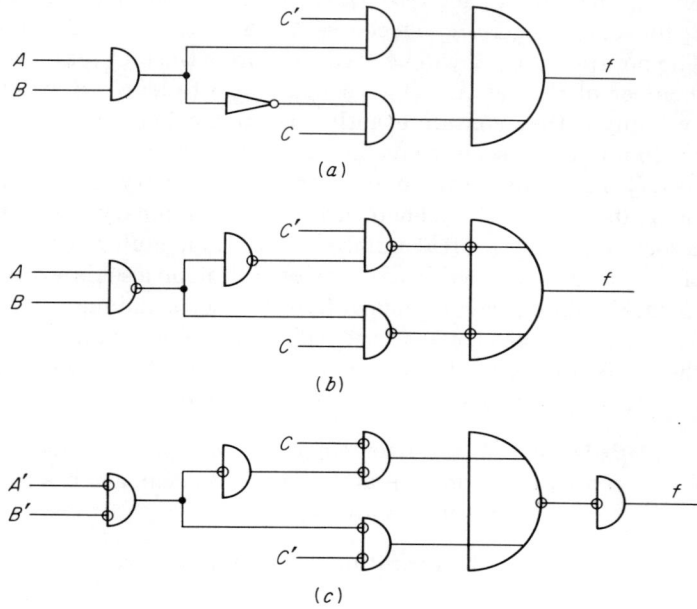

Fig. 3.5-9. *A network for $f(A,B,C,) = \Sigma(1,3,5,6)$ using AND and OR gates transformed into a network using AND-NOT gates and a network using OR-NOT gates. (a) AND gates, OR gates, and inverters; (b) AND-NOT gates; (c) OR-NOT gates.*

that the OR-NOT gate requires an additional gate at the output. This can be avoided by redesigning the original gate to have an AND gate as the output gate.

3.6 BOOLEAN ALGEBRA

The switching algebra developed in Sec. 3.1 is a particular example of a more general type of mathematical system called a *Boolean algebra*. Boolean algebras can be interpreted in terms of sets and logical propositions as well as switching variables. A general development of Boolean algebra will be presented next, and then the application to sets and logic will be discussed [4,5].

A set is a collection of objects or elements. The members of a set can be specified by listing them explicitly or by specifying some property such that all elements satisfying this property are members of the set, while all other objects are not members of the set. For example, the set

$X = \{a,b,c\}$, or the set $Y = \{$all integers x such that x can be expressed as $2y$ for some integer $y\}$. Of course the set Y consists of all the even integers. The notation $a \in X$ will be used to express the fact that the element a is a member of the set X. Two sets are said to be *equal*, written $X = Y$, if and only if they contain exactly the same elements.

In order to associate an algebraic structure with a set, it is necessary to specify one or more rules of combination or binary operations between the elements of the set. There are four such binary operations normally associated with the real numbers: addition, multiplication, subtraction, and division. Formally, a binary operation on a set is a rule which assigns to each ordered pair of elements of the set a unique element of the set. The fact that the element assigned to the ordered pair is also a member of the set is often called the *closure property*. The following formal definition of a Boolean algebra can now be stated:

Definition. A set B of elements $\{a,b,c, \ldots\}$ and two associated binary operations $+$ and $\cdot$ form a Boolean algebra if and only if the following postulates are satisfied:

(A1) The operations are commutative,

$$a + b = b + a \qquad \text{and} \qquad a \cdot b = b \cdot a$$

(A2) Each of the operations distributes over the other,

$$a(b + c) = a \cdot b + a \cdot c \qquad \text{and} \qquad a + b \cdot c = (a + b) \cdot (a + c)$$

(A3) There exist identity elements 0 and 1 for $+$ and $\cdot$ respectively,

$$0 + a = a \qquad \text{and} \qquad 1 \cdot a = a \qquad \text{for all } a \in B$$

(A4) For each $a \in B$ there exists an $a' \in B$ such that

$$a + a' = 1 \qquad \text{and} \qquad a \cdot a' = 0$$

From this set of postulates it is possible to prove the following theorems:

$a + a = a$	$a \cdot a = a$	(Idempotent)
$a + 1 = 1$	$a \cdot 0 = 0$	(Null elements)
	$(a')' = a$	(Involution)
$a + ab = a$	$a(a + b) = a$	(Absorption)
$a + (b + c) = (a + b) + c$	$a \cdot (b \cdot c) = (a \cdot b) \cdot c$	(Associative)
$(a + b)' = a' \cdot b'$	$(a \cdot b)' = a' + b'$	(DeMorgan's laws)

3.6 boolean algebra

These theorems are formally identical with the corresponding theorems proved for switching algebra. The remaining theorems of switching algebra can also be proved to follow from the postulates given above. It is not obvious that perfect induction can be used in proving these theorems, since the variables are not restricted to the two symbols 0 and 1.† In fact nothing in the definition restricts the set B to containing only a finite number of elements, and it is possible to have a Boolean algebra with an infinite set B. The proofs of the theorems will not be discussed here since they can easily be found in the literature [4, chap. 2].

This discussion shows that the switching algebra developed earlier is simply a Boolean algebra for which the set B consists of only two elements 0 and 1. An example of a Boolean algebra with more than two elements is given by $B = \{0,\alpha,\beta,1\}$, with $+$ and $\cdot$ defined by the accompanying tables.

$+$	0	α	β	1
0	0	α	β	1
α	α	α	1	1
β	β	1	β	1
1	1	1	1	1

$\cdot$	0	α	β	1
0	0	0	0	0
α	0	α	0	α
β	0	0	β	β
1	0	α	β	1

Propositional Logic

Another interpretation of a Boolean algebra with two elements in the set B is in terms of propositional logic. This interpretation will be discussed next. A proposition in this context is the content or meaning of a declarative sentence for which it is possible to determine the truth or falsity. Thus a sentence must be free of ambiguity and must not be self-contradictory in order to qualify as a proposition. The statements

"The sun is shining"
"All men have three heads"

are propositions. The first is sometimes true, while the second is always false. The statement "All men are tall" is not a proposition, since it is ambiguous, and the statement "The statement you are now reading is false" is not a proposition, since it is self-contradictory.

The lower case letters $p, q, r, s, t, \ldots$ will be used to represent arbitrary or unspecified propositions, i.e., propositional variables. Two propositional constants are necessary also: one for a proposition that is

† The fact that perfect induction is applicable in this general case is proved in 4, p. 36].

always false, which will be represented by the symbol F, and one for a proposition that is always true, which will be represented by the symbol T. Two propositions p and q are said to be equal if whenever one is true the other is true, and vice versa. Thus the expression $p = $ F represents a statement to the effect that the proposition p is always false. If p represents the proposition "x is an even integer" and q represents the proposition "x can be expressed as $2y$, with y an integer" then it follows that $p = q$.

Corresponding to any given proposition p, it is possible to form another proposition which asserts that p is false. This new proposition is called the *denial* (or complement, or negative) of p and is written as $\bar{p}$ (or p' or $\sim p$ or $-p$). It is clear that, if p is true, $\bar{p}$ is false, and vice versa. This can be stated formally by means of the accompanying table. If p is the

p	$\bar{p}$
T	F
F	T

statement "It is raining," then $\bar{p}$ is the statement "It is false that it is raining" or "It is not raining."

It is also possible to combine two propositions to form a new proposition. Thus, if p and q are two propositions, it is possible to form a new proposition called the *conjunction* of p and q, written $p \wedge q$ (or $p \& q$ or $p \cdot q$), which is true only when both p and q are true and is false otherwise. The table expressing this relationship is shown herewith.

p	q	$p \wedge q$
F	F	F
F	T	F
T	F	F
T	T	T

A rule for combining two propositions to form a new proposition is called a *logical connective*. The logical-connective conjunction corresponds to the common usage of "and" to combine two statements. Thus if p represents the proposition "It is raining" and q represents the proposition "The sun is shining," $p \wedge q$ represents the proposition "It is raining, and the sun is shining."

Another common means for combining two statements involves the use of "or." The usage in this case is not entirely unambiguous. This is illustrated by the following two statements:

1. "To get a degree, a student must take an economics course or a history course"
2. "Each student is to leave the room by the front door or the side door"

The meaning of the first sentence is clearly economics or history or both, and thus the usage corresponds to an "inclusive or." In the second sentence it is presumably impossible to leave by more than one door; so an "exclusive or" is intended—one door or the other but *not* both. In propositional logic it is necessary to distinguish clearly between these two usages of "or."

The disjunction of two propositions p and q, written $p \vee q$, is defined as a proposition which is true when either p or q or *both* are true. Thus disjunction corresponds to the inclusive or. The table for disjunction is shown herewith. Of course, it is also possible to define a logical connective which corresponds to the exclusive or.

p	q	$p \vee q$
F	F	F
F	T	T
T	F	T
T	T	T

The system of propositional logic which has just been described can be shown to be a Boolean algebra with a set B consisting of the two propositional constants T and F. The two binary operations are conjunction corresponding to $\cdot$ and disjunction corresponding to $+$. To show that propositional logic does form a Boolean algebra with this interpretation for B, $+$, and $\cdot$, it is necessary to verify that the four postulates are satisfied:

(P1) $p \vee q = q \vee p$ and $p \wedge q = q \wedge p$
(P2) $p \wedge (q \vee r) = (p \wedge q) \vee (p \wedge r)$
$p \vee (q \wedge r) = (p \vee q) \wedge (p \vee r)$
(P3) $F \vee p = p$ and $T \wedge p = p$
(P4) $p \vee \bar{p} = T$ and $p \wedge \bar{p} = F$

Switching Algebra

The correctness of the P1 postulate follows directly from the definitions of conjunction and disjunction. The remaining three postulates can be verified by using truth tables, such as those used to define conjunction and disjunction, in much the same fashion as tables of combination were used to verify the theorems of switching algebra in Sec. 3.4 (perfect induction).

It follows from the fact that propositional logic is a Boolean algebra that no additional logical connectives are necessary in order to be able to express any arbitrary propositional function. In fact, because of DeMorgan's theorem, $p \vee q = \overline{(\bar{p} \wedge \bar{q})}$, so that only conjunction and negation are really necessary. It is convenient to have more than this absolute minimum set of connectives available. An additional logical connective, called material implication, is of considerable importance in propositional logic. This connective corresponds to the compound statements "p implies q" or "If p, then q" in everyday usage. The formal definition of *material implication*, written $p \rightarrow q$, is that $p \rightarrow q$ is false only when p is true and q is false†. This is shown in the accompanying table. For the proposition $p \rightarrow q$ to be true, it is necessary only that q

p	q	$p \rightarrow q$
F	F	T
F	T	T
T	F	F
T	T	T

be true whenever p is true. If p is false, the proposition $p \rightarrow q$ is true independent of the truth value of q. Thus if p is the statement "The moon is a planet" and q is the statement "The moon is made of blue cheese," then $p \rightarrow q$ is true since p is false. It is possible to relate the material-implication connective to disjunction and negation: $p \rightarrow q = \bar{p} \vee q$.

A switching algebra can be developed for gate networks and contact networks directly in terms of propositional logic. This is done by associating propositions with circuit elements, as is shown in the following table. If leads i and j are inputs to an AND gate and lead k is the output lead of the AND gate, then it follows that $x_k = x_i \wedge x_j$ since there will be a high voltage on lead k only when leads i and j both have high voltages.

† This is also sometimes written as $p \supset q$.

3.6 boolean algebra

Element	Proposition	Symbol
Relay X	"Relay X is operated"	X
Make contact on relay X	"A closed path exists between the two terminals of each make contact on relay X"	x
Break contact on relay X	"A closed path does not exist between the two terminals of each make contact on relay X"	$\bar{x}$
Lead i in a gate network	"A high voltage is present on lead i"	x_i

Algebra of Sets

Another system whose structure is formally identical with Boolean algebra is the algebra of sets. This algebra is concerned mainly with the ways in which sets can be combined. Thus the *union* of two sets X and Y is written as $X \cup Y$ and is defined as the set containing all elements which are in either X or Y or both. If $X = \{0,1,2,3\}$ and $Y = \{0,2,4\}$, then $X \cup Y = \{0,1,2,3,4\}$. The *intersection* of two sets X and Y is written as $X \cap Y$ and is defined as the set containing all elements which are in both X and Y. For the particular X and Y given above, $X \cap Y = \{0,2\}$.

If two sets have exactly the same members, they are said to be equal. If a set X has only elements which are also elements of another set Y, X is said to be a *subset* of Y or X is said to be *included in* Y, written $X \subseteq Y$. If Y contains elements which are not also in X, then X is a *proper subset* of Y, written $X \subset Y$. This relation of inclusion has the following properties:

$$X \subseteq X \quad \text{(Reflexive)}$$
$$\text{If} \quad X \subseteq Y \quad \text{and} \quad Y \subseteq X \quad \text{then} \quad X = Y$$
$$\text{(Antisymmetric)}$$
$$\text{If} \quad X \subseteq Y \quad \text{and} \quad Y \subseteq Z \quad \text{then} \quad X \subseteq Z$$
$$\text{(Transitive)}$$

A relation such as $\subseteq$ which has these three properties is said to be a *partial ordering*.

If the two sets X and Y happen to have no elements in common, their intersection $X \cap Y$ will be empty. It is customary to define a special set, the *null*, or *empty*, *set*, which contains no elements. The null set is represented by the symbol ϕ, so that $X \cap Y = \phi$ if X and Y have no common elements. The other special set which is necessary is the *universal*

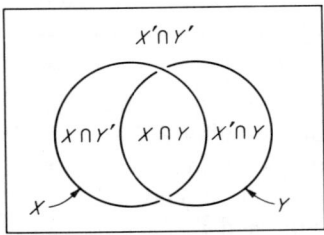

Fig. 3.6-1. *Venn diagram.*

set, written U, that by definition contains all elements under discussion. It is then possible to associate with each set X another set X', called the complement of X, which contains all members of U which are not in X. If only integers are being discussed so that U is the set of all integers, and X is the set of all even integers it follows that X' is the set of all odd integers. The set inclusion property can be related to intersection and complementation since $X \subseteq Y$ if and only if $X \cap Y' = \phi$. There can be no elements of X which are not in Y.

From the definitions just given it can be easily shown that the following rules hold:

(S1) $X \cup Y = Y \cup X$ and $X \cap Y = Y \cap X$
(S2) $X \cap (Y \cup Z) = (X \cap Y) \cup (X \cap Z)$
      $X \cup (Y \cap Z) = (X \cup Y) \cap (X \cup Z)$
(S3) $\phi \cup X = X$ and $U \cap X = X$
(S4) $X \cup X' = U$ and $X \cap X' = \phi$

This shows that the algebra of sets is formally identical to Boolean algebra if the elements of B are chosen to be the subsets of some universal set U and the connectives $\cap$ and $\cup$ take the place of $\cdot$ and $+$. In fact it has been shown [4, p. 42] that for any abstract Boolean algebra there is a corresponding set U such that the algebra of the subsets of U has the same structure as that of the abstract Boolean algebra. In other words, there is no abstract Boolean algebra which cannot be interpreted as the algebra of subsets of some universal set U.

Visualization of the combination of sets is often possible by using a *Venn diagram* in which each set is represented by a certain area. Specifically the universal set U is represented by the interior of a rectangle, and each other set is represented by the points inside a circle or closed region inside the rectangle. This is illustrated in Fig. 3.6-1.

Boolean Rings

Another important interpretation of a Boolean algebra is as a Boolean ring. The theory of rings is an important topic in abstract algebra.

Definition. A set R of elements $\{a,b,c, \ldots\}$ and two associated binary operations $\oplus$ and $\cdot$ form a ring if and only if the following postulates are satisfied:

(R1) $a \oplus (b \oplus c) = (a \oplus b) \oplus c$ (Addition is associative)
(R2) $a \oplus b = b \oplus a$ (Addition is commutative)
(R3)   The equation $a \oplus X = b$ has a solution x in R
(R4) $a(bc) = (ab)c$ (Multiplication is associative)
(R5) $a(b \oplus c) = ab \oplus ac$, $(b \oplus c)a = ba \oplus bc$
        (Distributivity)

If in addition to these properties it is also true that $a \cdot a = a$ for all elements of R (idempotent law), then the ring is said to be a *Boolean ring* [5,15]. Any Boolean algebra forms a Boolean ring if the same operation is used for multiplication in both systems and if ring addition is defined by

$$a \oplus b = a'b + ab'$$

Since the theory of switching circuits does not make any substantial use of Boolean rings, they will not be discussed further here.

Problems

1. For the networks of Fig. P3-1:
   (a) Write the output function.
   (b) Fill in the table of combinations.
   (c) Write the decimal specifications.

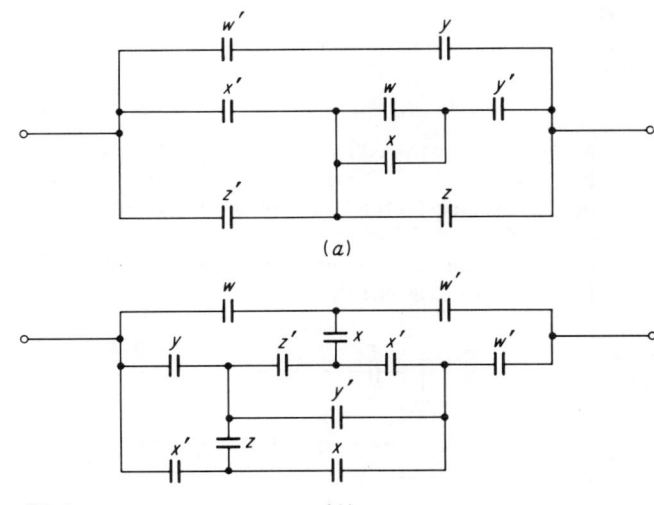

Fig. P3-1

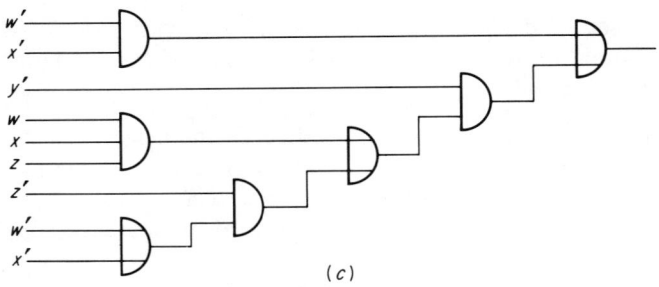

(c)

Fig. P3-1

2. For the network of Fig. P3-2 write the algebraic expressions for each of the outputs f_a, f_b, f_c, f_d in terms of the inputs x and y. Assume that the symbol 1 has been assigned to E_H, the high input-voltage level.

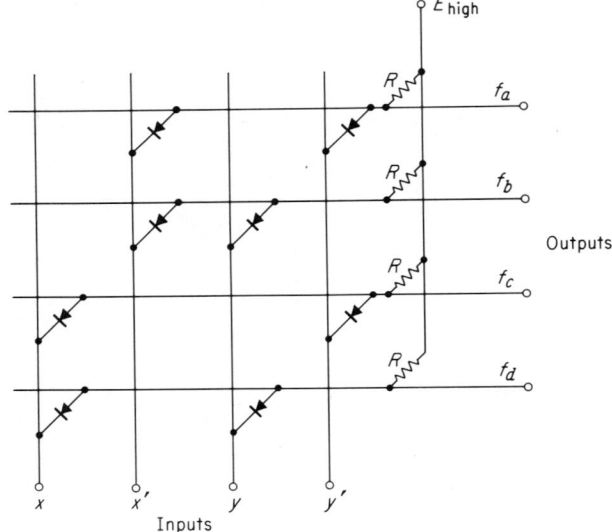

Fig. P3-2

3. A relay X is controlled by a contact network containing contacts from relays W, Y, Z as well as X itself. The network is shown in Fig. P3-3.
   (a) What must be the status of the $W, Y,$ and Z relays before relay X can operate?
   (b) Once operated, under what conditions can relay X be deenergized? Explain your answer fully.

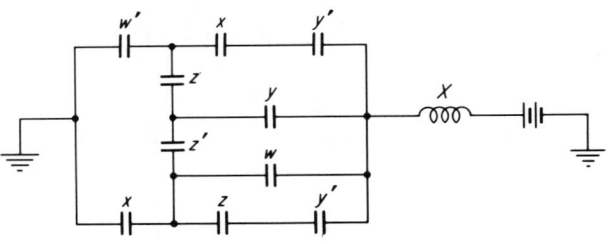

Fig. P3-3

4. Figure P3-4 shows a network consisting of a diode and contacts from relays W, X, Y, and Z. This network controls the operation of relay A.
   (a) When will relay A operate?
   (b) Redraw the network with the diode replaced with a contact from one of the relays W, X, Y, Z so that the conditions of operation of relay A are unchanged.

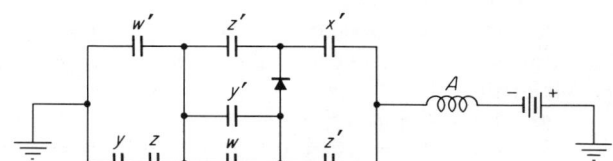

Fig. P3-4

5. The single-error-correcting code in the tabulation is to be used to transmit messages over a noisy channel. When the message bits are received, they

Message	Code word
	x_1 x_2 x_3 x_4 x_5 x_6
1	0 1 1 1 0 0
2	1 0 1 0 1 0
3	1 1 0 0 0 1
4	1 1 0 1 1 0
5	1 0 1 1 0 1
6	0 1 1 0 1 1
7	0 0 0 0 0 0
8	0 0 0 1 1 1

must be "decoded" in order to determine which message was sent. Write the decimal specification for the decoded output which corresponds to message 1 being sent.

6. A lock circuit is to be designed with 10 toggle switches, $x_0, x_1, \ldots, x_9$ as inputs. The lock is to be open only when the switches are alternately up and down. Write the decimal specification for the corresponding switching function.

7. An indicator circuit is to be designed for a room which has two swinging doors D_1 and D_2. Associated with each door there are two switches, e_1 and x_1 for D_1, e_2 and x_2 for D_2. The e_i switch is closed only when the corresponding door D_i is open *in*, and the x_i switch is closed only when the corresponding door D_i is open *out*. An indicator lamp is to be lit whenever there is a clear path through the room (one door open in and the other door open out).
   Fill out a table of combinations for the function corresponding to the indicator light.

8. Write the canonical sum and the canonical product for each of the following functions:
   (a) $f(x,y,z) = \Sigma(0,3,6)$
   (b) $f(x,y,z) = \Pi(1,2,7)$

9. Prove the following identities *without* using perfect induction:
$$ab' + bc' + ca' = a'b + b'c + c'a$$
$$ab + a'c + bcd = ab + a'c$$

10. Find the complements of the following functions:
$$f = a + bc$$
$$f = (a + b)(a'c + d)$$
$$f = ab + b'c + ca'd$$
Prove that your answers are correct by showing that $f \cdot f' = 0$ and $f + f' = 1$.

11. Prove whether or not the following identities are valid. Do not use perfect induction.
 (a) $ab + c'd' + a'bcd' + ab'c'd = (a + d')(b + c')$
 (b) $(a + b')(b + c')(c + a') = (a' + b)(b' + c)(c' + a)$
 (c) $(a + b)(b + c)(c + a) = (a' + b')(b' + c')(c' + a')$
 (d) $ab + a'b'c = (c + a)(c + b)$

12. (a) Write the transmission function for the network of Fig. P3-12.
  (i) By use of P sets (tie sets)
  (ii) By use of S sets (cut sets)
 (b) Show that the expressions obtained in (i) and (ii) are equivalent.

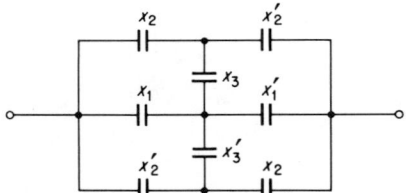

Fig. P3-12

13. Redesign the networks of Fig. P3-13, using as few gates or contacts as possible.

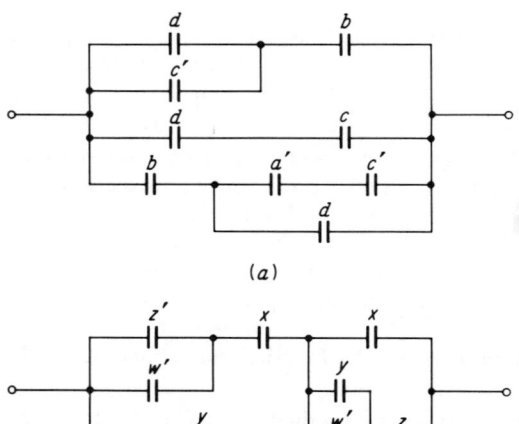

Fig. P3-13

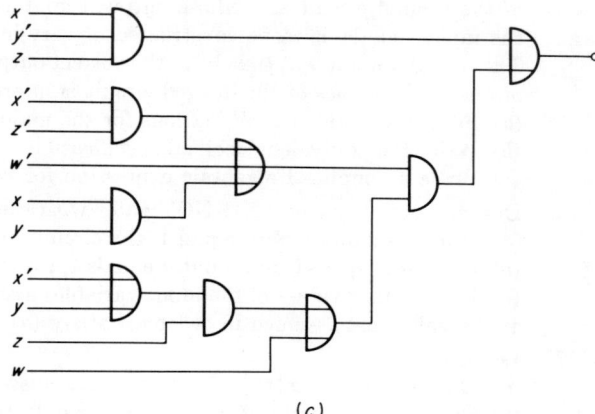

Fig. P3-13   (c)

14. Design a contact network whose transmission is the complement of the transmission of the network of Fig. P3-14.

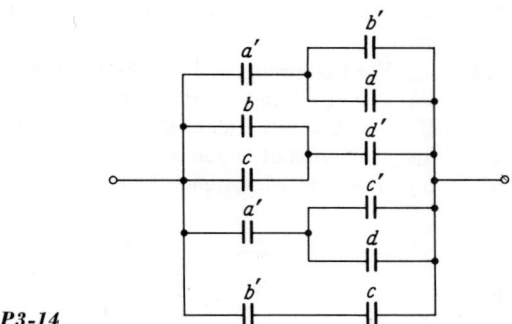

Fig. P3-14

15. In order to increase the reliability of a combinational circuit constructed of AND and OR gates, three copies of the desired circuit are to be built. An additional circuit, whose inputs are the outputs of the three copies

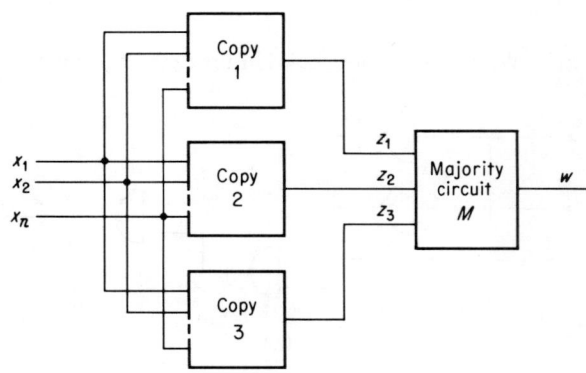

Fig. P3-15

of the desired circuit and whose output signal agrees with the majority of the input signals, is to be constructed of very-high-reliability components. The overall circuit will thus have the correct output even though the output of one of the copies of the desired circuit is in error (see Fig. P3-15).
   (a) Form the table of combinations for the majority circuit M.
   (b) Write the equivalent decimal specification.
   (c) Write a simplified algebraic expression for w as a function of z_1, z_2, z_3.
16. Design a circuit using AND-NOT gates with the following performance:
   (a) If inputs x and y both equal 1, the circuit output z is also equal to 1.
   (b) If input x equals 0 and input w equals 1, the circuit output z is to equal 0.
   (c) For all other values of the input variables w, x, and y the output *remains* at the value last assumed in order to satisfy (a) or (b).
17. Let $x \downarrow y = x'y'$.
   (a) Does $(x \downarrow y) \downarrow z = x \downarrow (y \downarrow z)$? Prove your answer.
   (b) Express the function $f = x'y'z'$, using only the $\downarrow$ connective.
18. Show that the operation $f = x'y' = x \downarrow y$, which is called *joint denial*, forms a complete set.
19. Does the operation $f = x' + y = x \rightarrow y$, which is called implication, form a complete set? If it does not, what must be added in order to form a complete set?
20. (a) For the circuit of Fig. P3-20 write algebraic expressions for A and B.
   (b) The box S represents a circuit for which $y = 0$ and $y' = 1$ if A equals 0. If $A = 1$, $y = 1$ unless $B = 1$, in which case $y = 0$. Write an expression for y in terms of w and x.
   (c) Give a word statement for the dependence of y on w and x.

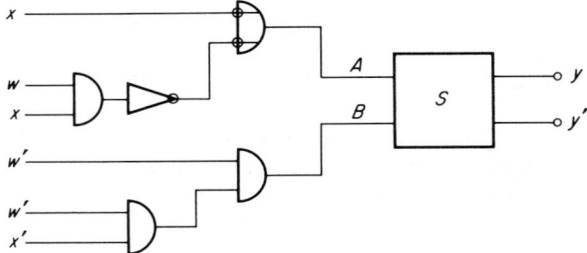

Fig. P3-20

21. Write the output function for each of the networks of Fig. P3-21.

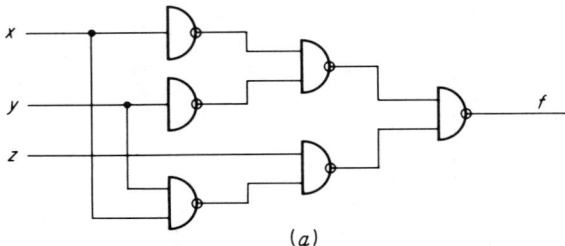

(a)

Fig. P3-21

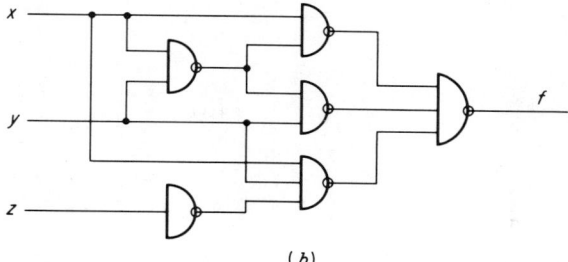

Fig. P3-21   (b)

22. Analyze the circuit in Fig. P3-22.
    (a) Write an algebraic expression for z.
    (b) Express z as a sum of products (this can be done with five products).

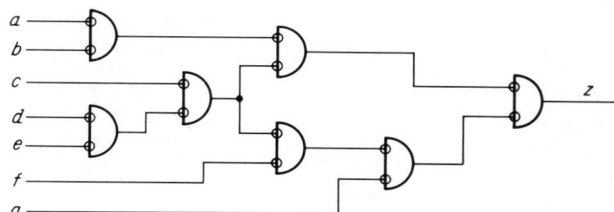

Fig. P3-22

23. (a) If $f = x \oplus y$, express f in terms of x, x', y, y', $+$, and $\cdot$.
    (b) Prove that $(x \oplus y) \oplus z = x \oplus (y \oplus z)$.
    (c) Prove that if $x \oplus y = z$
    then
    $$x \oplus z = y$$
    and
    $$x \oplus y \oplus z = 0$$
    (d) Prove that $x \oplus y = x + y$ if $xy = 0$.
    (e) Prove that $(a + b) \oplus (a + c) = a'(b \oplus c)$.

24. Analyze the network of Fig. P3-24, showing each step of your work clearly.

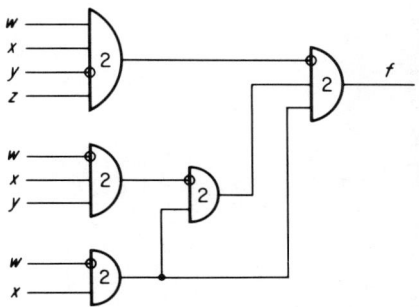

Fig. P3-24

Summary of Postulates and Theorems for Switching Algebra

Postulates

(P1) $X = 0$ if $X \neq 1$ (P1') $X = 1$ if $X \neq 0$
(P2) $0' = 1$ (P2') $1' = 0$
(P3) $0 \cdot 0 = 0$ (P3') $1 + 1 = 1$
(P4) $1 \cdot 1 = 1$ (P4') $0 + 0 = 0$
(P5) $1 \cdot 0 = 0 \cdot 1 = 0$ (P5') $0 + 1 = 1 + 0 = 1$

Single-variable theorems

(T1) $X + 0 = X$ (T1') $X \cdot 1 = X$ (Identities)
(T2) $X + 1 = 1$ (T2') $X \cdot 0 = 0$ (Null elements)
(T3) $X + X = X$ (T3') $X \cdot X = X$ (Idempotency)
(T4) $(X')' = X$ (Involution)
(T5) $X + X' = 1$ (T5') $X \cdot X' = 0$ (Complements)

Two- and three-variable theorems

(T6) $X + Y = Y + X$ (T6') $XY = YX$ (Commutative)
(T7) $X + XY = X$ (T7') $X(X + Y) = X$ (Absorption)
(T8) $(X + Y')Y = XY$ (T8') $XY' + Y = X + Y$
(T9) $X + Y + Z = (X + Y) + Z = X + (Y + Z)$ (T9') $XYZ = (XY)Z = X(YZ)$ (Associative)
(T10) $XY + XZ = X(Y + Z)$ (T10') $(X + Y)(X + Z) = X + YZ$ (Distributive)
(T11) $(X + Y)(X' + Z)(Y + Z) = (X + Y)(X' + Z)$ (T11') $XY + X'Z + YZ = XY + X'Z$ (Consensus)
(T12) $(X + Y)(X' + Z) = XZ + X'Y$

n-Variable theorems

(T13) $(X + Y + Z + \cdots)' = X'Y'Z' \cdots$ (T13') $(X \cdot Y \cdot Z \cdots)' = X' + Y' + Z' + \cdots$ (DeMorgan's theorems)
(T14) $f(X_1, X_2, \ldots, X_n, +, \cdot)' = f(X_1', X_2', \ldots, X_n', \cdot, +)$ (generalized DeMorgan's theorem)

(T15) $f(X_1, X_2, \ldots, X_n) = X_1 f(1, X_2, \ldots, X_n) + X_1' f(0, X_2, \ldots, X_n)$ (Expansion theorem)
(T15') $f(X_1, X_2, \ldots, X_n) = [X_1 + f(0, X_2, \ldots, X_n)] \cdot [X_1' + f(1, X_2, \ldots, X_n)]$

REFERENCES

1. Shannon, C. E.: A Symbolic Analysis of Relay and Switching Circuits, *Trans. AIEE*, vol. 57, pp. 713–723, 1938.
2. ———: The Synthesis of Two-terminal Switching Circuits, *Bell System Tech. J.*, vol. 28, pp. 59–98, January, 1949.
3. Keister, W., A. E. Ritchie, and S. H. Washburn: "The Design of Switching Circuits," D. Van Nostrand Company, Inc., Princeton, N.J., 1951.
4. Whitesitt, J. E.: "Boolean Algebra and Its Applications," Addison-Wesley Publishing Company, Inc., Reading, Mass., 1961.
5. Sikorski, R.: "Boolean Algebras," Springer-Verlag OHG, Berlin, 1960.
6. Slepian, D.: On the Number of Symmetry Types of Boolean Functions of n-variables, *Can. J. Math.*, vol. 5, no. 2, pp. 185–193, February, 1954.
7. Grea, R., and R. Higonnet: Etudes logiques des circuits électriques et des systèmes binares, Editions Berger-Levrault, Paris, 1955.
8. Staff of the Harvard University Computation Laboratory: Synthesis of Electronic Computing and Control Circuits," Harvard University Press, Cambridge, Mass., 1951.
9. Birkhoff, G., and S. MacLane: "A Survey of Modern Algebra," The Macmillan Company, New York, 1955.
10. Quine, W. V.: Two Theorems about Truth Functions, *Bol. Soc. Math. Mex.*, vol. 10, nos. 1, 2, pp. 64–70, 1953.
11. Gilbert, E. N.: Lattice Theoretic Properties of Frontal Switching Functions, *J. Math. Phys.*, vol. 33, pp. 57–67, April, 1954.
12. Markov, A. A.: On the Inversion Complexity of a System of Functions, *J. ACM*, vol. 5, no. 4, pp. 331–334, October, 1958.
13. McCluskey, E. J.: Logical Design Theory of NOR Gate Networks with No Complemented Inputs, Switching Circuit Theory and Logical Design, *Proc. Fourth Ann. Symposium*, Chicago, Ill., Oct. 28–30, September, 1963, pp. 105–116. Published by the IEEE, New York, *Spec. Publ.* S-156.
14. Washburn, S. H.: An Application of Boolean Algebra to the Design of Electronic Switching Circuits, *Trans. AIEE*, vol. 72, pt. 1, pp. 380–388, September, 1953.
15. Bartee, T. C., I. L. Lebow, and I. S. Reed: "Theory and Design of Digital Machines," McGraw-Hill Book Company, New York, 1962.

4 SIMPLIFICATION OF SWITCHING FUNCTIONS

In Chap. 3 it was shown how a combinational switching circuit can be specified by means of a table of combinations, how an algebraic expression (the canonical expansion) can be derived from the table of combinations, and how the canonical expansion can be simplified by using the switching algebra theorems. Algebraic simplification is satisfactory for only the simplest functions; for other functions more powerful methods are necessary. It is the object of this chapter to present these methods, which will be used throughout the remainder of the book.

4.1 THE MAP METHOD [1,2]

The simplest sum-of-product-terms form of a function will be called a *minimal sum*.† The precise definition will be discussed later, but for the present the sum-of-products form which has the fewest terms will be taken as the minimal sum. If there is more than one sum-of-products form having the minimum number of terms, and if these forms do not all contain the same total number of literals, then only the form(s) with the fewest literals will be called the minimal sum(s). For example, the function $f = x'yz + xyz + xyz'$ can be written as $f = yz + xyz'$, $f = x'yz + xy$, and $f = yz + xy$. Each of these forms contains two terms, but only the third form is a minimal sum, since it contains four literals, while the other two forms contain five literals each.

The minimal sum corresponds to a diode gate circuit in which the circuit inputs are connected to AND gates and the outputs of the AND gates form the inputs to an OR gate whose output is the

---
† This is called minimal rather than minimum, since there may be more than one such form.

circuit output. Such a circuit is called a *two-stage circuit*, since there are two gates connected in series between the circuit inputs and output. It is also possible to have two-stage circuits in which the circuit inputs are connected to OR gates and the circuit output is obtained from an AND gate. The minimal sum just defined corresponds to the two-stage circuit in which the output is derived from an OR gate and which contains the minimum number of gates. The basic method for obtaining the minimal sum is to apply the theorem $XY + X'Y = Y$ to as many terms as possible and then to use the theorem $XY + X'Z + YZ = XY + X'Z$ to eliminate as many terms as possible.

Example 4.1-1

$f = x'y'z' + x'y'z + xy'z + xyz$
$\quad x'y'z' + x'y'z = x'y'$
$\quad\quad xy'z + xyz = xz$
$\underline{f = x'y' + xz \quad\quad \textit{Minimal sum}\dagger}$

Example 4.1-2

$f = w'x'y'z + w'x'yz + w'xy'z + w'xyz + wxy'z'$
$\quad\quad\quad\quad\quad\quad\quad\quad\quad\quad\quad\quad\quad + wxy'z + wx'y'z' + wx'y'z$
$\quad w'x'y'z + w'x'yz = w'x'z \quad w'xy'z + w'xyz = w'xz$
$\quad wxy'z' + wxy'z = wxy'$
$\quad wx'y'z' + wx'y'z = wx'y'$
$\quad w'x'z + w'xz = w'z$
$\quad wxy' + wx'y' = wy'$
$\underline{f = w'z + wy' \quad\quad \textit{Minimal sum}\dagger}$

Example 4.1-3

$f = xyz + x'yz + xy'z$
$xyz = xyz + xyz$
$f = (xyz + x'yz) + (xyz + xy'z)$
$\underline{f = yz + xz \quad\quad \textit{Minimal sum}\dagger}$

Example 4.1-2 illustrates the fact that it may be necessary to apply the theorem $XY + X'Y = Y$ several times, the number of literals in the terms being reduced each time. A single term may be paired with more than one other term, as shown in Example 4.1-3.

The process of comparing pairs of terms to determine whether or not the theorem $XY + X'Y = Y$ applies can become very tedious for large functions. This comparison process can be simplified by using an n-cube map.

---

† It is possible to prove that this is a minimal sum. This will be discussed subsequently.

Simplification of Switching Functions

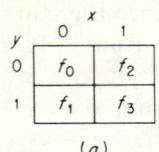 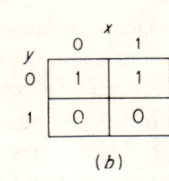

Fig. 4.1-1. Two-variable map. (a) General form; (b) map for $f = x'y' + xy' = \Sigma(0,2) = \pi(1,3)$.

Maps for Two, Three, and Four Variables

A map for a function of two variables, as shown in Fig. 4.1-1, is a square of four cells or a 2-cube map. The value 0 or 1 which the function is to equal when $x = 1, y = 0$ (the entry in the 10 location or 2 row of the table of combinations) is placed in the cell having coordinates $x = 1, y = 0$. In general, the scheme for filling in the map is to place a 1 in all cells whose coordinates form a binary number which corresponds to one of the fundamental products included in the function and to place a 0 in all cells whose binary numbers correspond to fundamental products not included in the function. This is done very simply by writing a 1 in each cell whose decimal designation (decimal equivalent of the binary number formed by the coordinates) occurs in the decimal specification of the function and writing 0's in the remaining cells.

The maps for functions of three and four variables are direct extensions of the two-variable map and are shown in Figs. 4.1-2 and 4.1-3. Discus-

	xy			
z	00	01	11	10
0	f_0	f_2	f_6	f_4
1	f_1	f_3	f_7	f_5

(a)

	xy			
z	00	01	11	10
0	1	1	0	0
1	0	0	1	0

(b)

Fig. 4.1-2. Three-variable map. (a) General form; (b) map for $f = x'y'z' + x'yz' + xyz$; $f = \Sigma(0,2,7) = \pi(1,3,4,5,6)$.

	wx			
yz	00	01	11	10
00	f_0	f_4	f_{12}	f_8
01	f_1	f_5	f_{13}	f_9
11	f_3	f_7	f_{15}	f_{11}
10	f_2	f_6	f_{14}	f_{10}

(a)

	wx			
yz	00	01	11	10
00	1	0	0	0
01	1	0	0	0
11	0	1	1	0
10	0	1	1	0

(b)

Fig. 4.1-3. Four-variable map. (a) General form; (b) map for $f = w'x'y'z' + w'x'y'z + w'xyz' + w'xyz + wxyz' + wxyz$; $f = \Sigma(1,6,7,14,15) = \pi(2,3,4,5,8,9,10,11,12,13)$.

4.1 the map method

sion of maps for more than four variables will be postponed temporarily, since such maps are more complex.

Prime Implicants

Two fundamental products can be "combined" by means of the theorem $XY + XY' = X$ if their corresponding binary numbers differ in only 1 bit. For the fundamental products $wxyz$ and $wxyz'$,

$$wxyz + wxyz' = wxy$$

The corresponding binary numbers are 1111 and 1110, which differ only in the lowest-order bit position. The fundamental products $wxyz$ and $w'xyz'$ cannot combine, and their corresponding numbers, 1111 and 0110, differ in the first and last bit positions.

In terms of the distance concept introduced in Chap. 2, two fundamental products combine only if the corresponding points on an n-cube or n-cube map are distance 1 apart. Points distance 1 apart on an n-cube become adjacent cells on a map, so that cells which represent fundamental products which can be combined can be determined very quickly by inspection. In carrying out this inspection process it must be remembered that cells such as f_4 and f_6 or f_1 and f_9 in Fig. 4.1-3 must be considered to be adjacent.

In a four-variable map each cell is adjacent to four other cells corresponding to the four bit positions in which two binary numbers can differ. In inspecting a map to determine which fundamental products can be combined, only cells with 1 entries (1 cells) need be considered, since these correspond to the fundamental products included in the function. Figure 4.1-4 shows a four-variable map with adjacent 1 cells encircled. Notice that the 0111 cell is adjacent to two 1 cells. The rule for writing down the algebraic expression corresponding to a map is that there will be one product term for each pair of adjacent 1 cells and a fundamental product for each 1 cell which is not adjacent to any other 1 cell. The fundamental products are written down according to the rule given in Chap. 3: any variable corresponding to a 0 in the binary number formed by the coordinates of the corresponding 1 cell is primed; the variables corresponding to 1's are left unprimed. The product terms corresponding to pairs of adjacent 1 cells are obtained by the same rule, with the excep-

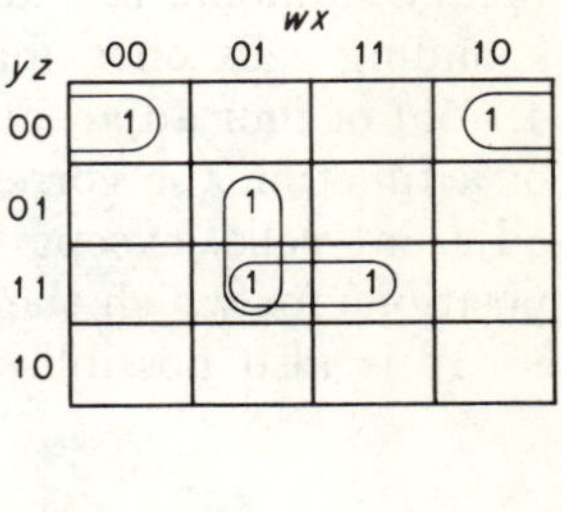

Fig. 4.1-4. Four-variable map with adjacent 1 cells encircled. $f = \Sigma(0,5,6,8,15)$ $= x'y'z' + w'xz + xyz$.

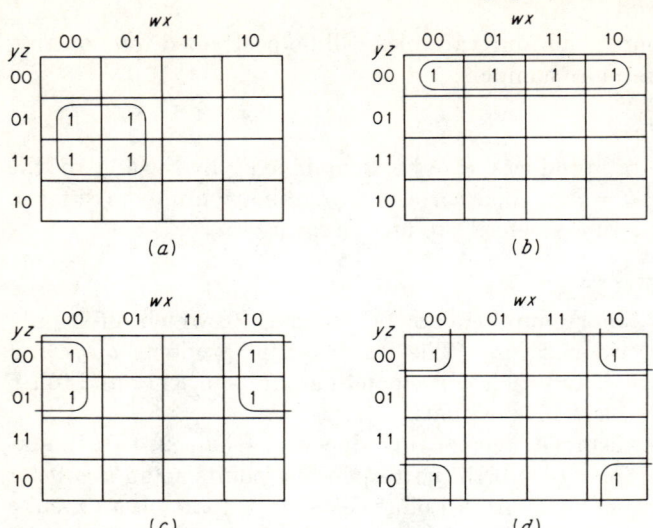

Fig. 4.1-5. *Four-variable maps showing sets of cells corresponding to four fundamental products which can be combined.* (a) $f = w'z$; (b) $f = y'z'$; (c) $f = x'y'$; (d) $f = x'z'$.

tion that one variable is not included in the product. The variable excluded is that corresponding to the bit position in which the coordinates of the two 1 cells differ (see Fig. 4.1-4).

The situation where it is possible to combine two of the terms obtained from pairs of the fundamental terms as in Example 4.1-2 must be considered next. In such a situation four of the fundamental products can be combined into a single product term by successive applications of the $XY + XY' = X$ theorem. A function which is the sum of four such fundamental products is $f = wxyz + wxyz' + wxy'z + wxy'z'$. Application of the theorem to this function yields

$$f = (wxyz + wxyz') + (wxy'z + wxy'z') = wxy + wxy' = wx$$

The characteristic property of four fundamental products which can be combined in this fashion is that all but two of the variables are the same (either primed or unprimed) in all four terms. The corresponding four binary numbers are identical in all but two bit positions. The corresponding cells on a map form "squares" (Fig. 4.1-5a) or "lines" (Fig. 4.1-5b) of four adjacent cells. For such a group of four cells on the map of a function the corresponding product term is written just as for two adjacent cells, except that two variables corresponding to the two bit positions for which the cell coordinates change must be omitted.

It is also possible that eight of the fundamental products can be

4.1 the map method

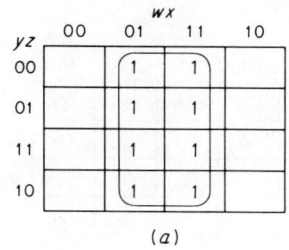
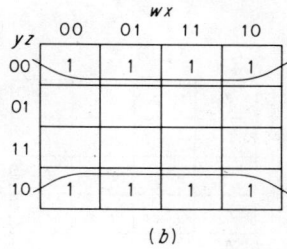

Fig. 4.1-6. *Four-variable maps showing sets of cells corresponding to eight fundamental products which can be combined.* (a) $f = x$; (b) $f = z'$.

Fig. 4.1-7. *Map showing prime implicants.*

$f = \Sigma\,(0, 5, 7, 8, 11, 13, 15) = xz + x'y'z' + wyz$

combined. In this case all but three of the variables are identical (either primed or unprimed) in all eight terms. Figure 4.1-6 shows some sets of eight cells on a map which have all but three coordinates fixed. The general rule is that, if in 2^i fundamental products all but i of the variables are identical (primed or unprimed), then the 2^i products can be combined and the i variables which change can be dropped.

In searching for a minimal sum for a function by means of a map, the first step is to encircle all sets of cells corresponding to fundamental products which can be combined (see Fig. 4.1-7). If one such set is contained in a larger set, only the larger set is encircled.† In Fig. 4.1-7 the set (0101,0111) is not encircled. The encircled sets and the corresponding product terms will be called *prime implicants*.‡ These are exactly the terms which would result from repeated applications of the theorem $XY + XY' = X$. The terms appearing in the minimal sum will be some or all of the prime implicants.

Maps for Five and Six Variables

While the map is most useful for functions of four variables, it is also helpful for five- and six-variable functions. A five-variable map is formed

† This corresponds to using the theorem $X + XY = X$.
‡ This term was introduced by W. V. Quine [3]. It is derived from the terminology of mathematical logic, but it has received widespread use in connection with switching theory.

Simplification of Switching Functions

v = 0

yz \ wx	00	01	11	10
00	f_0	f_4	f_{12}	f_8
01	f_1	f_5	f_{13}	f_9
11	f_3	f_7	f_{15}	f_{11}
10	f_2	f_6	f_{14}	f_{10}

v = 1

yz \ wx	00	01	11	10
00	f_{16}	f_{20}	f_{28}	f_{24}
01	f_{17}	f_{21}	f_{29}	f_{25}
11	f_{19}	f_{23}	f_{31}	f_{27}
10	f_{18}	f_{22}	f_{30}	f_{26}

Fig. 4.1-8. Five-variable map.

by using 2 four-variable maps (Fig. 4.1-8), and a six-variable map is composed of 4 four-variable maps (Fig. 4.1-9).

In the five-variable map, one of the four-variable maps represents all rows of the table of combinations for which $v = 0$, and the other four-variable map represents all rows for which $v = 1$. Similarly one of the four-variable maps making up the six-variable map represents all rows of the table of combinations for which $u = 0$ and $v = 0$, and another four-variable map represents all rows for which $u = 0$, $v = 1$, etc.

The basic rule for combining cells for five- or six-variable maps is the same as for four-variable maps: it is possible to combine any set of 1 cells for which some of the coordinates remain fixed while the remaining coordinates take on all possible combinations of values. For cells on the same four-variable map, the patterns of sets of cells which can be combined are the same patterns discussed in connection with four-variable maps. Two cells which are on different four-variable maps can be combined only

u = 0, v = 0

yz \ wx	00	01	11	10
00	f_0	f_4	f_{12}	f_8
01	f_1	f_5	f_{13}	f_9
11	f_3	f_7	f_{15}	f_{11}
10	f_2	f_6	f_{14}	f_{10}

u = 0, v = 1

yz \ wx	00	01	11	10
00	f_{16}	f_{20}	f_{28}	f_{24}
01	f_{17}	f_{21}	f_{29}	f_{25}
11	f_{19}	f_{23}	f_{31}	f_{27}
10	f_{18}	f_{22}	f_{30}	f_{26}

u = 1, v = 0

yz \ wx	00	01	11	10
00	f_{32}	f_{36}	f_{44}	f_{40}
01	f_{33}	f_{37}	f_{45}	f_{41}
11	f_{35}	f_{39}	f_{47}	f_{43}
10	f_{34}	f_{38}	f_{46}	f_{42}

u = 1, v = 1

yz \ wx	00	01	11	10
00	f_{48}	f_{52}	f_{60}	f_{56}
01	f_{49}	f_{53}	f_{61}	f_{57}
11	f_{51}	f_{55}	f_{63}	f_{59}
10	f_{50}	f_{54}	f_{62}	f_{58}

Fig. 4.1-9. Six-variable map.

4.1 the map method

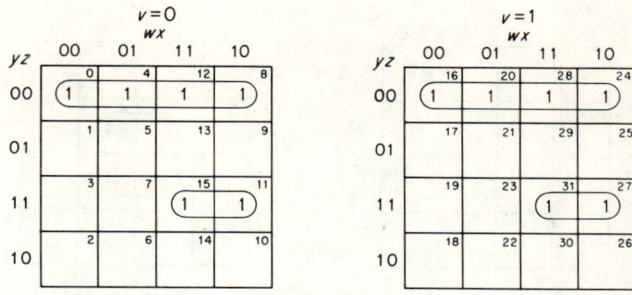

Prime implicants: (0, 4, 8, 12, 16, 20, 24, 28); (11, 15, 27, 31)

Fig. 4.1-10. *Five-variable map for* $f = \Sigma(0,4,8,11,12,15, 16,20,24,27,28,31) = y'z' + wyz$.

if they occupy the same relative position on their respective four-variable maps. In Fig. 4.1-8 the cells containing f_4 and f_{20} can be combined, but it is not permissible to combine the cells containing f_4 and f_{16}. For the six-variable map, only cells from two maps which are horizontally or vertically adjacent can be combined—a cell from the map labeled $u = 0$, $v = 0$ cannot be combined with a cell from the map labeled $u = 1, v = 1$, since the two cells differ in two coordinates rather then in one, as required. Four cells, such as those labeled f_5, f_{21}, f_{37}, f_{53} (in Fig. 4.1-9), which all occupy the same position in their individual four-variable maps can all be combined.

The first step in the procedure for picking the minimal sets on a five- or six-variable map is to determine the prime implicants for each of the individual four-variable maps (Fig. 4.1-10). Each prime implicant must now be compared with the prime implicants of the (horizontally and vertically) adjacent maps. If there is an identical prime implicant in an

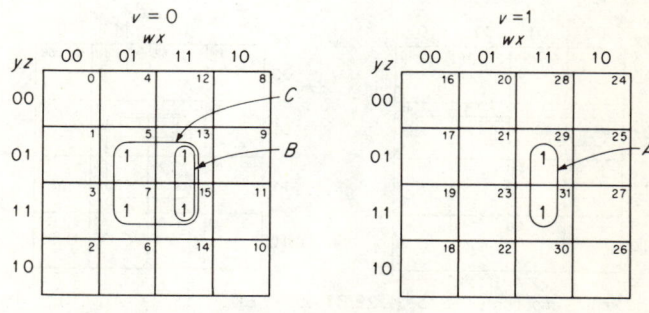

Prime implicants: (5, 7, 13, 15); (13, 15, 29, 31)

Fig. 4.1-11. *Five-variable map for* $f = \Sigma(5,7,13,15,29,31) = wxz + v'xz$.

Simplification of Switching Functions

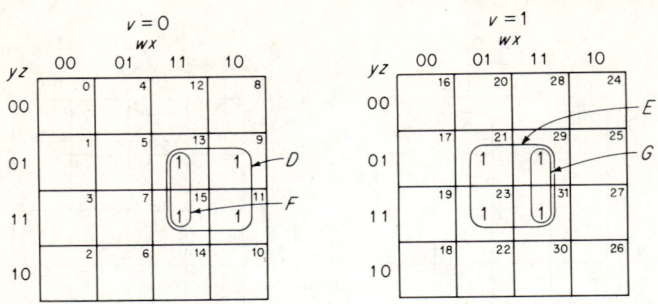

Prime implicants: (9,11,13,15); (21,23,29,31); (13,15,29,31)

Fig. 4.1-12. Five-variable map for $f = \Sigma(9,11,13,15,21,23, 29,31) = v'wz + vxz$.

adjacent map, the two prime implicants are combined into one prime implicant (Fig. 4.1-10).

If, in one four-variable map, there is a prime implicant (such as A in Fig. 4.1-11) which is identical with a subset (B, Fig. 4.1-11) of a prime implicant (C, Fig. 4.1-11) in an adjacent map, a new prime implicant is

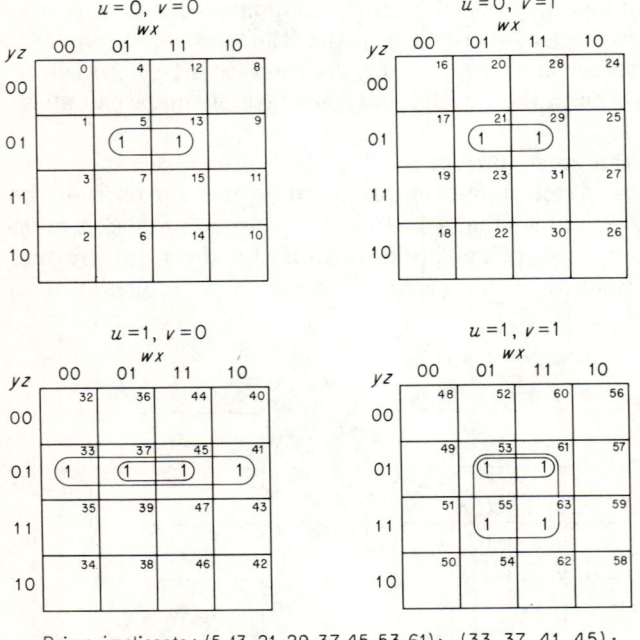

Prime implicants: (5,13,21,29,37,45,53,61); (33,37,41,45); (53,55,61,63)

Fig. 4.1-13. Six-variable map for $f = \Sigma(5,13,21,29,33,37, 41,45,53,55,61,63) = xy'z + uv'y'z + uvxz$.

formed from the original prime implicant A and the subset B. In such a case, the original prime implicant A is no longer a prime implicant, since it is included in the larger prime implicant A, B. The set C which included B is still a prime implicant. One further situation must be considered: there may be two prime implicants D, E (Fig. 4.1-12) in two different four-variable maps which are not identical and for which neither is identical to a subset of the other, but which both have identical subsets F, G (Fig. 4.1-12). The two identical subsets can be combined to form a new prime implicant. Both the original prime implicants remain as prime implicants. For a six-variable map, it is also necessary to consider prime implicants made up of a four-variable prime implicant which is identical with prime implicants or subsets of prime implicants in all the three remaining four-variable maps (see Fig. 4.1-13).

Formation of Minimal Sums

As was shown in Chap. 3, it is possible to express a function as a sum of the fundamental products which correspond to rows of the table of combinations for which the function is to equal 1 (the canonical sum). It is also possible to express any function as the sum of all its prime implicants. This form of the function will be called the *complete sum*. This is a correct representation for the function, since it is possible to derive the complete sum from the canonical sum by use of the theorems $XY + XY' = X$ and $X = X + X$. Moreover, just as there is only one canonical sum for any function and only one function corresponding to a given canonical sum, there is only one complete sum for each function, and vice versa.

Usually it is a minimal sum rather than the complete sum that is desired. As will be shown in the following section, the minimal sum always consists of a sum of the prime implicants. For some functions all the prime implicants must be included, and for these functions the minimal sum and the complete sum are identical. For most functions it is not necessary to include all prime implicants, since some of them can be removed by use of the theorem $XY + X'Z + YZ = XY + X'Z$. The minimal sum can be obtained from the complete sum by using this theorem to remove as many prime implicants as possible. There usually are several orders in which prime implicants can be eliminated, and some of these orders of elimination will result in minimal sums and others may not.

Example 4.1-4. For the function $f(w,x,y,z) = \Sigma(2,3,5,6,7,9,11,13)$ the order in which terms are eliminated from the complete sum determines whether or not the minimal sum is obtained.

Complete sum:

$$f = w'y + xy'z + w'xz + wy'z + x'yz + wx'z$$

Simplification of Switching Functions

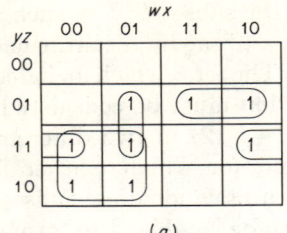
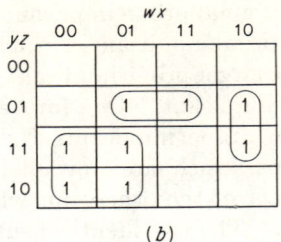

Fig. 4.1-14. *Maps for $f = \Sigma(2,3,5,6,7,9,11,13)$ showing two simplified forms of f. (a) $f = w'y + w'xz + wy'z + x'yz$; (b) $f = w'y + xy'z + wx'z$.*

First order of elimination:

$xy'z + w'xz + wy'z = x'yz + wx'z$
$wy'z + x'yz + wx'z = wy'z + x'yz$
$\underline{f = w'y + w'xz + wy'z + x'yz}$

No further eliminations are possible.

Second order of elimination:

$w'xz + w'y + xy'z = w'y + xy'z$
$wy'z + xy'z + wx'z = xy'z + wx'z$
$x'yz + w'y + wx'z = w'y + wx'z$
$\underline{f = w'y + xy'z + wx'z}$ *Minimal sum*

The maps corresponding to these two simplified forms of f are shown in Fig. 4.1-14.

A sum-of-products form from which no term or variable(s) can be deleted without changing the value of the expression is called an *irredundant sum*. Both the simplified expressions of Example 4.1-4 are irredundant sums.

Once the prime implicants have been formed, the minimal sum can be determined directly from the map. The rule that must be followed in choosing the prime implicants which are to correspond to terms of the minimal sum is that each 1 cell must be included in at least one of the chosen prime implicants. The problem of obtaining a minimal sum is equivalent to that of selecting the fewest prime implicants such that each 1 cell is included in at least one of these prime implicants. This rule is based on the fact that, for each combination of values of the input variables for which the function is to equal 1, the minimal sum must equal 1 and therefore at least one of its terms must equal 1. More simply, the map corresponding to the minimal sum must have the same 1 cells as the map of the original function.

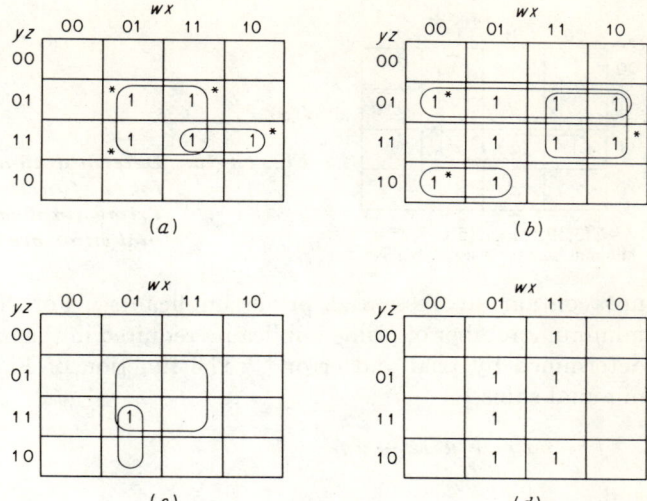

Fig. 4.1-15. *Determination of minimal sums.* (a) $f = \Sigma(5,7,11,13,15)$, *minimal sum:* $f = xz + wyz$; (b) $f = \Sigma(1,2,5,6,7,9,11,13,15)$, *essential prime implicants;* (c) *Fig. 4.2-15b after removal of the essential prime implicants;* (d) $f = \Sigma(5,6,7,12,13,14)$, *no essential prime implicants present.*

A procedure for determining the minimal sum is first to determine whether any 1 cells are included in only one prime implicant. In Fig. 4.1-15 an asterisk has been placed in each 1 cell which is included in only one prime implicant. A 1 cell which is included in only one prime implicant is called a *distinguished* 1 *cell*.

A prime implicant which includes a 1 cell which is not included in any other prime implicant is called an essential prime implicant and must be included in the corresponding minimal sum.† In Fig. 4.1-15a both the prime implicants are essential and must be included in the minimal sum. A minimal sum does not always consist only of essential prime implicants. In Fig. 4.1-15b, only the essential prime implicants are shown. Cell 7 is not included in any of these; so another prime implicant which includes cell 7 must be present in the minimal sum. Figure 4.1-15c shows the function of Fig. 4.1-15b after removal of the essential prime implicants. One of the two prime implicants shown must be included in the minimal sum, and the larger is chosen because the corresponding term contains fewer literals. The final minimal sum is $f = y'z + wz + w'yz' + xz$.

There are some functions, such as that shown in Fig. 4.1-15d, which do

† Actually an essential prime implicant must be included in all irredundant sums.

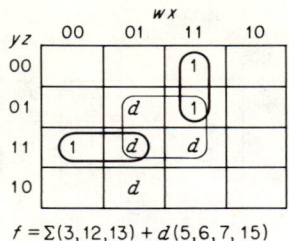

Fig. 4.1-16. Determination of minimal sum for a function with d terms. Prime implicants used in minimal sums are shown darkened.

not contain any essential prime implicants. For such functions the minimum number of prime implicants required in the minimal sum can be determined by trial and error.† The function of Fig. 4.1-15d has two minimal sums,

$$f = wxy' + w'xz + xyz'$$

and

$$f = wxz' + xy'z + w'xy$$

Incompletely Specified Functions

The addition of d terms does not introduce any extra complexity into the procedure for determining minimal sums. Any d terms which are present are treated as 1 terms in forming the prime implicants, *with the exception that no prime implicants containing only d terms are formed. The d terms are disregarded in choosing terms of the minimal sum.* No prime implicants are included in order to ensure that each d term is contained in at least one prime implicant of the minimal sum. The explanation of this procedure is that the d terms are used to make the prime implicants as large as possible so as to include the maximum number of 1 cells and to contain as few literals as possible. No prime implicants need be included in the minimal sum because of the d terms, for it is not required that the function equal 1 for the d terms. An example of a function with d terms is given in Fig. 4.1-16.

It is often convenient to avoid determining all the prime implicants. This can sometimes be done by searching for 1 cells which are contained in only one prime implicant and thus determining the essential prime implicants. A 1 cell is selected, and the prime implicant or prime implicants which include the 1 cell are determined. If there is only one prime implicant, it is essential and must be included in the minimal sum. This procedure is continued until all the 1 cells are included in prime implicants of the minimal sum.

† Systematic procedures for such functions will be discussed in connection with tabular methods.

4.1 the map method

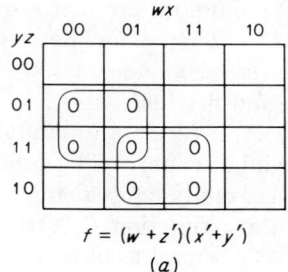
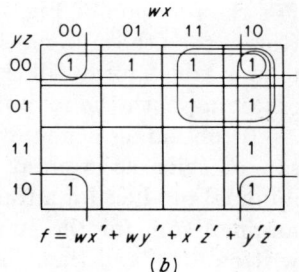

Fig. 4.1-17. *Derivation of minimal product and minimal sum for $f = \Sigma(0,2,4,8,9,10,11,12,13)$.* (a) *Minimal product;* (b) *minimal sum.*

Minimal Products

It is also possible to express a function as a product of factors. The simplest such form will be called a *minimal product*. The definition of minimal product is analogous to the definition of minimal sum: the minimal product is the product-of-factors form which contains the fewest factors. If there is more than one such form, only the form or forms also containing the fewest literals are minimal products. Minimal products are obtained by using the theorems $(X + Y)(X + Y') = X$ and $(X + Y)(X' + Z)(Y + Z) = (X + Y)(X' + Z)$ to simplify the canonical product.

Example 4.1-5

$f = (x + y' + z')(x + y' + z)(x + y' + z)(x' + y' + z)$
$(x + y' + z')(x + y' + z) = (x + y')$
$(x + y' + z)(x' + y' + z) = (y' + z)$
$\underline{f = (x + y')(y' + z)} \quad Minimal\ product$

By combining 0 cells rather than 1 cells, a map can be used for minimal products in exactly the same way as it is used for minimal sums. The encircled sets and the corresponding sum factors will be called *prime implicates*. The algebraic expressions are obtained from the map in the same way as the fundamental sums are obtained from the table of combinations. A variable corresponding to a 0 is left unprimed, and a variable corresponding to a 1 is primed. The rule for omitting variables is the same for prime implicates as for prime implicants. An example of the formation of a minimal product by means of a map is given in Fig. 4.1-17a.

The minimal product corresponds to the two-stage diode gate circuit in which the output is derived from an AND gate and which contains the minimum number of gates. For some functions the minimal sum leads to a more economical circuit, and for other functions the converse is true.

For the function of Fig. 4.1-17 the minimal product requires fewer gates (three) than the minimal sum (five). There is no known method for determining which form will lead to a more economical circuit without actually obtaining both the minimal sum and product.

In obtaining a minimal product, d terms are handled in exactly the same fashion as in obtaining a minimal sum. The process of forming a minimal product for a function f is exactly equivalent to that of writing a minimal sum for the complementary function f'. If a minimal sum is written for f' and then Theorem 14† is used to obtain f, the result will be the minimal product for f.

Additional techniques using maps are described in [4 and 5].

4.2 GENERAL PROPERTIES OF MINIMAL SUMS AND PRODUCTS [6]

There are several other definitions of minimal sum and minimal product besides the definitions given in the preceding section. Two factors affect the definition chosen: (1) the devices and circuits to be used and (2) the difficulty of obtaining an expression satisfying the definition. The definitions for minimal sums and minimal products used in the preceding section correspond to a two-stage diode gate circuit which contains the minimum number of gates. This is appropriate for circuits constructed out of standard gate packages which are prewired for a fixed number of inputs. In such circuits, it is customary to minimize the total number of packages. When the circuits are to be built by interconnecting diodes and resistors, the common practice is to minimize the total number of diodes. For this, the desired expression is one in which the sum of the number of literals plus the number of terms (or factors) containing more than a single literal is a minimum. The number of terms (or factors) which contain more than a single literal is equal to the number of gates whose outputs are connected as inputs to the output gate. The number of literals is equal to the number of gate inputs which are connected to circuit inputs. Since the number of diodes equals the number of gate inputs, this is the proper quantity to minimize. These remarks apply only to two-stage circuits. Other types of circuits will be discussed in a later section. Another possible definition of the minimal sum would involve obtaining the expression containing the fewest literals.

For the vast majority of functions, these three definitions of the minimal sum all result in the same final expression. Because of this, the exact definition used is relatively unimportant. Since the definition of the minimal sum given in Sec. 4.1 is ordinarily easier to use than the other definitions, this is the one which will be used here.

† $f(X_1, X_2, \ldots, X_n, +, \cdot)' = f(X_1', X_2', \ldots, X_n', \cdot, +)$.

4.2 general properties of minimal sums and products

In general, a rule is given for associating a number called the *cost* with each sum-of-products expression, and the minimal sum is defined as an expression for which this cost takes on its minimum value. The three definitions of the minimal sum given previously can all be restated in terms of this cost function. If the total number of literals is to be minimized, then the cost is just equal to the number of literal appearances in an expression. When the number of gate inputs is to be minimized, the cost is defined as the number of literals plus the number of terms (or factors) containing more than one literal. If the definition of Sec. 4.1 is used, the appropriate cost is equal to the number of literals plus a large multiple of the number of terms containing more than one literal. This is necessary so that the cost will always be lower for an expression containing more literals but fewer terms than some other expression.

Example 4.2-1
$f = w'y + xy'z + wx'y$
N_l = number of literals = 8
N_t = number of terms containing more than 1 literal = 3
$(\text{Cost})_A = N_l = 8$
$(\text{Cost})_B = N_l + N_t = 8 + 3 = 11$
$(\text{Cost})_C = N_l + 10^6 N_t = 8 + 3 \times 10^6$

These three definitions of cost all have the property that the cost increases if a literal is added to any expression. The following theorem shows that the corresponding minimal sums must always contain only prime implicants.

It is usually true that the total number of inputs that any single gate can have is restricted. To be strictly accurate, this restriction should be taken into account in the cost function. Since there has been only limited success in including such considerations in the theory of combinational-circuit design [7], no further attention will be devoted to them here.

Prime Implicant Theorem

> **Theorem 4.2-1.** A minimal sum must always consist of a sum of prime implicants if any definition of cost is used in which the addition of a single literal to any expression increases the cost of the expression [3].

In order to prove this theorem, a more formal definition of prime implicant is required. It is convenient first to give the following preliminary definition:

> **Definition.** One function $f(x_1, x_2, \ldots, x_n)$ is said to *include* another function $g(x_1, x_2, \ldots, x_n)$, written $f \supseteq g$ or $g \subseteq f$, if, for each combination of values of the variables for which $g = 1$, it is also

Simplification of Switching Functions

true that $f = 1$. This does not exclude the case where $f = g$. In this case $f \supseteq g$ and $g \supseteq f$.

Thus, if $f \supseteq g$, then f has a 1 in every row of the table of combinations in which g has a 1. In terms of the decimal specification, $f \supseteq g$ if every number which appears in the decimal specification of g as a canonical sum also appears in the decimal specification of f as a canonical sum. It follows from this definition that, if

$$f(x_1, x_2, \ldots, x_n) = f_1(x_1, x_2, \ldots, x_n) + f_2(x_1, x_2, \ldots, x_n)$$

then $f \supseteq f_1$ and $f \supseteq f_2$, since $f = 1$ whenever f_1 or f_2 equals 1.

Definition. A prime implicant of a function $f(x_1, x_2, \ldots, x_n)$ is a product of literals $x_{i_1}^* x_{i_2}^* \cdots x_{i_m}^*$, $m \leq n$, which is included in f—$f \supseteq x_{i_1}^* x_{i_2}^* \cdots x_{i_m}^*$—and which has the property that, if any literal is removed from the product, the remaining product is not included in f.

The prime implicants which were discussed in connection with the n-cube maps (Sec. 4.1) satisfy this definition. The product terms derived from the map equal 1 only for combinations of values of the variables which correspond to 1 cells of the map. Removing a literal from a product term corresponds to picking a larger set of 1 cells on the map, but this contradicts the rule of Sec. 4.1 that prime implicants correspond to sets of 1 cells which are not included in any larger set of 1 cells.

PROOF OF PRIME IMPLICANT THEOREM. Suppose that, for some function f, a minimal sum exists in which at least one of the product terms is not a prime implicant. Let $f = P + R$, where P is a term which is not a prime implicant and R is equal to the remaining terms of f. Then $f \supseteq P$, as discussed previously. Since P is not a prime implicant, it must be possible to remove a literal from P, forming $Q (P = x_i^* Q)$, and have $f \supseteq Q$. Since Q equals 1 whenever $P = 1$, $f = Q + R$. This is a sum-of-products expression which contains the same number of terms but one fewer literal than the given minimal sum, $f = P + R$. For any definition of cost which decreases with the removal of a literal, this proves that the original expression could not be a minimal sum and that any minimal sum must contain only prime implicants.

This theorem does not apply to a situation in which only the gate cost is important and the number of gate inputs has no effect on the cost of the circuit. It is possible to prove a more general theorem which does apply to this situation and which shows it is always possible to obtain a minimal sum by considering only sums of prime implicants.

Generalized Prime Implicant Theorem

Theorem 4.2-2. When a definition of circuit cost is used such that the cost does not increase when a literal is removed from the corresponding sum-of-products expression, there is at least one minimal sum which is a sum of prime implicants.

PROOF. Suppose that a minimal sum exists which is not a sum of prime implicants. This means that there must be some product terms which are not prime implicants, because some of their literals can be removed without changing the fact that they are included in the original function. Consider the expression which results from removing all such literals. It must be a sum of prime implicants. Further it must be a minimal sum, since the removal of the literals does not increase the cost associated with the expression.

4.3 MULTIPLE-OUTPUT NETWORKS

Very often combinational circuits are desired which have several outputs rather than just one. The design specifications for such a multiple-output network typically consist of several functions—$f_1(w,x,y,z)$, $f_2(w,x,y,z)$, . . . , $f_m(w,x,y,z)$—of the same set of input variables. Each function refers to one of the output leads and specifies the relationship between the condition on this output lead and the conditions of the input leads. The methods described in the preceding sections for single-output networks can be extended to the multiple-output case, but certain modifications will be required [8,9].

Perhaps the most obvious technique to try for the multiple-output case is to find minimal sums for each of the output functions separately and then to construct a separate network for each output function. While this technique has the advantage of simplicity, it unfortunately does not lead to the most economical (two-stage) multiple-output network. The truth of this statement can be demonstrated by means of some simple examples. If the functions $f_1(x,y,z) = \Sigma(1,3,7)$ and $f_2(x,y,z) = \Sigma(3,6,7)$ are minimized separately, the network of Fig. 4.3-1b results. A more economical two-stage network is shown in Fig. 4.3-1c. For this circuit it is fairly obvious that it is not necessary to include two AND gates to form the yz term appearing in both outputs. A less obvious example is shown in Fig. 4.3-2 for the functions $f_1(x,y,z) = \Sigma(1,3,7)$ and $f_2(x,y,z) = \Sigma(2,6,7)$. In the most economical circuit for these functions, use is made of the term xyz, which is not a prime implicant of either of the output functions. For a single-output circuit, only the prime implicants need be considered in determining the minimal two-stage circuit. This example shows that it is not sufficient to consider only the

Simplification of Switching Functions

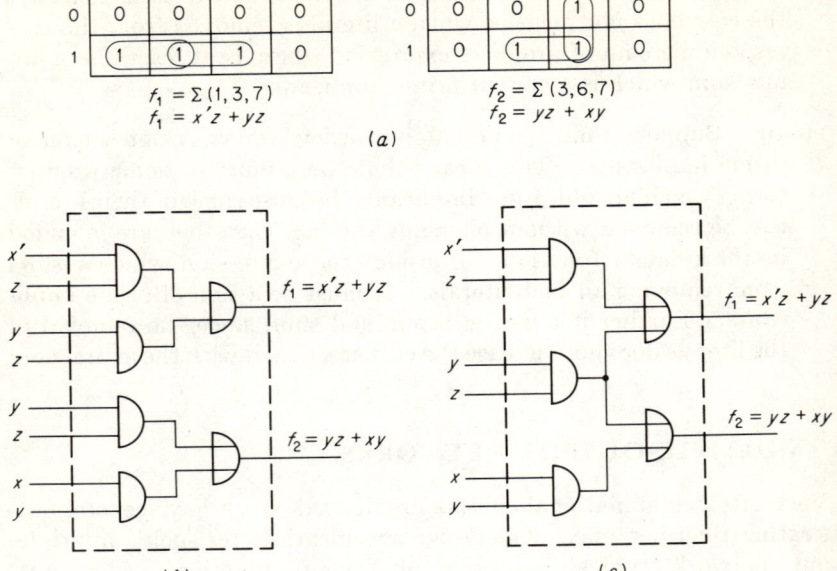

Fig. 4.3-1. *An example of a multiple-output circuit.* (a) *Output functions;* (b) *circuit obtained from minimal sums for f_1 and f_2;* (c) *minimal two-stage circuit for f_1 and f_2.*

prime implicants of the output functions in designing multiple-output networks. A more general study of the multiple-output problem will be necessary before a synthesis technique can be arrived at.

Multiple-output Prime Implicants

The general form of a two-stage three-output circuit is shown in Fig. 4.3-3. Each of the input gates must directly drive output gates because of the two-stage requirement. The only freedom which exists is in the number of output gates which are directly driven by a given input gate. The numbers inside the AND gates indicate to which outputs they are connected. Thus, the AND gates labeled 1 are connected only to output f_1, those labeled 23 are connected only to outputs f_2 and f_3, etc. The AND gates which "drive" a single output will be considered first. Whenever all the inputs to one of these gates are equal to 1, the output of the AND gate will be equal to 1 and the output of the OR gate to which it is connected will also be equal to 1. Thus the product $x_{i_1}^* x_{i_2}^* \cdots x_{i_k}^*$ realized by an AND gate labeled j must be included in the function f_j (Whenever this product is 1, the function is 1.) The products realized by

4.3 multiple-output networks

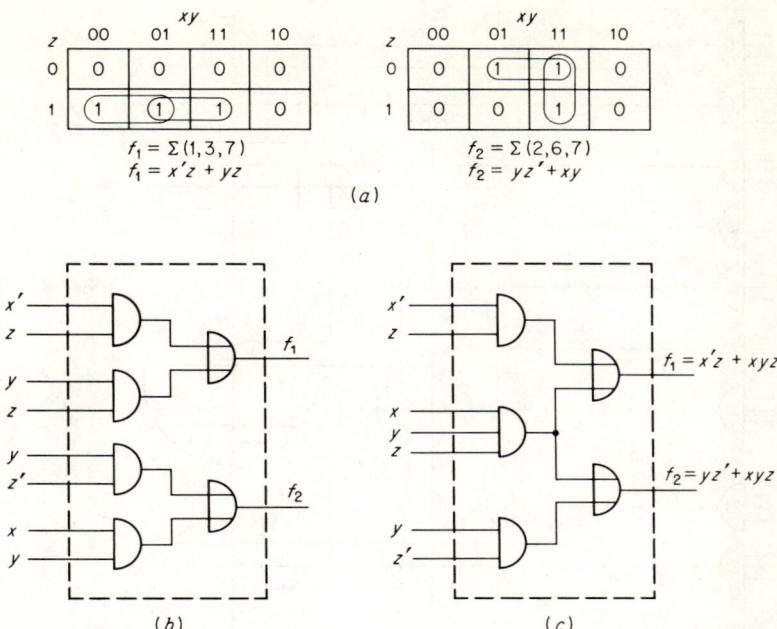

Fig. 4.3-2. *An example of a multiple-output circuit.* (a) *Output functions;* (b) *circuit obtained from minimal sums for f_1 and f_2;* (c) *minimal two-stage circuit for f_1 and f_2.*

the gates labeled j therefore satisfy the first requirement on the prime implicants of the function f_j. For the usual definitions of circuit cost it would be uneconomical to include any more inputs to these gates than necessary. Under these conditions, the AND gates with single labels must also satisfy the second prime-implicant requirement—none of the inputs to a gate labeled j can be removed without changing the function f_j. Throughout the remainder of this section it will be assumed that *the cost criterion used is one for which the addition of an input to any gate increases the circuit cost*. It follows from this that *the AND gates which are connected only to output f_j must correspond to prime implicants of f_j*. Input variables which are connected directly to output gates can be thought of as input AND gates with only one input.

The situation for input gates which are connected to two outputs is more complicated. Whenever all the inputs to one of these gates are equal to 1, the gate output will be equal to 1 and *both* the circuit outputs to which this input gate is connected will be equal to 1. Thus the product $x_{i_1}^* x_{i_2}^* \cdots x_{i_p}^*$ realized by a gate labeled jk is included not only in f_j and f_k but also in the function $f_j \cdot f_k$, which is 1 only when both f_j and f_k

Simplification of Switching Functions

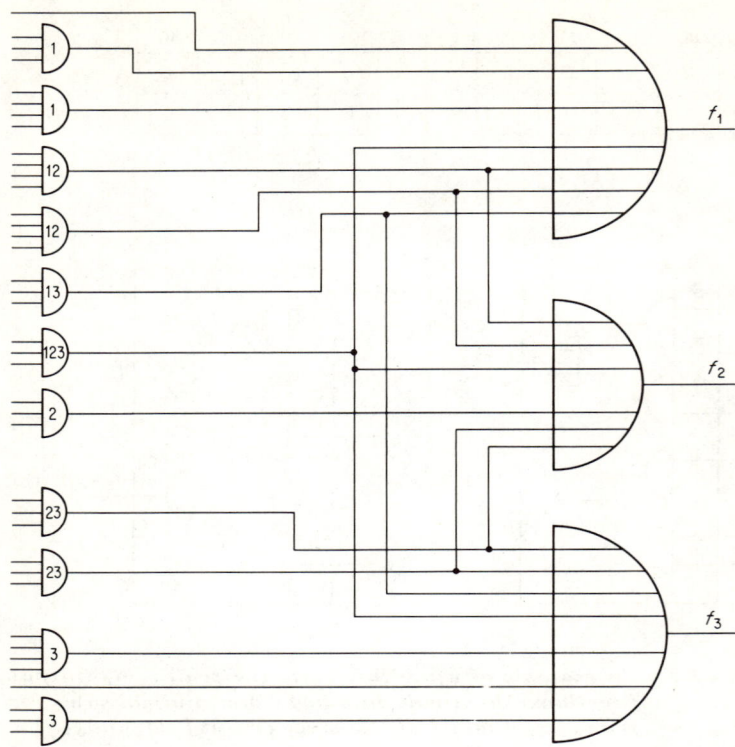

Fig. 4.3-3. *General form of two-stage three-output circuit.*

are equal† to 1. If the circuit is minimal, removing one of the inputs to a gate labeled jk must change either f_j or f_k or both. Thus the product realized by this gate satisfies both the requirements for prime implicants for the function $f_j \cdot f_k$. From this it follows that *the AND gates which are connected to both outputs f_j and f_k must correspond to prime implicants of the function $f_j \cdot f_k$*. These AND gates need not correspond to prime implicants of f_j, for example, since removing one of the input leads need not change f_j if f_k is changed. Thus the second requirement for prime implicants of f_j need not be satisfied. Removal of a different input lead might change f_j but not f_k so that the product would be a prime implicant of $f_j \cdot f_k$ but not a prime implicant of f_j or of f_k. This situation is illustrated by the functions f_1 and f_2 of Fig. 4.3-2. The product xyz is prime implicant of $f_1 \cdot f_2 = \Sigma(7)$ but is not a prime implicant of either (because, for f_1, the x can be removed, leaving yz) or f_2 (because, for f_2, the z can be removed, leaving xy).

† If $f_j(x,y,z) = \Sigma(1,3,7) = x'z + yz$ and $f_k(x,y,z) = \Sigma(2,6,7) = yz' + xy$, then $f_j \cdot f_k = \Sigma(7) = (x'z + yz)(yz' + xy) = xyz$.

4.3 multiple-output networks

The situation for gates which are connected to more than two outputs is a direct extension of the two-output case. The entire situation can be summarized as follows: *an AND gate which is connected to all the outputs $f_1, f_2, \ldots, f_r$ must correspond to a prime implicant of the function $f_1 \cdot f_2 \cdots f_r$.*

In the technique which was developed for single-output networks all the prime implicants were determined, and then a selection was made of those prime implicants which should appear in the minimal circuit. In this case the set of prime implicants corresponds to all the input gates which could possibly appear in a minimal circuit. For the multiple-output problem, the set of possible input gates corresponds to all the prime implicants of $f_1 \cdot f_2, f_1 \cdot f_3, f_2 \cdot f_3, \ldots, f_1 \cdot f_2 \cdot f_3, f_1 \cdot f_2 \cdot f_4, \ldots$. In designing a multiple-output network it is necessary to generate the prime implicants of each of the individual output functions plus the prime implicants of the functions which are equal to all possible products of two output functions, of three output functions, etc. This collection of prime implicants will be called the *multiple-output prime implicants*. The algebraic expressions corresponding to the minimal multiple-output circuits will be called *multiple-output minimal sums*. The selection of those prime implicants to be used in the minimal circuit is similar to the single-output technique. Naturally, all these remarks about multiple-output networks with OR gates as output gates apply equally well to networks using AND gates as output gates, provided that the obvious changes in terminology are made. The preceding discussion of multiple-output minimal sums can be summarized formally as follows:

Definition. A *multiple-output prime implicant* of a set of functions $\{f_1(x_1, x_2, \ldots, x_n), f_2(x_1, x_2, \ldots, x_n), \ldots, f_m(x_1, x_2, \ldots, x_n)\}$ is a product of literals $x_{i_1}^* x_{i_2}^* \cdots x_{i_h}^* (h \leq n, 1 \leq i_j \leq n)$, which is:

1. Either a prime implicant of one of the functions

$$f_k(x_1, x_2, \ldots, x_n)$$

2. Or a prime implicant of one of the product functions

$$f_{j_1}(x_1, x_2, \ldots, x_n) \cdot f_{j_2}(x_1, x_2, \ldots, x_n) \cdots f_{j_l}(x_1, x_2, \ldots, x_n)$$

The fact that the multiple-output prime implicants are the only product terms that need be considered in designing a minimum-cost two-stage multiple-output network is demonstrated by the following theorem:

Theorem 4.3-1. For any definition of network cost such that the cost does not increase when a gate or gate input is removed, there exists at least one minimum-cost two-stage network in which the corresponding expressions for the output functions f_j are all sums of multiple-output prime implicants. All the product terms which

Simplification of Switching Functions

occur only in the expression for f_j are prime implicants of f_j; all the product terms which occur in both the expressions for f_j and f_k but in no other expressions are prime implicants of $f_j \cdot f_k$, etc.

Essential Multiple-Output Prime Implicants and Maps

The first step in obtaining a multiple-output minimal sum is to determine the multiple-output prime implicants. One technique for doing this is to form all the appropriate product functions and then for each of these to obtain the prime implicants. This procedure is illustrated for a three-output example in Fig. 4.3-4.

Once the multiple-output prime implicants have been determined, a selection must be made of the prime implicants to be used in constructing the minimal circuit. Just as in the case of single-output circuits these are essential prime implicants which *must* be included in any minimal multiple-output circuit. For example, in the map of Fig. 4.3-4 for f_1 the essential

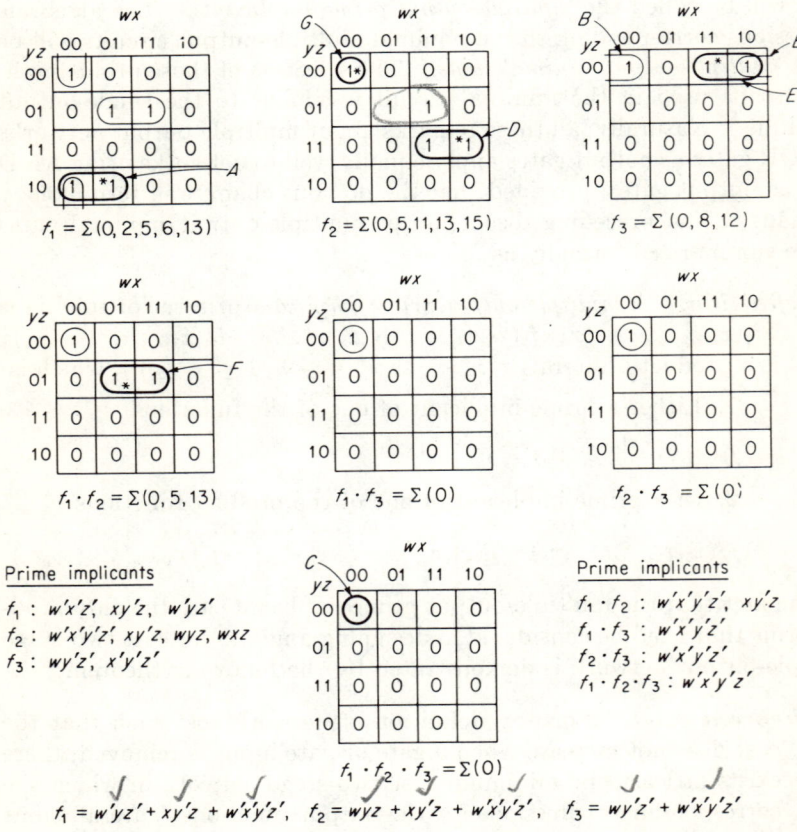

Fig. 4.3-4. Determination of multiple-output prime implicants for $f_1 = \Sigma(0,2,5,6,13)$, $f_2 = \Sigma(0,5,11,13,15)$, and $f_3 = \Sigma(0,8,12)$.

prime implicant $A(w'yz')$ is shown darkened, since it is the only prime implicant which includes the distinguished fundamental product $w'xyz'$. On the other hand, the prime implicant $B(x'y'z')$ of f_3 is *not* essential even though it is the only prime implicant of f_3 which includes the fundamental product $w'x'y'z'$. The reason why this prime implicant is *not* essential is that the fundamental product $w'x'y'z'$ is also included in the prime implicant $C(w'x'y'z')$ of $f_1 \cdot f_2 \cdot f_3$. Basically, the reason why B is not essential is that there is a choice possible between using B or using C to ensure that f_3 will be equal to 1 when $w = x = y = z = 0$ ($w'x'y'z' = 1$). In general, a multiple-output prime implicant is essential for a function f_i if there is a fundamental product of f_i that is included in only the one multiple-output prime implicant.

Definition. Let $\{f_1(x_1,x_2, \ldots ,x_n), f_2(x_1,x_2, \ldots ,x_n), \ldots ,f_m(x_1, x_2, \ldots ,x_n)\}$ be a set of output functions, and let $\{P_1, P_2, \ldots , P_t\}$ be the corresponding set of multiple-output prime implicants. Then, a fundamental product of one of the output functions of the set, f_i, is a *distinguished fundamental product* if and only if the fundamental product is included in only one product of literals which is a multiple-output prime implicant of f_i or of any of the product functions involving f_i ($f_i \cdot f_j$, $f_i \cdot f_j \cdot f_k$, etc.).

Definition. A multiple-output prime implicant of a set $\{f(x_1,x_2, \ldots ,x_n), f_2(x_1,x_2, \ldots ,x_n), \ldots ,f_m(x_1,x_2, \ldots ,x_n)\}$ is essential *for an output function f_i* if and only if it includes a distinguished fundamental product of f_i.

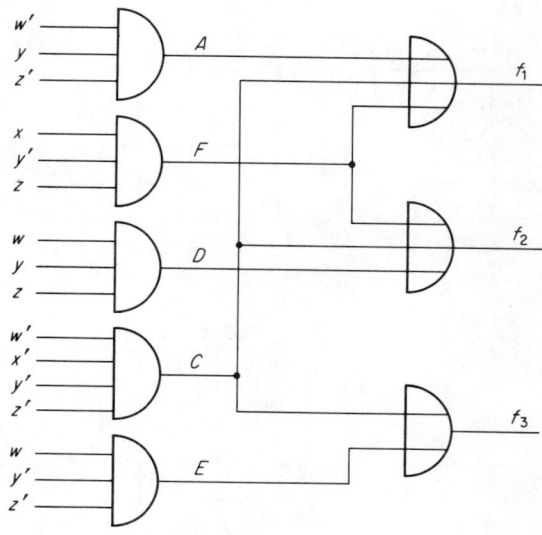

Fig. 4.3-5. Minimal circuit for the functions of Fig. 4.3-4.

Theorem 4.3-2. Let $\{f_1(x_1,x_2,\ldots,x_n), f_2(x_1,x_2,\ldots,x_n),\ldots, f_m(x_1,x_2,\ldots,x_n)\}$ be a set of output functions, and let $\{E_1(x_1,x_2,\ldots,x_n), E_2(x_1,x_2,\ldots,x_n),\ldots, E_m(x_1,x_2,\ldots,x_n)\}$ be a set of multiple-output minimal sums corresponding to these functions. Then a multiple-output prime implicant which is essential for function f_i must appear in the corresponding minimal-sum expression E_i.

In Fig. 4.3-4 the distinguished fundamental products are marked with an asterisk, and the corresponding essential prime implicants are shown darkened. The prime implicant G is a little different from the others. It is an essential prime implicant of f_2 but is also a prime implicant of $f_1 \cdot f_2 \cdot f_3$. What this means is that it is necessary that G be used for f_2 and that it may also be used for f_1 and f_3. For example, it is possible to use B rather than G for f_3. It is not possible to reduce the total number of gates by *not* using G for f_1 and f_3 as well as f_2; the only possibility is that

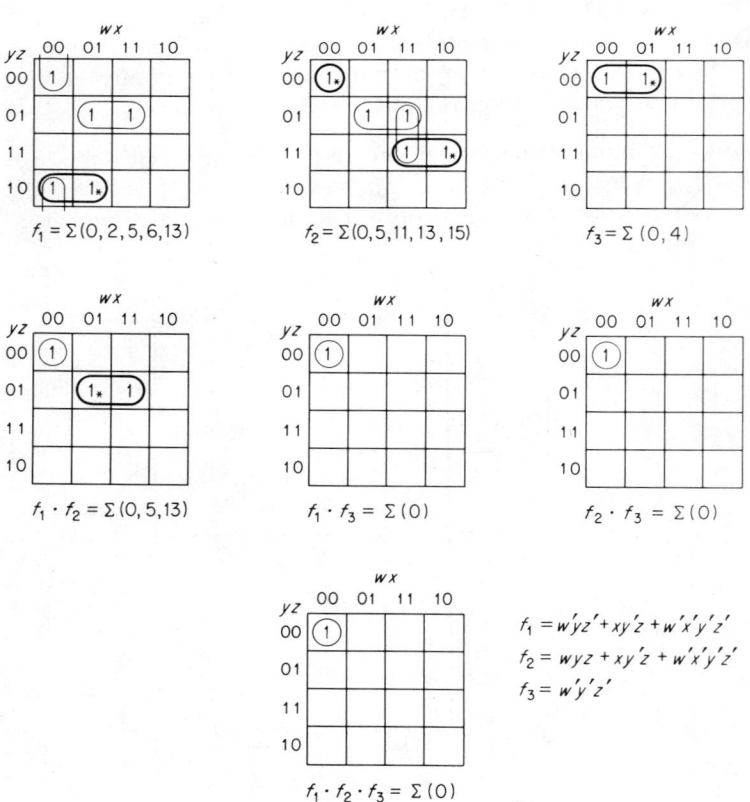

Fig. 4.3-6. Determination of multiple-output minimal sums for $f_1 = \Sigma(0,2,5,6,13)$, $f_2 = \Sigma(0,5,11,13,15)$, and $f_3 = \Sigma(0,4)$.

some inputs can be saved on the f_1 or f_3 OR gates. This will be illustrated in a later example. G is an example of a prime implicant of $f_1 \cdot f_2 \cdot f_3$ which is essential only for f_2; F is an example of a prime implicant of $f_1 \cdot f_2$ which is essential for both f_1 and f_2.

The prime implicants A, D, E, G, and F of Fig. 4.3-4 have been determined as essential. Only two fundamental products which have not been included in prime implicants remain: $w'x'y'z'$ of f_1 and $w'x'y'z'$ of f_3. The obvious solution is to connect G to f_1 and f_3 as well as to f_2, since this requires no additional gates. The resulting circuit is shown in Fig. 4.3-5.

Figure 4.3-6 shows a three-output problem in which there is a prime implicant $w'x'y'z'$ of $f_1 \cdot f_2 \cdot f_3$ which is used only for f_1 and f_2 in the minimal circuit. The fundamental product $w'x'y'z'$ is included in f_3, but it is not necessary to connect the $w'x'y'z'$ gate output to f_3, since this fundamental product is included in the $w'y'z'$ prime implicant which is essential for f_3. By not using the $w'x'y'z'$ gate output it is possible to avoid using any OR gate in forming f_3. The circuit corresponding to Fig. 4.3-6 is shown in Fig. 4.3-7.

If a realization is desired in which the outputs are derived from AND gates rather than OR gates (minimal products rather than minimal sums), the easiest procedure is to work with $f_1', f_2', \ldots, f_m'$ to get minimal sums and then obtain minimal products by using DeMorgan's theorem to obtain f_i from f_i'.

Unspecified output conditions are treated in exactly the same fashion as for single-output problems—the d entries are treated as 1's in forming the prime implicants and are then disregarded in forming the minimal sums.

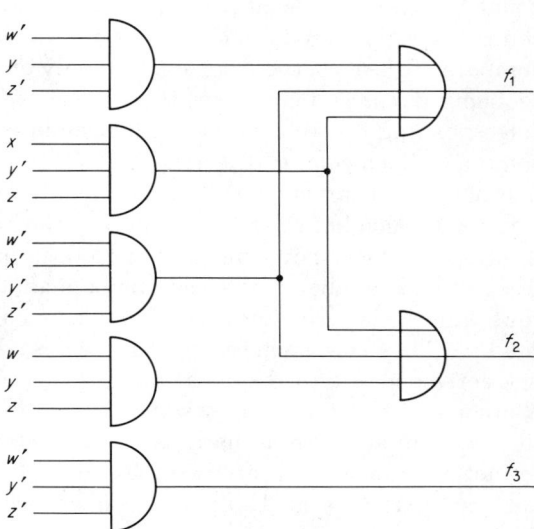

Fig. 4.3-7. Minimal circuit for the functions of Fig. 4.3-6.

4.4 TABULAR DETERMINATION OF PRIME IMPLICANTS

The map techniques described in the preceding sections are usually quite satisfactory for functions of four or fewer variables and are useful for some functions of five or six variables. In order to handle functions of larger numbers of variables, different techniques must be developed. What would be most desirable would be an algorithm for obtaining minimal sums which could be used for hand calculation and also could be programmed on a digital computer. An algorithm satisfying these criteria will be described first for single-output functions [10] and will then be extended to multiple-output functions.

Binary-character Method†

In order to avoid any limitation as to the number of variables which can be handled, it will be necessary to avoid any attempt to rely on geometric intuition and to work directly with the fundamental products. Actually it will be more convenient to work with binary characters corresponding to algebraic product terms than to work with the algebraic expressions themselves. Thus each fundamental product will be represented by the corresponding row of the table of combinations; for example, the fundamental product $w'xy'z$ will be represented by the binary character 0101. The first step in this minimization procedure is to list in a column the binary characters corresponding to the fundamental products of the function for which the minimal sum is desired. This amounts to listing the rows of the table of combinations for which the function is to equal 1. Each row is labeled with its decimal equivalent. Such a list is shown in Table 4.4-1a. Each pair of these binary characters is then compared to see whether they differ in only one coordinate (corresponding to being distance 1 apart on the n-cube). The fundamental products corresponding to two characters differing in only one coordinate could be combined by means of the theorem $XY + XY' = X$. Thus, for each pair of binary characters which differ in only one position, a new character is formed which has the same value as both the original characters in each position in which they agree and has a dash in the position in which they disagree. The label of the new row is made up of the labels of the two rows from which it is formed. These new characters are listed in Table 4.4-1b. They correspond to product terms which have one variable (the one corresponding to the dash) missing. A check is placed next to each character of the first column which is used in forming a character of the second column, since a character which can be combined with another cannot correspond to a prime implicant.

The characters in Table 4.4-1a have been arranged in a particular format in order to facilitate the comparison process. They have been

†Additional examples are available in the discussion of the Quine-McCluskey method on pages 145-156 of [4].

Table 4.4-1. Determination of Prime Implicants for
$f = \Sigma(0,2,4,6,7,8,10,11,12,13,14,16,18,19,29,30)$

(a)

	v	w	x	y	z	
(0)	0	0	0	0	0	✓
(2)	0	0	0	1	0	✓
(4)	0	0	1	0	0	✓
(8)	0	1	0	0	0	✓
(16)	1	0	0	0	0	✓
(6)	0	0	1	1	0	✓
(10)	0	1	0	1	0	✓
(12)	0	1	1	0	0	✓
(18)	1	0	0	1	0	✓
(7)	0	0	1	1	1	✓
(11)	0	1	0	1	1	✓
(13)	0	1	1	0	1	✓
(14)	0	1	1	1	0	✓
(19)	1	0	0	1	1	✓
(29)	1	1	1	0	1	✓
(30)	1	1	1	1	0	✓

(b)

	v	w	x	y	z	
(0,2)	0	0	0	–	0	✓
(0,4)	0	0	–	0	0	✓
(0,8)	0	–	0	0	0	✓
(0,16)	–	0	0	0	0	✓
(2,6)	0	0	–	1	0	✓
(2,10)	0	–	0	1	0	✓
(2,18)	–	0	0	1	0	✓
(4,6)	0	0	1	–	0	✓
(4,12)	0	–	1	0	0	✓
(8,10)	0	1	0	–	0	✓
(8,12)	0	1	–	0	0	✓
(16,18)	1	0	0	–	0	✓
(6,7)	0	0	1	1	–	
(6,14)	0	–	1	1	0	✓
(10,11)	0	1	0	1	–	
(10,14)	0	1	–	1	0	✓
(12,13)	0	1	1	0	–	
(12,14)	0	1	1	–	0	✓
(18,19)	1	0	0	1	–	
(13,29)	–	1	1	0	1	
(14,30)	–	1	1	1	0	

(c)

	v	w	x	y	z	
(0,2,4,6)	0	0	–	–	0	✓
(0,2,8,10)	0	–	0	–	0	✓
(0,2,16,18)	–	0	0	–	0	
(0,4,8,12)	0	–	–	0	0	✓
(2,6,10,14)	0	–	–	1	0	✓
(4,6,12,14)	0	–	1	–	0	✓
(8,10,12,14)	0	1	–	–	0	✓

(d)

	v w x y z
(0,2,4,6,8,10,12,14)	0 – – – 0

Simplification of Switching Functions

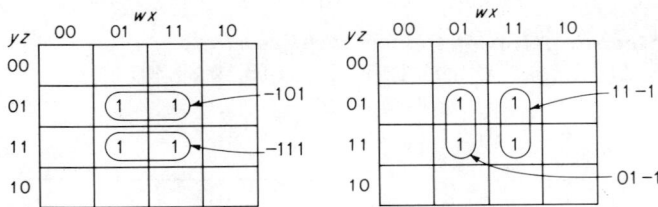

Fig. 4.4-1. Maps for Table 4.4-2.

arranged in groups so that all the characters in one group contain the same number of 1 entries. Since two characters that differ in only one position must also differ by exactly 1 in the number of 1's which they contain, it is necessary only to compare each character with each of the characters in the next lower group. Comparison with the next upper group is not necessary, since this would cause each comparison to be carried out twice. A simple rule is that a new character is formed when a character has 1's wherever the character (from the next upper group) with which it is being compared has 1's.

The characters in the second column are now compared in the same fashion, and a new character is formed whenever two characters are found which differ in only one position and which both have their dashes in the same position. Again the characters which are used in forming a new character have check marks placed next to them. This procedure corresponds to combining two product terms which both have the same variable missing and which are identical in all the other variables except one. This procedure will result in a new column such as Table 4.4-1c. The same procedure is repeated for each of the new columns (always requiring that the dashes "line up" in two characters which can combine) until no further combinations are possible. Of course, one character can combine with several other characters to form new characters. The unchecked characters which remain after no further combinations are possible correspond to the prime implicants.

One phenomenon should be pointed out in connection with this process. Each of the characters with more than one dash will be formed in more than one way. This is illustrated in Table 4.4-2, and the corresponding map is shown in Fig. 4.4-1. One way to avoid this repetition is to compare only pairs of characters whose labels form an increasing sequence of decimal numbers. If this is done, care must be taken to place check marks not only next to all characters used in forming new characters but also next to each character which has all the numbers of its label occurring in another character's label. Thus, in Table 4.4-2 the characters (5,13) and (7,15) would not be compared, since their labels do not form an increasing sequence, but they would both receive check marks

4.4 tabular determination of prime implicants

Table 4.4-2. An Example of the Two Ways of Forming a Character with Two Dashes

	w	x	y	z			w	x	y	z			w	x	y	z
(5)	0	1	0	1	(5,7)		0	1	–	1	(5,7,13,15)		–	1	–	1
					(5,13)		–	1	0	1	[(5,13,7,15)]		–	1	–	1]
(7)	0	1	1	1												
(13)	1	1	0	1	(7,15)		–	1	1	1						
					(13,15)		1	1	–	1						
(15)	1	1	1	1												

because their labels are included in the (5,7,13,15) label. The other possibility is to ignore the labels and form each character several times as a check on the calculations but not to write down the repetitions of the characters.

Use of Octal Numbers

For hand calculation it is usually easier to work with octal numbers rather than the binary characters [11]. This can be done by making use of the fact that, if two binary numbers differ in only 1 bit, the two corresponding octal numbers must differ in only 1 octal digit (since each octal digit corresponds directly to 3 binary digits) and that the difference in the single octal digit must be a power of 2 (1, 2, or 4). This is illustrated in Table 4.4-3, which lists all the combinations possible for 3 binary digits, along with the corresponding differences in the octal equivalents. Table 4.4-4 shows the use of octal numbers to determine the prime implicants for the function of Table 4.4-1.

The octal equivalents of the binary numbers corresponding to the fundamental products are listed in a column according to the number of 1's in the binary numbers (Table 4.4-4a). Each number of this column is then compared with each number of the next lower group; and if two numbers differ in only one of their octal digits, the difference being a power of 2, an entry is made in the next column (Table 4.4-4b). This entry is a double entry consisting of the smaller of the pair of octal numbers which differ by a power of 2 and also the power of 2 by which they differ. Thus the two octal numbers 15 and 35 would give rise to a new entry 15,20. The smaller octal number is called the *base label* (15), and the power of 2 is called the *difference label* (20). These entries are partitioned into groups according to the number of 1's in their base labels.†

† The easiest way to do this is to note that two entries will have the same number of 1's in their base labels only if they were formed from entries from the same pair of groups of the preceding column.

Simplification of Switching Functions

Table 4.4-3. **Example Showing that Octal Equivalents of Two Binary Numbers Which Combine Always Differ by a Power of 2**

	x	y	z
0	0	0	0
1	0	0	1
2	0	1	0
4	1	0	0
3	0	1	1
5	1	0	1
6	1	1	0
7	1	1	1

		x	y	z
0,1	(1)	0	0	–
0,2	(2)	0	–	0
0,4	(4)	–	0	0
1,3	(2)	0	–	1
1,5	(4)	–	0	1
2,3	(1)	0	1	–
2,6	(4)	–	1	0
4,5	(1)	1	0	–
4,6	(2)	1	–	0
3,7	(4)	–	1	1
5,7	(2)	1	–	1
6,7	(1)	1	1	–

↑—Difference in octal labels

Each entry of the second column (Table 4.4-4b) is then compared with each entry of the next lower group having the *same difference label*. A new entry in the third column is formed whenever two entries of the second column are found which have the same difference label and which differ by a power of 2 in their base labels. The new entry has the same base label as the smaller of the two octal numbers from which it is formed and has a difference label made up of the difference label of the original pair of entries and also of the power of 2 by which the base labels of the two entries differ. Thus, the two second-column entries 02; 04 and 12; 04 will produce the third-column entry 02; 04, 10.

This process is repeated, the rule always being continued that entries can be combined only if their difference labels are identical, until no further combinations are possible. The rule about the difference labels arises from the fact that they specify the locations of the dashes in the corresponding binary characters. The rule for assigning check marks is the same as in using binary characters, and the same remarks about the formation of an entry in more than one way also still apply.

Table 4.4-5 illustrates the procedure for converting from the octal

4.4 tabular determination of prime implicants

Table 4.4-4. Determination of Prime Implicants for the Function of Table 4.4-1 by Use of Octal Numbers

(a)		(b)			(c)		
00	✓	Base	Difference		Base	Difference	
		00	02	✓	00	02,04	✓
02	✓	00	04	✓	00	02,10	✓
04	✓	00	10	✓	00	02,20	G
10	✓	00	20	✓	00	04,10	✓
20	✓						
		02	04	✓	02	04,10	✓
06	✓	02	10	✓	04	02,10	✓
12	✓	02	20	✓	10	02,04	✓
14	✓	04	02	✓			
22	✓	04	10	✓		(d)	
		10	02	✓			
07	✓	10	04	✓	00	02,04,10	H
13	✓	20	02	✓			
15	✓						
16	✓	06	01	A			
23	✓	06	10	✓			
		12	01	B			
35	✓	12	04	✓			
36	✓	14	01	C			
		14	02	✓			
		22	01	D			
		15	20	E			
		16	20	F			

entries representing the prime implicants to the corresponding prime implicants. This is done by converting the base label to the equivalent binary number and then replacing by dashes the binary digits corresponding to the powers of 2 appearing in the difference labels. The binary numbers corresponding to the fundamental products included in a given prime implicant are obtained by replacing the dashes in the binary character representing the prime implicant by all possible combinations of 0's and 1's. For example, the binary numbers corresponding to – 0 1 – are 0 0 1 0, 0 0 1 1, 1 0 1 0, 1 0 1 1. Schemes have also been devised for working directly with the decimal equivalents [12]

Simplification of Switching Functions

Table 4.4-5. Formation of the Prime Implicants from the Unchecked Entries of Table 4.4-4

(a)			(b) v w x y z	(c)
A	06	01	0 0 1 1 —	v' w' x y
B	12	01	0 1 0 1 —	v' w x' y
C	14	01	0 1 1 0 —	v' w x y'
D	22	01	1 0 0 1 —	v w' x' y
E	15	20	— 1 1 0 1	w x y' z
F	16	20	— 1 1 1 0	w x y z'
G	00	02,20	— 0 0 — 0	w' x' z'
H	00	02,04,10	0 — — — 0	v' z'

rather than the octal equivalents, but these decimal techniques are more involved than the octal method and are less generally useful.

The techniques just described will result in a list of all the prime implicants. Additional techniques are required for selecting the prime implicants to be used in the minimal sum.

4.5 PRIME IMPLICANT TABLES

The basic requirement which the terms of the minimal sum must satisfy is that each fundamental product of the function must be included in at least one of the terms of the minimal sum. The relation between the fundamental products and the prime implicants can be specified most conveniently by means of a *prime implicant table* [10] such as Table 4.5-1. Each row of this table corresponds to a prime implicant, and each column corresponds to a fundamental product. An × is placed at the intersection of a row and column if the corresponding prime implicant includes the corresponding fundamental product. In terms of the table the basic requirement on the minimal-sum terms becomes that *each column must have an × in at least one of the rows which correspond to minimal-sum terms.*

Essential Rows

If any column contains only a single ×, the column corresponds to a distinguished fundamental product (it is included in only one prime implicant) and the row in which the × occurs corresponds to an essential prime implicant. Rows and columns corresponding to essential prime implicants and distinguished fundamental products will be called *essential rows* and *distinguished columns*, respectively.

4.5 prime implicant tables

Table 4.5-1. Prime Implicant Table for the Function of Table 4.4-1

	0	2	④	⑧	⑯	6	10	12	18	⑦	⑪	13	14	⑲	㉙	㉚	
1	×	×	×	×		×	×	×					×				*
2	×	×			×				×								*
3													×			×	*
4														×	×		*
5									×					×			*
6						×						×					
7							×				×						*
8									×	×							*

Minimal sum: $f = (0,2,4,6,8,10,12,14) + (0,2,16,18) + (14,30) + (13,29)$
$+ (18,19) + (10,11) + (6,7)$
$f = v'z' + w'x'z' + wxyz' + wxy'z + vw'x'y + v'wx'y + v'w'xy$

The first step in obtaining a minimal sum from a prime implicant table is to determine the distinguished columns and essential rows (if any exist). In Table 4.5-1 the essential rows are marked with an asterisk, and the labels of the distinguished columns are encircled. The next step is to draw a line through each column which contains an × in any of the essential rows, since inclusion of the essential rows in the solution will guarantee that these columns contain at least one ×. The result of doing this for Table 4.5-1 would be a table in which all the columns were "lined out." Thus, for this table, the essential prime implicants include all the fundamental products, and the minimal sum is just the sum of all the essential prime implicants. This function represents a special case, since for most functions the essential prime implicants do not cover all the fundamental products. Usually, after the essential rows have been discovered and the corresponding columns have been lined out, a reduced table in which each column has at least two ×'s will result.

Dominance

A prime implicant table for which the essential prime implicants do not include all the fundamental products is shown in Table 4.5-2. The essential rows are marked with an asterisk, and the labels of the distinguished columns are encircled. The reduced table which results when the essential rows and the columns in which they have ×'s are removed is shown in Table 4.5-3 (row L has also been removed, since it has no ×'s in the remaining columns). Each of the columns of this reduced table contains at least two ×'s. It is not possible to select any more rows which *must* be included in the minimal sum. However, certain rows can be eliminated, since it can be shown that at least one minimal sum exists which does not include the prime implicants corresponding to these rows.

Simplification of Switching Functions

Table 4.5-2. Prime Implicant Table for $f = \Sigma(1,4,5,7,8,9,11,$
$13,14,15,18,19,20,21,23,24,25,26,27,28,29,30)$

	①	④	⑧	5	9	⑱	20	24	7	11	13	14	19	21	25	26	28	15	23	27	29	30
*A	×		×	×					×													
*B		×	×			×								×								
*C			×	×			×								×							
D				×				×	×									×				
E				×				×							×				×			
F				×						×				×								×
G					×				×	×								×				
H					×						×										×	
I					×					×				×							×	
*J						×							×			×				×		
K							×								×		×				×	
L							×								×	×				×		
M							×								×		×				×	
N																×	×					×
O													×				×					
P													×									×
Q																			×				

Table 4.5-3. Table Which Results after Removal of Essential Rows and Corresponding Columns from Table 4.5-2

	7	11	14	28	15	23	29	30
D	×				×			
*E	×					×		
F								×
*G		×			×			
H		×						
I						×		
*K				×			×	
M				×			×	
N				×				×
O*			×	×				
P		×						×
Q						×		

$F = I$ $K = M$ remove I and M

4.5 prime implicant tables

For example, in Table 4.5-3, row F is identical with row I, and row K is identical with row M. Also, the gates corresponding to the rows F, I, K, M all have the same number of gate inputs. (This can be seen in Table 4.5-2 by noting that each of these rows contains four $\times$'s.) Thus, these four rows correspond to gates which all have the same cost regardless of whether gates or gate inputs are being minimized. It follows from this that the cost will be the same for two circuits which are identical, except that one circuit contains a gate corresponding to row F and the other contains instead a gate corresponding to row I. Since rows I and F include the same set of fundamental products, it makes no difference if one of these rows is chosen rather than the other. A decision to exclude one of these rows from the minimal sum being sought cannot prevent the discovery of at least *one* minimal sum. If a minimal sum exists which contains row F, then another minimal sum containing I instead of F *must* exist. Thus, it is possible to remove rows I and M from Table 4.5-3 and work with the reduced table shown in Table 4.5-4. This table still contains at least two $\times$'s in each column so that further reduction must be attempted before any rows can be chosen for inclusion in the minimal sum.

Rows K and F of Table 4.5-4 both correspond to gates having the same cost, and row F has an $\times$ in column 29, while row K has $\times$'s in column 29 and also in column 28. Thus any minimal sum which contains row F will also be a minimal sum if row F is replaced by row K. The converse is not true, since replacing row K by row F could cause column 28 to be left without any $\times$ in a row of the "minimal sum." From this it follows that row F can be removed from Table 4.5-4 without preventing the obtaining of a minimal sum. The reasoning is analogous to that used for equal

Table 4.5-4. Table Which Results from Table 4.5-3 after Removal of Rows I and M

	7	11	14	28	15	23	29	30
D	×				×			
E	×					×		
F							×	
G		×			×			
H		×						
K				×			×	
N			×					×
O			×		×			
P			×					×
Q						×		

$K \supset F$, $G \supset H$, $E \supset Q$ remove F, H, Q,

Simplification of Switching Functions

Table 4.5-5. Table Which Results from Table 4.5-4 after Removal of Dominated Rows

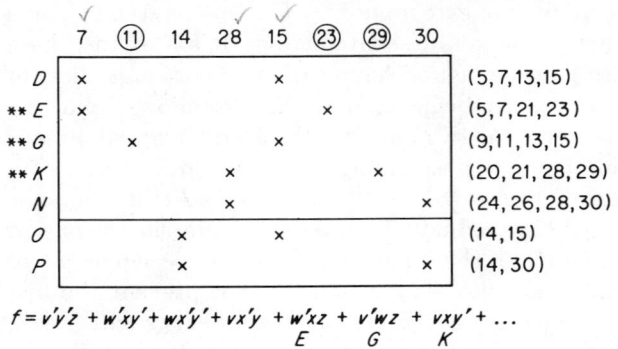

$$f = v'y'z + w'xy' + wx'y' + vx'y + w'xz + v'wz + vxy' + \ldots$$
$$ E G K$$

rows. In Table 4.5-4, row G contains $\times$'s in all the columns where row H has $\times$'s. Since rows G and H do not differ in their corresponding costs, row H can be removed from the table. Row Q has an $\times$ only in column 23, while row E has $\times$'s in columns 23 and 7. These two rows do not necessarily correspond to gates of equal cost. From Table 4.5-2 it can be seen that row Q corresponds to a gate with four inputs, while row E corresponds to a gate with three inputs.† In Table 4.5-4 row E includes all the fundamental products which row Q includes and also corresponds to a gate of equal (if gates are being minimized) or smaller (if gate inputs are being minimized) cost than row Q. Row Q can be removed from the table, since replacement of Q by E in a minimal sum cannot increase (and may actually decrease) the corresponding circuit cost. The table which results from removal of rows F, H, and Q from Table 4.5-4 is shown in Table 4.5-5. Before proceeding with Table 4.5-5, the reductions discussed in connection with Table 4.5-4 will be summarized in general terms.

Definition. Two rows I and J of a prime implicant table which have $\times$'s in exactly the same columns are said to be *equal* (written $I = J$).

Definition. A row K of a prime implicant table is said to *dominate* another row L of the same table (written $K \supset L$) if row K has $\times$'s in all the columns in which row L has $\times$'s and if, in addition, row K has at least one $\times$ in a column in which row L does not have an $\times$.

---

† Row Q contains two (2^1) $\times$'s, indicating that it corresponds to the combination of two fundamental products and therefore to a product having one fewer variable than a fundamental product. Row E contains four (2^2) $\times$'s. It corresponds to the combination of four fundamental products and thus to a product term containing two fewer variables than a fundamental product.

4.5 prime implicant tables

Table 4.5-6. Table Which Results from Table 4.5-5 after Removal of Essential Rows and Corresponding Columns

	14	30	
D			
N		×	N ⊂ P
O	×		O ⊂ P
** P	×	×	

Minimal sum
$f = v'y'z + w'xy' + wx'y' + vx'y + w'xz$
$\quad + v'wz + vxy' + wxyz'$

Theorem 4.5-1. A row I of a prime implicant table can be removed and at least one minimal sum can still be obtained from the reduced table (with row I missing) if (1) there is another row J of the table which is equal to row I and which does not have a higher cost than row I† or (2) there is another row K of the table which dominates row I and which does not have a higher cost than row I.

Table 4.5-5 contains three columns which have only a single ×. The corresponding rows must therefore be chosen. These rows are marked with a double asterisk on the figure and are called *secondary essential rows*. The reason for the qualifier "secondary" is that an essential row is one which must appear in *all* minimal sums (in fact it must appear in all irredundant sums), while a secondary essential row may not appear in some of the minimal sums. For example, if the secondary essential row was one of a pair of equal rows (for example, K and M), another minimal sum will exist which includes the other member of the pair. After removal of the secondary essential rows and corresponding columns from Table 4.5-5, Table 4.5-6 is obtained. In this table row P dominates rows N and O. Row O can be removed, since it has the same cost as row P. With row O removed, row P is secondary essential. Selection of row P completes the process of obtaining the minimal sum, since row P has ×'s in both columns of Table 4.5-6.

The relation of dominance between rows has been presented and shown to be useful for "solving" prime implicant tables. There is a similar relation between columns of a prime implicant table which is also useful.

Definition. Two columns i and j of a prime implicant table which have ×'s in exactly the same rows are said to be equal (written $i = j$).

† Of course, only one of a pair of equal rows can be removed, and the row of higher cost is the one which should be removed. Since the theorem applies to the removal of rows one at a time, this is automatically taken into account.

Simplification of Switching Functions

Table 4.5-7. A Table to Illustrate Column Dominance

	1	2	3	4
A	×	×	×	
B	×	×		×
C		×	×	×

Definition. A column i of a prime implicant table is said to *dominate* another column j of the same table (written $i \supset j$) if column i has ×'s in all the rows in which column j has ×'s and if, in addition, column i has at least one × in a row in which column j does not have an ×.

In Table 4.5-7, column 2 dominates column 1. Column 1 requires that either row A or row B must be selected, while column 2 requires that row A or row B or row C must be selected. If the requirement of column 1 is satisfied, that of column 2 will automatically be satisfied (column 1 will cause row A or row B to be selected and this selection will satisfy the column 2 requirement). Removal of column 2 from this table will thus have no influence on the final minimal sum. It is sometimes possible first to use column dominance to eliminate some columns, and then eliminate rows by row dominance, and then have some additional column dominance develop so that more columns can be eliminated, etc. This will be illustrated in an example in the section on Multiple-output Prime Implicant Tables.

Theorem 4.5-2. A column i of a prime implicant table can be removed without affecting the minimal sum being sought if (1) there is another column j of the table which is equal to column i or (2) there is another column h of the table which is dominated by column i.

Note that, for rows, the *dominated row* is removed and that, for columns, the *dominating column* is removed. It is not always true that removing essential rows, dominated rows, and dominating columns will suffice for the "solution" of a prime implicant table. It can happen that a table results in which each column contains at least two ×'s and no rows or columns can be removed. Such a table is called a *cyclic* table.

Cyclic Prime Implicant Tables

Table 4.5-8 shows a prime implicant table which contains one essential row and one secondary essential row. When these and the dominated rows (there are no dominating columns) are removed, the cyclic table of Table 4.5-9 results. This table cannot have any rows or columns removed from

4.5 prime implicant tables

Table 4.5-8. Prime Implicant Tables for
$f = \Sigma(0,4,12,16,19,24,27,28,29,31)$

(a) Removal of essential row G and dominated row I

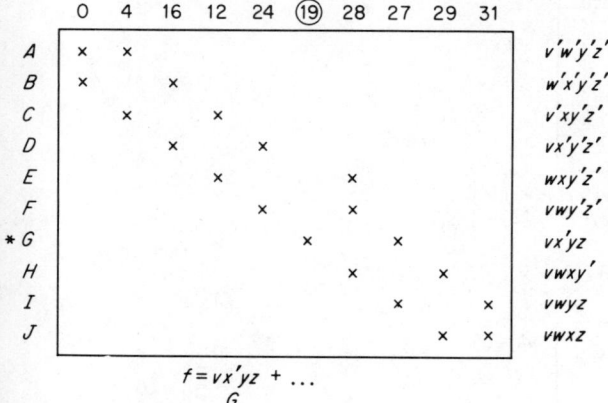

	0	4	16	12	24	⑲	28	27	29	31	
A	×	×									$v'w'y'z'$
B	×		×								$w'x'y'z'$
C		×		×							$v'xy'z'$
D				×	×						$vx'y'z'$
E				×			×				$wxy'z'$
F					×		×				$vwy'z'$
*G					×				×		$vx'yz$
H							×		×		$vwxy'$
I								×		×	$vwyz$
J									×	×	$vwxz$

$$f = \underset{G}{vx'yz} + \ldots$$

(b) Removal of secondary essential row J and dominated row H

	0	4	16	12	24	28	29	31
A	×	×						
B	×		×					
C		×		×				
D				×	×			
E				×		×		
F					×	×		
H						×	×	
**J							×	×

$$f = \underset{G}{vx'yz} + \underset{J}{vwxz} + \ldots$$

$H \subset F$

Table 4.5-9. The Cyclic Prime Implicant Table Which Results from Table 4.5-8

	✓	✓	✓	✓	✓	✓
	0	4	16	12	24	28
A	×	×				
B	×		×			
C		×		×		
D				×	×	
E				×		×
F					×	×

Table 4.5-10. The Branching Method Applied to Table 4.5-9

(a) After selection of row A

	16	12	24	28	
B	×				$B \subset D$
C		×			
** D	×		×		
E		×		×	
F			×	×	

$$f = vx'yz + vwxz + v'w'y'z' + vx'y'z' + wxy'z'$$
$$\;G \quad\quad J \quad\quad\; A \quad\quad\;\; D \quad\quad\; E$$

(b) After selection of row B

	4	12	24	28	
A	×				$A \subset C$
** C	×	×			
D			×		
E		×		×	
F			×	×	

$$f = vx'yz + vwxz + w'x'y'z' + v'xy'z' + vwy'z'$$
$$\;G \quad\quad J \quad\quad\; B \quad\quad\;\; C \quad\quad\; F$$

it so that a new technique is required for selecting the remaining terms of the minimal sum.

One method for "solving" cyclic tables consists in arbitrarily selecting one row for inclusion in the minimal sum and then using the reduction techniques to remove rows and columns from the table which results after removal of this row. This entire process must then be repeated for each row which could replace the original selected row, and the final minimal sum is obtained by comparing the costs of the expressions which result from each arbitrary choice of a selected row. This process is commonly called the *branching method* [10, sec. 6]. In Table 4.5-9 column 0 requires that either row A or row B must appear in the minimal sum. If row A is arbitrarily chosen, Table 4.5-10a results. In this table row B is dominated by row D. After removal of row B, row D is secondary essential, and after selection of row D, rows C and F are dominated by row E and row E must be selected. There is no guarantee that the rows selected from Table 4.5-10a actually correspond to a minimal sum. It is necessary to determine also the result of arbitrarily selecting row B instead of row A (column 0 ensures that either A or B must be in the minimal sum). If row B is selected, Table 4.5-10b results, and the corresponding sum turns out to be of the same cost as the sum for Table 4.5-10a. For this particular function, there happen to be two minimal sums. This is *not* true in general, and, for this reason, all alternative arbitrary selections must be checked in detail.

The choice of column 0 in this example is purely arbitrary. Any other column could equally well have been used. If column 12 was chosen, first row C and then row E would have to be selected and the resulting "minimal" sums determined. Usually it is convenient to choose a column with only two X's. If a column with more than two X's is chosen, then more than two alternative solutions will have to be determined.

It is possible for another cyclic table to result after the arbitrary selection of a row and the resulting reduction of the table. If this happens, another arbitrary selection must be made and all alternative "minimal" sums determined.

Cyclic tables can be "solved" by means of another method which produces all the irredundant sums [10, sec. 9].† The costs of the irredundant sums are then compared in order to choose the minimal sum. In this method a binary variable is associated with each row of the prime implicant table. This variable is set equal to 1 if the corresponding row is selected and is set equal to 0 if the row is not selected. Since these are binary variables, it is possible to interpret them as switching variables and to specify a new switching function which is equal to 1 only when each column has an X in at least one of the selected rows. This function will be called a *prime implicant function*,‡ or *p function*, and will be represented by p. A p function could be specified by a table of combinations, but this would not usually be satisfactory because of the large number of variables involved. Another possibility is to write an algebraic expression for p directly from the prime implicant table. This can be done for Table 4.5-9 by observing that the p function for this table must be equal to 0 when the variables A and B are both equal to 0. This must be true because, if A and B are both 0, neither of these rows is selected and no selected row can have an X in column 0. If the p function has the form $p = (A + B)q$, where q is a function of A, B, C, D, E, F which is as yet undetermined, p will be 0 when both A and B are 0. In a similar fashion, column 4 requires that $p = 0$ when $A = 0$ and $C = 0$. This means that p must also have $(A + C)$ as a factor and must be of the form

$$p = (A + B)(A + C)r$$

Either the condition $A = B = 0$ or $A = C = 0$ will thus make p equal to 0. By similar reasoning, each column will contribute a factor to p which contains those variables which correspond to rows in which the column has X's. It is also true that this product of factors completes the speci-

† This technique is sometimes called *Petrick's method*.
‡ It should be remembered that this is a new function in which each *variable* corresponds to a prime implicant of the original function.

fication of p. The p function should equal 0 if and only if there is some column of the prime implicant table which does not have an × in any selected row. The product of factors corresponding to the columns will equal 0 only when some factor is equal to 0, and a factor can equal 0 only if there are no selected rows having ×'s in the corresponding column. The p function for Table 4.5-9 is

$$p = (A + B)(A + C)(B + D)(C + E)(D + F)(E + F)$$

The preceding remarks will be summarized before the procedure for determining the irredundant sums is presented.

> **Definition.** The *prime implicant function*, or *p function*, corresponding to a given prime implicant table is a switching function which is equal to 1 only when each column of the table has an × in at least one row for which the corresponding switching variable is equal to 1 (selected rows).

> **Theorem 4.5-3.** The p function for a prime implicant table can be expressed as a product of factors. Each factor corresponds to one column in the table and is equal to the sum of the variables which correspond to rows in which the column has ×'s.

If the total number of gates is to be minimized, the minimal sum corresponds to the fewest variables which when set equal to 1 will cause p to equal 1. It is difficult to discover these variables when p is expressed as a product of factors. If the factors are "multiplied out" by using the theorems of switching algebra, a sum-of-product-terms expression will result. When all the variables of any of these product terms are equal to 1, the function will equal 1. Thus the product terms involving the smallest number of variables correspond to the minimal sums. The p function for Table 4.5-9 is multiplied out as follows:

$$p = (A + B)(A + C)(B + D)(C + E)(D + F)(E + F)$$
$$p = (A + BC)(D + BF)(E + CF)$$
$$p = (AD + ABF + BCD + BCF)(E + CF)$$
$$p = ADE + ACDF + ABEF + ABCF + BCDE + BCDF$$
$$+ BCEF + BCF$$
$$p = ADE + ACDF + ABEF + BCDE + BCF$$

This shows that the minimal sums contain the prime implicants corresponding either to rows A, D, and E or to rows B, C, and F. This agrees with the results of Table 4.5-10. The other product terms of p correspond to other irredundant sums.

If a different cost criterion is used, the procedure is to associate a cost with each of the rows of the prime implicant table. The cost corresponding to each term of the associated p function is found by adding the costs for each of the variables in the product term. The minimal sum then corresponds to the term with lowest cost. This will be illustrated by an example in connection with multiple-output minimal sums.

4.6 TABULAR METHODS FOR MULTIPLE-OUTPUT CIRCUITS

For multiple-output problems which are too large or complicated to be solved by map techniques, it is necessary to turn to a tabular method for either hand computation or digital-computer usage [8,9]. The most obvious way to extend the single-output technique would be to form each of the product functions—$f_1 \cdot f_2 \cdot f_3 \cdot \cdot \cdot$ —and then to determine the prime implicants for all the original functions and all the product functions. Although this approach is straightforward, it is unnecessarily lengthy. All the required prime implicants can be obtained without ever forming the product functions explicitly. This is done by forming a binary character for each fundamental product that appears in any of the output functions. This binary character is made up of two parts—the *identifier*, which is the same as the single-output binary character and identifies the corresponding fundamental product, and the *tag*, which specifies which of the output functions include the fundamental product specified by the identifier portion of the character. Each symbol of the identifier corresponds to one of the variables and is 0, 1, or – depending on whether the variable is primed, unprimed, or missing. Each symbol of the tag corresponds to one of the output functions and is either 0 or – depending on whether the corresponding fundamental product is not included in the output function or is included in the output function. These multiple-output characters are shown in Table 4.6-1a. They have been ordered according to the number of 1's in their identifiers. The first step in forming the multiple-output prime implicants is to compare each pair of characters to determine whether or not the identifier portions differ in only one coordinate.† If two characters whose identifiers satisfy this condition are found, a new character is formed. The identifier of the new character is formed in the same way as in the single-output technique. *The tag portion of the new character will have 0's in all coordinates in which* either *of the original characters has 0's* and dashes in the remaining coordinates (Table 4.6-1b). The reasoning behind this tag-formation rule is as follows: The new character corresponds to a product term which is included in those functions which include *both* the fundamental products used in forming the product term. This is why there are dashes in the new tag only where there are dashes in *both* the original characters. This is illustrated in Fig. 4.6-1, in

† The rules for this comparison are the same as in the single-output case. Only characters which differ by 1 in their total number of 1's need be compared.

Simplification of Switching Functions

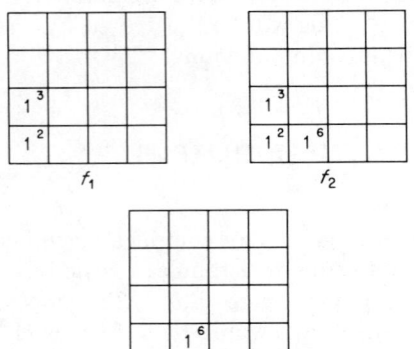

Fig. 4.6-1. *Illustration of the rule for formation of binary-character tags.*

which three maps are shown corresponding to the three functions of Table 4.6-1. Only the entries corresponding to the 2, 3, and 6 fundamental products are shown. The (2,3) character of Table 4.6-1b has dashes corresponding to f_1 and f_2, since both the 2 character and the 3 character of Table 4.6-1a have dashes for f_1 and f_2. In Fig. 4.6-1 it is evident that the 2 cell and the 3 cell can be combined on both the f_1 and f_2 maps. On the other hand, the (2,6) character of Table 4.6-1b has a dash only for f_2. In Table 4.6-1a, the 2 character has dashes for f_1 and f_2, while the 6 character has dashes for f_2 and f_3. The only function for which *both* these characters have dashes is f_2. In Fig. 4.6-1, the 2 and 6 cells can be combined only on the f_2 map.

Just as in the single-output technique, it is necessary to check off some of the characters, since they do not all correspond to prime implicants. The rule for this is as follows: A binary character is checked when (1) it is used in the formation of a new binary character and (2) the new character has dashes in the same positions as the character to be checked. Thus, in Table 4.6-1, when the (2,3) character is formed, both the 2 character and the 3 character are checked. However, when the (2,6) character is formed, neither the 2 nor the 6 character should be checked. The reasoning behind this rule can be seen by considering Fig. 4.6-1. Even though the 6 cell is combined with the 2 cell for f_2, the 6 cell itself is still a prime implicant of $f_2 \cdot f_3$.

The multiple-output prime implicants are obtained by continuing the process of comparing binary characters, using the single-output rule for forming new identifiers, and using the rule just stated for forming new tag portions. The characters which remain unchecked after the completion of this process correspond to the multiple-output prime implicants. Of course, a binary character with an all-0 tag portion need not be written down, since it corresponds to a product which is not included in any of the output functions. Octal numbers can also be used for the identifiers, but

4.6 tabular methods for multiple-output circuits

Table 4.6-1. Determination of Multiple-output Prime Implicants for $f_1(w,x,y,z) = \Sigma(2,3,5,7,8,9,10,11,13,15)$, $f_2(w,x,y,z) = \Sigma(2,3,5,6,7,10,11,14,15)$, and $f_3(w,x,y,z) = \Sigma(6,7,8,9,13,14,15)$

(a)

	w	x	y	z	f_1	f_2	f_3	
2	0	0	1	0	–	–	0	✓
8	1	0	0	0	–	0	–	✓
3	0	0	1	1	–	–	0	✓
5	0	1	0	1	–	–	0	✓
6	0	1	1	0	0	–	–	✓
9	1	0	0	1	–	0	–	✓
10	1	0	1	0	–	–	0	✓
7	0	1	1	1	–	–	–	✓
11	1	0	1	1	–	–	0	✓
13	1	1	0	1	–	0	–	✓
14	1	1	1	0	0	–	–	✓
15	1	1	1	1	–	–	–	✓

Identifier | Tag

(b)

	w	x	y	z	f_1	f_2	f_3	
(2,3)	0	0	1	–	–	–	0	✓
(2,6)	0	–	1	0	0	–	0	✓
(2,10)	–	0	1	0	–	–	0	✓
(8,9)	1	0	0	–	–	0	–	
(8,10)	1	0	–	0	–	0	0	✓
(3,7)	0	–	1	1	–	–	0	✓
(3,11)	–	0	1	1	–	–	0	✓
(5,7)	0	1	–	1	–	–	0	
(5,13)	–	1	0	1	–	0	0	✓
(6,7)	0	1	1	–	0	–	–	✓
(6,14)	–	1	1	0	0	–	–	✓
(9,11)	1	0	–	1	–	0	0	✓
(9,13)	1	–	0	1	–	0	–	
(10,11)	1	0	1	–	–	–	0	✓
(10,14)	1	–	1	0	0	–	0	✓
(7,15)	–	1	1	1	–	–	–	
(11,15)	1	–	1	1	–	–	0	✓
(13,15)	1	1	–	1	–	0	–	
(14,15)	1	1	1	–	0	–	–	✓

(c)

	w	x	y	z	f_1	f_2	f_3	
(2,3,6,7)	0	–	1	–	0	–	0	✓
(2,3,10,11)	–	0	1	–	–	–	0	
(2,6,10,14)	–	–	1	0	0	–	0	✓
(3,9,10,11)	1	0	–	–	–	0	0	
(3,7,11,15)	–	–	1	1	–	–	0	
(5,7,13,15)	–	1	–	1	–	0	0	
(6,7,14,15)	–	1	1	–	0	–	–	
(9,11,13,15)	1	–	–	1	–	0	0	
(10,11,14,15)	1	–	1	–	0	–	0	✓

(d)

	w	x	y	z	f_1	f_2	f_3
(2,3,6,7,10,11,14,15)	–	–	1	–	0	–	0

(e) Prime Implicants

	w	x	y	z	f_1	f_2	f_3	(Cost)$_d$
(2,3,6,7,10,11,14,15)	–	–	1	–	0	–	0	1 ✓
(9,11,13,15)	1	–	–	1	–	0	0	3 ✓
(6,7,14,15)	–	1	1	–	0	–	–	3 ✓
(5,7,13,15)	–	1	–	1	–	0	0	3 ✓
(3,7,11,15)	–	–	1	1	–	–	0	3
(8,9,10,11)	1	0	–	–	–	0	0	3 ✓
(2,3,10,11)	–	0	1	–	–	–	0	3 ✓
(13,15)	1	1	–	1	–	0	–	4 ✓
(7,15)	–	1	1	1	–	–	–	4
(9,13)	1	–	0	1	–	0	–	4 ✓
(5,7)	0	1	–	1	–	–	0	4
(8,9)	1	0	0	–	–	0	–	4 ✓

159

Simplification of Switching Functions

the details of this technique will not be described, since they are so similar to the octal technique for single-output prime implicants.

Multiple-output Prime Implicant Tables [9]

The process of selecting those prime implicants which are used in forming the multiple-output minimal sums is carried out by means of a prime implicant table which is quite similar to the one used in the single-output case. There must be a column of this table for each fundamental product of each of the output functions. The table is partitioned into sets of columns so that all the fundamental products corresponding to one of the output functions are represented by a set of adjacent columns as in Table 4.6-2. If a fundamental product occurs in more than one of the functions it will be represented by more than one column of the table. (In Table 4.6-2, there are two columns labeled 2, one for f_1 and one for f_2.) Each row of the table corresponds to one of the multiple-output prime implicants. These are also partitioned into sets of rows by listing first the rows which correspond to prime implicants of f_1, then those for f_2, ... those for $f_1 \cdot f_2$, etc.

Just as in the single-output case enough rows must be selected so that there is at least one × in each column. Again, a column which contains only one × is a distinguished column, and the corresponding row represents an essential prime implicant. In Table 4.6-2 there are five distinguished columns—those with their labels encircled. The fact that the 2 column of f_1 is distinguished shows that the (2,3,10,11) prime implicant (row F) is essential *for* f_1. This prime implicant is a prime implicant of both f_1 and f_2; however, it is essential only for f_1. This is illustrated in Fig. 4.6-2. It has been determined that the solid connection from the $x'y$ gate to the f_2 gate (shown dotted) is also *possible*, but it has not yet been determined whether or not it should be present in the minimal circuit. Thus, <u>the fact that the 2 column of f_1 is distinguished shows that the portion of the F row must be selected</u>. This has been indicated on the table by darkening the ×'s in the f_1 portion of row F. The 2,3,10, and 11 columns of f_1 can now be removed from the table, since they will have × in the selected portion of a selected row. The 2,3,10, and 11 columns of f_2 *cannot* be removed, since this portion of row F has not been selected. Similar remarks apply to column 5 of f_2 and columns 6,8, and 14 of f_3. The

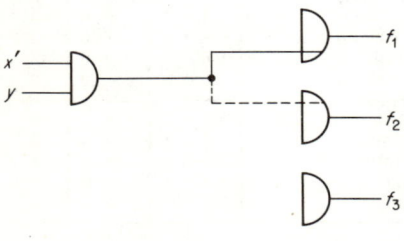

Fig. 4.6-2. *Illustration of the fact that $x'y$ is a prime implicant of $f_1 \cdot f_2$ but essential only for f_1.*

Table 4.6-2. Multiple-output Prime Implicant Table for the Functions of Table 4.6-1

		✓	✓	✓	✓f₁	✓	✓	✓	✓	✓	✓		✓	✓	✓	✓f₂	✓	✓	✓	✓		✓	✓	f₃	✓	✓	
		②	3	5	7	8	9	10	11	13	15		2	3	⑤	6	7	10	11	14	15		⑥	7	⑧	9	13	⑭	15
(9,11,13,15)	A						×		×	×	×																			f₁
(5,7,13,15)	B			×	×					×	×																			
(8,9,10,11)	C					×	×	×	×																					
(2,3,6,7,10,11,14,15)	D		×	×									×	×		×	×	×	×	×	×									f₂
(3,7,11,15)	E		×		×				×		×			×			×		×		×									
*(2,3,10,11)	F	**×**	**×**					×	**×**				×	×				×	×											f₁f₂
*(5,7)	G			×	×										**×**		**×**													
(13,15)	H									×	×																×		×	f₁f₃
(9,13)	I						×			×																×	×			
*(8,9)	J					×	×																		**×**	×				
*(6,7,14,15)	K															×	×			×	×		**×**	**×**				**×**	×	f₂f₃
(7,15)	L				×												×				×			×					×	f₁f₂f₃

$f_1 = (2,3,10,11) + D + I$
$f_2 = (5,7) + D$
$f_3 = (8,9) + (6,7,14,15) + H$

Simplification of Switching Functions

corresponding rows F,G,J,K have been marked with an asterisk to indicate that they are essential and have been selected. The table which results after removal of the appropriate columns is shown in Table 4.6-3.

Dominance can be used to remove rows and columns from Table 4.6-3. The same basic dominance rules apply as in the single-output case, but column dominance rules apply only to two columns from the same function. A dominating column can be removed only if the column it dominates refers to the same function. Since these rules depend on the costs corresponding to the rows, these costs have been listed in Table 4.6-3. Both the gate cost C_g and the gate-input cost C_d have been listed. The C_g for row A is 1, since one gate will have to be added if row A is selected. The C_d for row A is 3, since the prime implicant corresponding to row A contains two literals: two inputs will be required on the corresponding AND gate plus one input on the f_1 OR gate. The C_g for row D is 0, since the corresponding prime implicant has only one literal, and the C_d is 1, since only one input on the output f_2 gate would be required for row D. Row F also has a C_g equal to 0 and a C_d equal to 1, but for a different reason from row D. A gate is already necessary for row F, since it has already been determined that the corresponding prime implicant is essential for f_1. If row F is selected for f_2, it is not necessary to form an additional gate—just one additional gate input is required on the f_2 output gate. (In a sense, the costs listed are incremental costs.) Rows J and K have the same costs as row F, for the same reasons. Rows $E,H,I,$ and L have two values listed for C_d. This is because these rows can be used for either one or two functions. If they are used for two functions, one more output-gate input is required than if they are used for only one function.

Examination of Table 4.6-3 shows that row C is equal to row J ($C = J$). Since the costs for C are greater than the costs of J, row C is removed. Rows $F,K,$ and L can also be removed, for they are dominated by lower-cost rows D and E. Row G is dominated by row B, but row G cannot be removed, since it has lower costs than row B. Columns 7 and 9 of f_1 can be removed, since they dominate columns 5 and 8 of f_1. In f_2, column 2 equals column 10, and column 6 equals column 14, so that columns 10 and 14 can be removed. Also in f_2, columns 3, 11, and 15 can be removed because they dominate columns 2 and 6. The table which results after the removal of these rows and columns is shown in Table 4.6-4.

In Table 4.6-4a there are three distinguished columns (shown encircled) and two essential rows D and J. Selection of row D completes the formation of the minimal sum for f_2. Also, rows A, E, and I can be removed, for they are dominated by rows of equal cost, B and H. Row G is dominated by row B, but it cannot be removed, for row B is of higher cost than row G. Column 13 of f_1 dominates column 13 of f_3, but it cannot be removed, since the two columns refer to different functions. The table which results after the removal of these rows and columns is

Table 4.6-3. *The Table Which Results from Table 4.6-2 after Determination of Essential Rows*

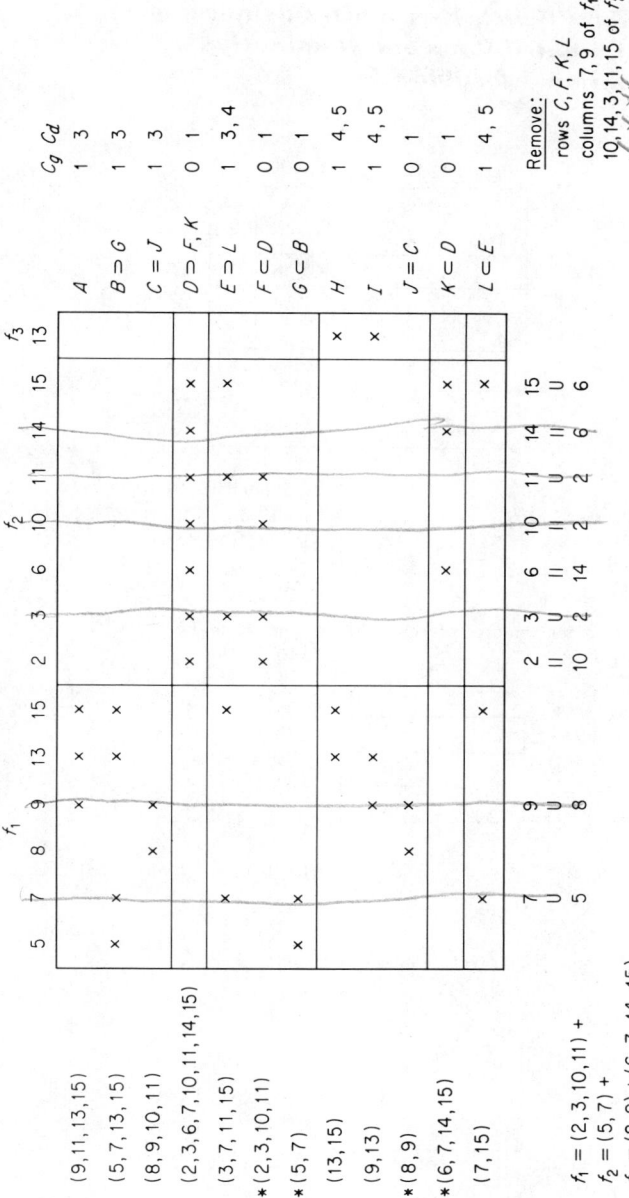

Simplification of Switching Functions

Table 4.6-4. Tables Which Result after Removal of Dominated Rows and Dominating Columns of Table 4.6-3

(a)

	f_1				f_2		f_3		C_g	C_d
	5	⑧	13	15	②	⑥	13			
(9, 11, 13, 15)			×	×				$A \subset B$	1	3
(5, 7, 13, 15)	×		×	×				B	1	3
**(2, 3, 6, 7, 10, 11, 14, 15)					×	×		D	0	1
(3, 7, 11, 15)			×					$E \subset A, B, H$	1	3
*(5, 7)	×							$G \subset B$	0	1
(13, 15)			×	×			×	H	1	4
(9, 13)			×				×	$I \subset H$	1	4
***(8, 9)		×						J	0	1

Remove rows A, E, I, D, J,
columns 8 of f_1
2, 6 of f_2

$f_1 = (2, 3, 10, 11) + (8, 9) +$
$f_2 = (5, 7) + (2, 3, 6, 7, 10, 11, 14, 15)$.
$f_3 = (8, 9) + (6, 7, 14, 15) +$

(b) Table which results after removal of rows and columns from (a)

	f_1			f_3		C_g	C_d
	5	13	15	⑬			
(5, 7, 13, 15)	×	×	×		B	1	3
(5, 7)	×				$G \subset B$	0	1
(13, 15)		×	×	×	H	1	4

Remove column 13 $f_3 = (8, 9) + (6, 7, 14, 15) + (13, 15)$

(c) Table which results from (b) after removal of column 13 of f_3

	f_1				C_g	C_d
	5	13	15			
(5, 7, 13, 15)	×	×	×	B	1	3
(5, 7)	×			$G \subset B$	0	1
(13, 15)		×	×	$H \subset B$	0	1

$$p = (B + G)(B + H) = \underset{(1,3)}{B} + \underset{(0,2)}{GH}$$

(d) Minimal sums

$f_1 = (2, 3, 10, 11) + (8, 9) + (5, 7) + (13, 15) = x'y + wx'y' + w'xz + wxz$
$f_2 = (5, 7) + (2, 3, 6, 7, 10, 11, 14, 15) \qquad = w'xz + y$
$f_3 = (8, 9) + (6, 7, 14, 15) + (13, 15) \qquad = wx'y' + xy + wxz$

shown in Table 4.6-4b. In this table, column 13 of f_3 is distinguished, and row H must be selected for f_3. This completes the formation of the minimal sum for f_3. The table which results after removal of column 13 of f_3 is shown in Table 4.6-4c. Rows G and H of this table are dominated by row B, but since row B has higher cost than rows G and H, no rows can be removed from the table, which is thus cyclic. The p function is shown in Table 4.6-4c. This p function has two product terms B and GH after it has been multiplied out. The costs for row B are (1,3), and the sums of the costs for rows G and H are (0,2). The minimal sum is thus obtained by choosing rows G and H since they are of lower cost. The multiple-output minimal sums are shown in Table 4.6-4d.

4.7 ITERATIVE CONSENSUS

The preceding methods for forming minimal sums all assume that the functions are originally specified by means of a canonical sum, table of combinations, or some equivalent form which specifies each fundamental product directly. A function can also be specified by means of an arbitrary sum-of-products expression which is not a minimal sum and which can contain product terms which are not prime implicants. As an example of a situation in which such a specification could arise, consider the problem of monitoring a chemical process involving four different raw materials. If the process is not constantly attended, it is necessary to have a circuit for stopping it whenever one of the raw materials is not being supplied. Many chemical processes are potentially explosive so that it may be necessary also to provide an emergency signal if particular partial combinations of the raw materials occur. The specification for this emergency signal might take a form such as the following:

Turn on the emergency signal if:

1. raw material z is being supplied and either
   raw material x is being supplied and raw material w is not supplied,
   or
   raw material w is being supplied and raw material y is not supplied,
   or
   raw material y is being supplied and raw material x is not supplied,
   or
2. raw materials z and w are not being supplied, and raw material y is supplied.

For this specification it is reasonable to write the expression

$$f = w'xz + x'yz + wy'z + w'yz'$$

Simplification of Switching Functions

as a definition of the circuit to be designed. This expression is neither a minimal sum nor a canonical sum. It is considered in more detail subsequently.

Since it is always possible to write the table of combinations which corresponds to such an expression, it is always possible to obtain a minimal sum by the methods already given. On the other hand, the resulting canonical sum can contain many more fundamental products than there were product terms in the original expression. It is possible to avoid this difficulty by deriving the complete sum (sum of all the prime implicants) directly from the original sum-of-products expression by means of the method to be presented next.

The Consensus Operation

Use was made of the theorems $XY + XY' = X$ and $X + XY = X$ in the methods for deriving the complete sum from the canonical sum. The theorems $XY + X'Z = XY + X'Z + YZ$ and $X + XY = X$ form the basis for the method to be described next.

Definition. Let $P = xy_1^* y_2^* \cdots y_n^*$ and $Q = x'z_1^* z_2^* \cdots z_m^*$, where it is possible that $y_i^* = z_j^*$ for some i and j. The *consensus* of P and Q, written $P \mathbin{\not{c}} Q$, is defined to be $y_1^* y_2^* \cdots y_n^* z_1^* z_2^* \cdots z_m^*$ (with any repeated literals removed) unless $y_i^* = (z_j^*)'$, in which case the consensus is said not to exist.

For example, $wxy'z \mathbin{\not{c}} w'xu'v = xy'zu'v$, while $wxy'z \mathbin{\not{c}} w'x'u'v$ does not exist. Figure 4.7-1 shows some examples of the formation of consensus terms plotted on the appropriate maps. For both cases shown in Fig. 4.7-1 the consensus term is included in the sum of the two terms from which it is derived. Moreover there is no other product of literals which contains fewer literals (more fundamental products) than the consensus term and which is not included in one of the original terms but is included in their sum. The following theorem shows that this property is not peculiar to the examples of Fig. 4.7-1 but is true in general.

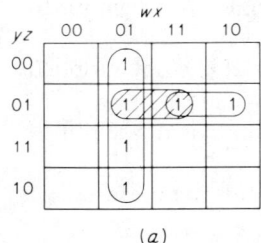
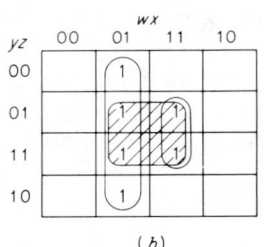

Fig. 4.7-1. *Some examples of consensus on maps.* (a) $w'x \mathbin{\not{c}} wy'z = xy'z$; (b) $w'x \mathbin{\not{c}} wxz = xz$.

Theorem 4.7-1. Let A, B, and C each represent a product of literals. If $A + B \supseteq C$, $A \not\supseteq C$, and $B \not\supseteq C$, then $A \not\mathrel{c} B \supseteq C$.

PROOF. Set all the literals of C equal to 1 so that C is equal to 1. Since $A \not\supseteq C$ and $B \not\supseteq C$, neither A nor B can be identically equal to 1. Let A/C be the product of those literals of A which are not present in C and B/C be the product of those literals of B which are not present in C. For example, if $A = w'x$, $B = wxz$, and $C = xy'z$, then $A/C = w'$ and $B/C = w$, which are the results of setting the literals of C equal to 1 in A and B. Then, since $A + B \supseteq C$, it must be true that $A/C + B/C = 1$, while $A/C \neq 1$ and $B/C \neq 1$. This is possible only if $A/C = (B/C)'$. Since A/C and B/C are both products of literals, it follows that A/C must be a single literal, say, w, and B/C must be w'. Thus the consensus of A and B, $A \mathrel{c} B$, must exist, since there is exactly one literal which occurs primed in one and unprimed in the other. Further, since neither A nor B can contain any literals other than w which are not present in C and since w and w' are absent from $A \mathrel{c} B$, it follows that $A \mathrel{c} B \supseteq C$. Thus $A \mathrel{c} B$ is the "largest" product of literals included in $A + B$ but not included in A or B. If $A = w'x$, $B = wxz$, and $C = xy'z$, then $A \mathrel{c} B = xz \supset xy'z$.

Complete Sums

It will be shown in the following that the successive addition of consensus terms to a sum-of-products expression and the removal of terms which are included in other terms $(X + XY = X)$ will result in a complete sum. This method is called *iterative consensus*. Before presenting a precise formal statement of the method and proving a theorem to the effect that a complete sum is obtained, it will be illustrated by means of an example.

Figure 4.7-2a shows the expression $f = w'xz + wy'z + x'yz + w'yz'$ plotted on a map. This is an irredundant sum for the function

$$f(w,x,y,z) = \Sigma(2,3,5,6,7,9,11,13)$$

It is not a minimal sum. The consensus terms which result from comparing all pairs of terms from this expression are shown in Fig. 4.7-2b. If these consensus terms are added to the original expression, the expression $f = w'xz + wy'z + x'yz + w'yz' + xy'z + w'yz + w'xy + wx'z + w'x'y$ results. Comparison of pairs of terms in this expression for consensus shows that $w'yz' \mathrel{c} w'yz = w'y$, as shown in Fig. 4.7-2c. Thus $w'y$ can be added to the expression, and since $w'y \supseteq w'yz'$, $w'y \supseteq w'yz$, $w'y \supseteq w'xy$, $w'y \supseteq w'x'y$, these terms can be removed from the expression, leaving $f = w'y + w'xz + wy'z + x'yz + xy'z + wx'z$, which is the complete sum for this function. Note that it is particularly simple to test whether or

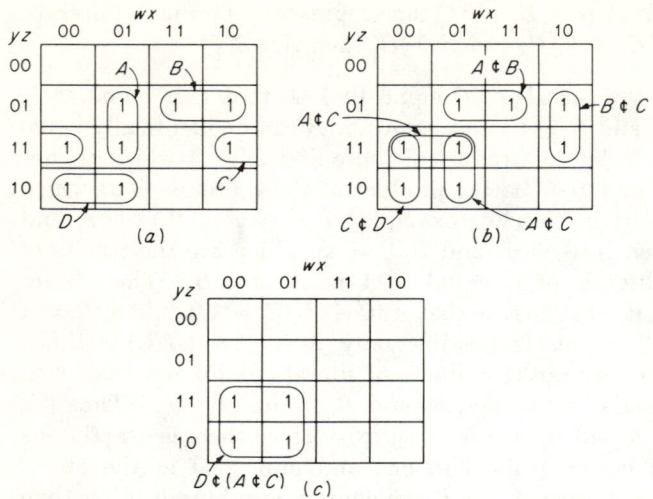

Fig. 4.7-2. *Application of iterated consensus to $f = w'xz + wy'z + x'yz + w'yz'$. (a) Map for $f = w'xz + wy'z + x'yz + w'yz'$; (b) map showing consensus terms derived from terms in (a); (c) map showing consensus term resulting from terms in (a) and (b).*

not one *product term* includes another *product term*, since the including product term must contain no literals which are not present in the included product term.† In order to show that the process illustrated in Fig. 4.7-2 always leads to a complete sum it is necessary to prove the following theorem [13]:

Theorem 4.7-2. A sum-of-products expression $E = P_1 + P_2 + \cdots + P_m$ for the function $f(x_1, x_2, \ldots, x_n)$ is a complete sum for f if and only if:

1. No product term includes any other product term, $P_i \not\supseteq P_j$ for any i and j, $i \neq j$.
2. The consensus of any two product terms, $P_i \mathbin{\rlap{\,/}{\mathrm{c}}} P_j$, either does not exist or is included in some other product term, $P_i \mathbin{\rlap{\,/}{\mathrm{c}}} P_j \subseteq P_k$.

There are two propositions contained in the statement of this theorem. The proposition "The sum-of-products expression

$$E = P_1 + P_2 + \cdots + P_m$$

is a complete sum" will be represented by the symbol α. The proposition "The sum-of-products expression $E = P_1 + P_2 + \cdots + P_m$ is such

---

† Some writers, notably Quine [13], refer to *subsumption* rather than inclusion and say that one product term is *subsumed* by another if it includes the other. Thus, if A and B are product terms and $A \supseteq B$, then B is said to subsume A.

4.7 iterative consensus

that (1) no product term includes any other product term, $P_i \not\supseteq P_j$, and (2) the consensus of any two product terms either does not exist or is included in some other product term" will be represented by the symbol β. The theorem states that α implies $\beta (\alpha \rightarrow \beta)$, that whenever α is true β must be true also, and that β implies $\alpha (\beta \rightarrow \alpha)$. Thus, in order to prove the theorem it is necessary to prove that $\alpha \rightarrow \beta$ and $\beta \rightarrow \alpha$.

The more difficult implication to prove is the one stating that $\beta \rightarrow \alpha$, and this will not be proved directly. Instead, its contrapositive, $\alpha' \rightarrow \beta'$, will be proved. Thus it is necessary to show that the fact that $E = P_1 + P_2 + \cdots + P_m$ is *not* a complete sum implies that either $P_i \supseteq P_j$ for some i and j or that some consensus $P_i \not\subset P_j$ exists which is not included in any P_k. The validity of proving the contrapositive $\alpha' \rightarrow \beta'$ in order to prove $\beta \rightarrow \alpha$ is discussed in [14, pp. 53–54] and is illustrated in the accompanying truth table.

α	β	$\beta \rightarrow \alpha$	$\alpha' \rightarrow \beta'$
F	F	T	T
F	T	F	F
T	F	T	T
T	T	T	T

Some simple examples of implications and the corresponding contrapositives are

Implication:
If x is an even integer, then x can be expressed as $2y$, with y an integer

Contrapositive:
If x cannot be expressed as $2y$, with y an integer, then x is not an even integer

Implication:
If John is married, then John has a spouse

Contrapositive:
If John has no spouse, then John is not married

It will now be assumed that $E = P_1 + P_2 + \cdots + P_m$ is a sum-of-products expression but not a complete sum for the function $f(x_1, x_2, \ldots, x_n)$. It will then be shown that there must be some P_i and P_j such that $P_i \supseteq P_j$ or that $P_i \not\subset P_j$ exists for some i and j and there is no P_k such that $P_k \supseteq (P_i \not\subset P_j)$. There are two possible reasons for E not being a complete sum. Either one of the P_i is not a prime implicant, or

Simplification of Switching Functions

some prime implicant of $f(x_1, x_2, \ldots, x_n)$ is missing from E. If P_i is not a prime implicant, there must be some prime implicant π of $f(x_1, x_2, \ldots, x_n)$ which includes P_i, $\pi \supseteq P_i$. If π occurs as one of the products, say, $P_j = \pi$, then it follows that $P_j \supseteq P_i$. If π does not occur as one of the P_j, then this is the situation where at least one of the prime implicants is missing from E.

It will thus be assumed next that there is some prime implicant π of $f(x_1, x_2, \ldots, x_n)$ which is missing from E. Since π is a prime implicant and is not identical with any of the P_i, it follows that $P_i \not\supseteq \pi$ for any i. It may be possible to add some literals to π, forming a product term $\hat{\pi}$ which still has the property that $P_i \not\supseteq \hat{\pi}$ for all i. For example, if the function of Fig. 4.7-2, $f(w,x,y,z) = \Sigma(2,3,5,7,9,11,13)$, is taken as f and the expression $w'xz + wy'z + x'yz + w'yz'$ is taken as E, then $\pi = w'y$ is a prime implicant which does not appear in E. Further the literal z can be joined to π to form $\hat{\pi} = w'yz$. Note that $w'yz$ is not included in any single term of $E = w'xz + wy'z + x'yz + w'yz'$. It is also possible to add x to π to form $\hat{\pi} = w'xy$. In general there may be several product terms satisfying the requirements placed on $\hat{\pi}$. In this case $\hat{\pi}$ is defined as one of those product terms which satisfies the requirements given and contains as many literals as any other product term satisfying these requirements. Note that if $xy'z$ rather than $w'y$ is chosen as π, then it is not possible to join an additional variable to π without having the resulting product included in one of the P_i. In a case such as this, $\hat{\pi}$ is defined to be π itself.

In summary $\hat{\pi}$ is defined as one of the product terms for which:

1. $\pi \supseteq \hat{\pi}$.
2. $P_i \not\supseteq \hat{\pi}$ for any i.
3. No product term exists having more literals than $\hat{\pi}$ and satisfying (1) and (2).
4. $\hat{\pi}$ contains no variables other than $x_1, x_2, \ldots, x_n$.

It follows from the definition of $\hat{\pi}$ that one of the x_i variables, say, x_h, will be missing from $\hat{\pi}$. If $\hat{\pi}$ had all the x_i variables appearing in it, then $\hat{\pi}$ would be a fundamental product. There is only one combination of values of the x_i for which a fundamental product is equal to 1. Since P_i does not include $\hat{\pi}$, P_i must equal 0 for this combination of values. This must be true for all the P_i so that E and thus f will equal 0 for this combination of values. However, π is a prime implicant of $f(x_1, x_2, \ldots, x_n)$ so that $f \supseteq \pi \supseteq \hat{\pi}$. It is thus not possible for f to equal 0 and $\hat{\pi}$ to equal 1 for the same combination of values for the x_i. Since the assumption that $\hat{\pi}$ includes all the x_i variables leads to a contradiction, it follows that at least one of the x_i, say, x_h must be missing from $\hat{\pi}$.

Next, the terms formed by joining x_h and x_h' to $\hat{\pi}$, $x_h\hat{\pi}$, and $x_h'\hat{\pi}$ must be considered. In the example with $E = w'xz + wy'z + x'yz + w'yz'$

$\pi = w'y$, and $\hat{\pi} = w'yz$ the variable x is missing from $\hat{\pi}$. Thus the terms to be considered are $xw'yz$ and $x'w'yz$. In general the terms $x_h\hat{\pi}$ and $x'_h\hat{\pi}$ both satisfy conditions 1 and 4. Unless $x_h\hat{\pi}$ and $x'_h\hat{\pi}$ fail to satisfy condition 2, it is clear that $\hat{\pi}$ could not have satisfied condition 3 as originally assumed. Thus $x_h\hat{\pi}$ and $x'_h\hat{\pi}$ must violate condition 2, and there must be some P_r and P_s such that $P_r \supseteq x_h\hat{\pi}$ and $P_s \supseteq x'_h\hat{\pi}$. In the case of $x_h\hat{\pi} = xw'yz$, $x'_h\hat{\pi} = x'w'yz$, these relations are $P_r = w'xz \supseteq xw'yz$ and $P_s = x'yz \supseteq x'w'yz$.

From the facts that $P_r \supseteq x_h\hat{\pi}$ and $P_r \not\supseteq \hat{\pi}$ it follows that the literal x_h must appear in P_r so that it must be possible to express P_r as $P_r = x_hQ_r$. Similarly it must be possible to express P_s as $P_s = x'_hQ_s$. In the example these expressions are $P_r = xQ_r = x(w'z)$ and $P_s = x'Q_s = x'(yz)$. It follows from the fact that $P_r = x_hQ_r \supseteq x_h\hat{\pi}$, that $Q_r \supseteq \hat{\pi}$† and similarly that $Q_s \supseteq \hat{\pi}$, so that $Q_r \cdot Q_s \supseteq \hat{\pi}$ and thus $Q_r \cdot Q_s$ is nonzero. In the example, $Q_r \cdot Q_s = w'yz \supseteq w'yz$. This shows that $P_r \not\subset P_s = Q_r \cdot Q_s$ exists and is nonzero. All that remains to be shown is that there exist no P_i such that $P_i \supseteq P_r \not\subset P_s = Q_r \cdot Q_s$. This is not possible, for if $P_i \supseteq Q_r \cdot Q_s$, then because $Q_r \cdot Q_s \supseteq \hat{\pi}$ it follows that $P_i \supseteq \hat{\pi}$, which contradicts the definition of $\hat{\pi}$. This completes the proof that if a sum-of-products expression $E = P_1 + P_2 + \cdots + P_m$ is not a complete sum then either $P_i \supseteq P_j$ or $P_i \not\subset P_j$ exists and there is no P_h such that $P_h \supseteq P_i \not\subset P_j$.

In order to complete the proof of the theorem, it is necessary to show that if $E = P_1 + P_2 + \cdots + P_m$ is the complete sum for the function $f(x_1, x_2, \ldots, x_n)$, then (1) no product term includes any other product term $(P_i \not\supseteq P_j)$ and (2) the consensus of any two product terms either does not exist or is included in some other product term $(\alpha \to \beta)$. Each of the P_i is a prime implicant by definition, and it follows from the definition of a prime implicant that it is not included in any other prime implicant of the same function. If the consensus $P_i \not\subset P_j$ exists, it is a product of literals which is included in $f(x_1, x_2, \ldots, x_n)$. By the definition of a prime implicant, any product of literals which is included in $f(x_1, x_2, \ldots, x_n)$ must be included in some prime implicant of $f(x_1, x_2, \ldots, x_n)$, and thus $P_i \not\subset P_j$ must be included in one of the P_k.

An Algorithm

In order to make use of the theorem just presented in forming a complete sum from a sum-of-products expression, it is necessary to convert the original expression into one in which the consensus of any two terms is included in some other term and in which no term includes any other. This is done by comparing each term with every other term in the expression and (1) removing any term which is included in another term and

---

† $x_hQ_r \supseteq x_h\hat{\pi}$ if and only if $(x_hQ_r)'(x_h\hat{\pi}) = (x'_h + Q'_r)(x_h\hat{\pi}) = Q'_rx_h\hat{\pi} = 0$. Since neither Q_r nor π contains x_h, it must be true that $Q'_r\hat{\pi} = 0$ and thus that $Q_r \supseteq \hat{\pi}$. If $Q_r \supseteq \hat{\pi}$, and $Q_s \supseteq \hat{\pi}$, then $Q'_r \cdot \hat{\pi} = Q'_s \cdot \hat{\pi} = 0$ and $(Q'_r + Q'_s) \cdot \hat{\pi} = 0$ so that $Q_r \cdot Q_s \supseteq \hat{\pi}$.

Simplification of Switching Functions

(2) adding the consensus of any two terms to the expression, provided that the consensus term is not included in some other term. A tabular technique for carrying out this process is shown in Table 4.7-1. The sum-of-products expression used in forming this table is

$$f = w'xz + wy'z + x'yz + w'yz'$$

The same rule is used for representing the product terms by 1's, 0's and —'s as was used in Table 4.4-1. The algorithm is carried out as follows:

1. Each row of the table is compared with each row above it in the table.
2. If any row is found to be included in another row (to have 1's wherever the other row has 1's and 0's wherever the other row has 0's), the included row is removed from the table.
3. If any two rows have a consensus, the consensus term is compared with all other rows of the table and then added at the bottom of the table if it is not included in any other row. Two rows have a consensus if there is only one column in which one row has a 1 and the other row has a 0. The consensus row has a dash in the column in which the two original rows differ and in any column in which both the original rows have dashes. It has a 0 in any column in which either of the original rows has a 0 and a 1 in any column in which either of the original rows has a 1.

Table 4.7-1. The Use of Iterative Consensus to Obtain a Complete Sum for $f = w'xz + wy'z + x'yz + w'yz'$

	w	x	y	z	
A	0	1	–	1	
B	1	–	0	1	Initial sum-of-products terms
C	–	0	1	1	
D	0	–	1	0	✓
B ¢ A	–	1	0	1	
C ¢ B	1	0	–	1	
C ¢ A	0	–	1	1	✓
D ¢ C	0	0	1	–	✓
D ¢ A	0	1	1	–	✓
(C ¢ A) ¢ D	0	–	1	–	

Complete sum: $f = w'y + wx'z + xy'z + x'yz + wy'z + w'xz$

4.7 iterative consensus

4. This process terminates when every row has been compared with all rows lower down in the table. The rows which remain in the table correspond to all the prime implicants. The consensus of any pair of rows either must appear as a row of the table or must be included in some row of the table. In Table 4.7-1, rows D, $C \not\subset A$, $D \not\subset C$, and $D \not\subset A$ are removed from the table because they are included in row $(C \not\subset A) \not\subset D$.

This iterative consensus algorithm can be directly extended to multiple-output circuits by using the identifier and tag representation described in Sec. 4.6. In carrying out the steps of the algorithm no distinction is made between the tag and identifier portions of the rows in checking for inclu-

Table 4.7-2. *An Example of the Use of Iterative Consensus to Obtain Multiple-output Complete Sums*

(a) Original specification

$$f_1 = w'x'y'z' + w'yz' + xy'z$$
$$f_2 = w'x'y'z' + wx'yz + wxz + w'xy'z$$
$$f_3 = w'x'y'z' + x'y'z' + wxy'z'$$

(b) Table

	w	x	y	z	f_1	f_2	f_3	
A	0	0	0	0	—	—	—	
B	0	—	1	0	—	0	0	
C	—	1	0	1	—	0	0	
D	1	0	1	1	0	—	0	$E \not\subset D$
E	1	1	—	1	0	—	0	
F	0	1	0	1	0	—	0	$F \not\subset E$
G	—	0	0	0	0	0	—	
H	1	1	0	0	0	0	—	$H \not\subset G$
$B \not\subset A$	0	0	—	0	—	0	0	
$E \not\subset D$	1	—	1	1	0	—	0	
$F \not\subset E$	—	1	0	1	0	—	0	
$C, F \not\subset E$	—	1	0	1	—	—	0	
$H \not\subset G$	1	—	0	0	0	0	—	

(c) Multiple-output complete sums

$$f_1 = w'x'y'z' + w'yz' + w'x'z' + xy'z$$
$$f_2 = w'x'y'z' + wxz + wyz + xy'z$$
$$f_3 = w'x'y'z' + x'y'z' + wy'z'$$

173

Simplification of Switching Functions

sion or consensus terms. One additional rule must be added for the multiple-output case:

5. If two rows have identical identifier portions and differ in their tag positions, a new row is formed having the same identifier and dashes wherever either of the original rows have dashes in their tags. The original rows are then removed from the table [8].

Table 4.7-2 shows an example of the use of iterative consensus for multiple-output complete sums. The next-to-the-last row of part *b* of the table is formed by use of rule 5.

It is always possible to form a prime implicant table from the complete sum and then to obtain a minimal sum. In order to form the table, a process equivalent to expanding the complete sum must be carried out. This is often undesirable, since there may be a very large number of fundamental products in the complete sum. Some results on techniques for obtaining a minimal sum without resorting to a prime implicant table have been published [15,16], but a satisfactory general method is not yet available.

Problems

1. Find minimal sums *and* minimal products for each of the following functions:
   - (a) $\Sigma(0,2,4,8,10,12)$
   - (b) $\Sigma(2,3,6,7,8,9,12,13)$
   - (c) $\Sigma(0,2,3,4,6,7,8,9,10,12,13,14)$
   - (d) $\Sigma(1,4,6,7,13)$
   - (e) $\Sigma(1,3,7,13,15)$
   - (f) $\Sigma(9,11,12,13,14,15,16,18,24,25,26,27)$
   - (g) $\Sigma(8,9,13,14,15,24,26,30)$
   - (h) $\Sigma(8,9,10,11,17,19,21,23,25,27,41,43,44,45,46,47,56,57,58,59)$
   - (i) $\Sigma(3,5,7,11) + d(6,15)$
   - (j) $\Sigma(3,5,7,11,12,29,31) + d(1,2,6,10,28)$
   - (k) $\Sigma(2,7,9,10,11,12,14,15)$

2. For each of the functions listed below:
   - (a) Determine the complete sum.
   - (b) Underline the essential prime implicants.
   - (c) Show (on a map) *one* minimal sum.
   - (d) Determine the number of different minimal sums.
   - (e) If there are any irredundant sums which are not minimal sums, display one of these irredundant sums on a map.

$$f_1(w,x,y,z) = \Sigma(1,5,6,7,11,12,13,15)$$
$$f_2(w,x,y,z) = \Sigma(2,3,5,7,8,10,12,13)$$
$$f_3(w,x,y,z) = \Sigma(0,2,5,6,7,8,9,12,13,15)$$
$$f_4(w,x,y,z) = \Sigma(0,1,4,5,6,7,9,10,13,14,15)$$
$$f_5(w,x,y,z) = \Sigma(3,5,6,7,9,10,11,12,13,14,15)$$
$$f_6(w,x,y,z) = \Sigma(0,1,2,3,5,6,7,8,9,10,12,13,14,15)$$

problems

3. For the function $f(v,w,x,y,z) = \Sigma(4,5,8,9,12,13,14,15,16,17,20,21,22,23,24,$
   $25,26,28,29,30,31)$:
   (a) Plot the function on a map, encircling the prime implicants.
   (b) How many prime implicants does this function have?
   (c) How many essential prime implicants does the function have?

4. For the function $f(w,x,y,z) = \Sigma(1,4,5) + d(2,3,6,7,8,9,12,13)$:
   (a) Determine a minimal sum.
   (b) Determine a minimal product.

5. For the circuit of Fig. P4-5:
   (a) Write an algebraic expression for f.
   (b) Write the decimal specification for f.
   (c) Design the analogous circuit for f, using OR-NOT gates having at most three inputs to a gate (nine gates are sufficient, no inverters).

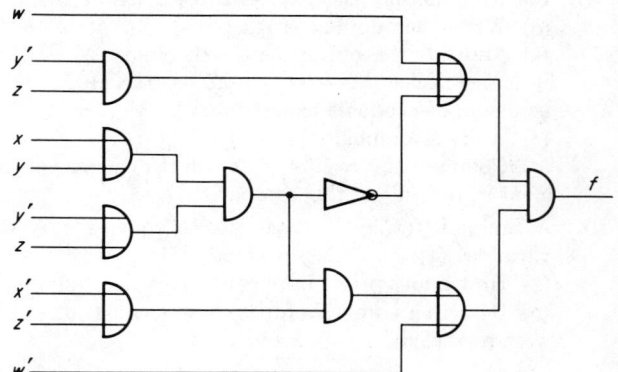

Fig. P4-5

6. For the circuit of Fig. P4-6:
   (a) What must g be in order that $f(w,x,y,z) = \Sigma(0,1,2,3,4,7,8,11)$?
   (b) Draw a map for g.
   (c) Draw a two-stage diode gate circuit for g. Use the minimum number of gates. (Four gates are sufficient.)

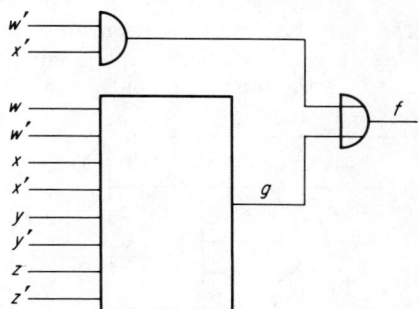

Fig. P4-6

7. Given that $x \oplus y = xy' + x'y$ and that $x \oplus (y \oplus z) = (x \oplus y) \oplus z$:
   (a) Prove that $x + y = x \oplus y \oplus xy$.
   (b) If the canonical sum for a function f is given by $f = p_{i_1} + p_{i_2} +$

175

Simplification of Switching Functions

$\cdots + p_{i_r}$, where the p_i are fundamental products, prove that

$$f = p_{i_1} \oplus p_{i_2} \oplus \cdots \oplus p_{i_r}.$$

(c) Given that $x \oplus x = 0$ and that $f = \Sigma(5,7,8,9,12,13,14,15)$, design a circuit in which the output is derived from a $\oplus$ gate and the inputs are connected to AND gates. (Use the minimum number of gates.)

8. A circuit is to be constructed for the function $f(v,w,x,y,z) = \Sigma(0,3,5,7,8,11,13,15,20,21,22,23,28,29,30,31)$. The gates which are to be used are OR gates and AND gates. Each gate has either two or four inputs, and gates with one, three, or five inputs are not available and *cannot be used*. Design a two-stage circuit for this function, using the minimum number of gates:
   (a) With an OR gate as the output gate
   (b) With an AND gate as the output gate

9. For the function $f(w,x,y,z) = \Sigma(0,1,2,3,4,5,7,10)$:
   (a) Write the minimal *product*.
   (b) Multiply this out, using the theorems $(X + Y)(W + Z) = WX + WY + YZ + XZ$ and $XX = X$, $XX' = 0$, $X + XY = X$. Your result should be a sum-of-products expression.
   (c) Write a minimal sum for f.
   (d) Compare the results of (b) and (c)—can you make any general statement as a result of this comparison?

10. A function $f(x_1, x_2, \ldots, x_{17})$ is to equal 1 only when exactly one of the variables $(x_1, \ldots, x_{17})$ is equal to 1.
    (a) How many prime implicants does this function have?
    (b) If a circuit for this function was constructed and then the input leads were relabeled

    x_1 lead changed to x_2 lead
    x_2 lead changed to x_3 lead
    . . . .
    x_{16} lead changed to x_{17} lead
    x_{17} lead changed to x_1 lead

    what function would the circuit with the new labels realize?

11. (a) Write an expression for the output of the circuit shown in Fig. P4-11.
    (b) Design a two-stage circuit using only OR-NOT gates which realizes the same function as the circuit of Fig. P4-11.

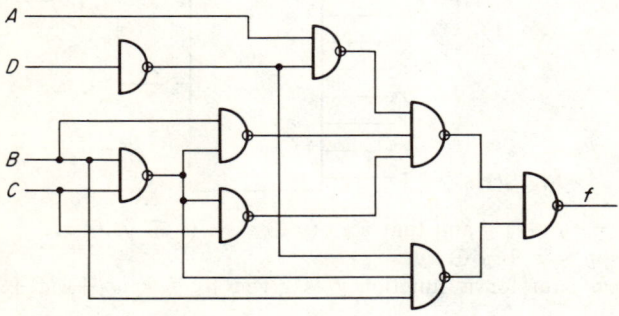

Fig. P4-11

problems

12. Show that the following definitions of *inclusion* are equivalent to those given in Sec. 4.2:
    (a) $f \supseteq g$ if and only if $f'g = 0$.
    (b) $f \supseteq g$ if and only if $f + g' = 1$.

13. Design a minimal multiple-output two-stage (AND-OR or OR-AND) circuit for the following functions:
    $$f_1 = \Sigma(0,4,5)$$
    $$f_2 = \Sigma(0,2,3,4,5)$$
    $$f_3 = \Sigma(0,1,2)$$

14. Determine the multiple-output minimal sums for each of the following sets of functions:
    (a) $f_1 = wxz + w'y'z$
    $f_2 = x'z' + wxz + xy'z$
    $f_3 = x'z' + wxyz + w'x'y'$
    (b) $f_1 = yz + wz$
    $f_2 = yz + wz + wxy$
    $f_3 = wy'z + w'xyz'$
    (c) $f_1 = wx' + x'z + xy'z'$
    $f_2 = wz + xz' + xy + w'x'z$
    (d) $f_1 = y'z + w'z$
    $f_2 = w'xy' + wy'z + wx'y + x'yz'$
    $f_3 = xyz' + wx'z$

15. Design a minimal-gate multiple-output two-stage network (using AND gates, OR gates) for the following functions:
    $$f_1(w,x,y,z) = \Sigma(1,4,5,7,13) + d(3,6)$$
    $$f_2(w,x,y,z) = \Sigma(3,5,7) + d(6)$$
    $$f_3(w,x,y,z) = \Sigma(3,4,11,13,15) + d(9,14)$$

16. For each of the following specifications, plot functions satisfying these specifications on *n*-cube maps. (Use variables v, w, x, y, z.)
    (a) A five-variable function containing no essential prime implicants
    (b) A five-variable function for which all the prime implicants are essential
    (c) A four-variable function which is unchanged when the variables y and z are interchanged
    (d) Two four-variable functions for which the multiple-output minimal sums contain none of the prime implicants of the product function
    (e) A four-variable function having at least two terms in its minimal sum and having no variables primed in the minimal sum
    (f) A function having at least two different irredundant sums

17. For the function $f(x_1,x_2,x_3,x_4,x_5,x_6,x_7) = \Sigma(0,5,16,21,32,37,45,48,53,61,64,69, 80,96,112,117,125)$ use a tabular method to find
    (a) All prime implicants
    (b) A minimal sum

18. Use a tabular method to determine a minimal multiple-output two-stage circuit (AND-OR) for the following functions:
    $$f_1(w,x,y,z) = \Sigma(1,5,7,8,12,13,14,15)$$
    $$f_2(w,x,y,z) = \Sigma(0,1,5,8,14)$$
    $$f_3(w,x,y,z) = \Sigma(0,2,3,6,8,9,10,11,14)$$

19. Use the iterated consensus technique to obtain a complete sum for each of the following functions:
(a) $f(v,w,x,y,z) = vy + v'y' + w'z + vxy + wxy' + vwxz$
(b) $f(t,u,v,w,x,y,z) = t'u'w'y'z' + t'w'y'z + t'uw'x + u'vx'y + tu'v'w'xy + tuvwy'z' + t'u'vwyz + tu'v'wz$

20. (a) Prove that $A \not\mathrel{\not{c}} (A \not\mathrel{\not{c}} B)$ never exists.
(b) The consensus operation can be interpreted as a binary connective. Is this connective commutative? Associative? Does it distribute over addition? Multiplication?

21. Write factored algebraic expressions which contain as few literals as possible for each of the following functions:
(a) $f(x,y,z) = \Sigma(0,1,2,5,6)$
(b) $f(w,x,y,z) = \Sigma(0,3,12)$
(c) $f(w,x,y,z) = \Sigma(0,1,2,13,14)$
(d) $f(w,x,y,z) = \Sigma(0,1,2,5,14)$

22. Design a circuit to realize the function $f = x_1 x_2 x_3 + x_1' x_2 + x_2' x_4$:
(a) Using only AND-NOT gates
(b) Using only OR-NOT gates

REFERENCES

1. Veitch, E. W.: A Chart Method for Simplifying Truth Functions, *Proc. ACM*, Pittsburgh, Pa., May 2, 3, 1952, pp. 127–133.
2. Karnaugh, M.: The Map Method for Synthesis of Combinational Logic Circuits, *Trans. AIEE*, vol. 72, pt. I, pp. 593–598, 1953.
3. Quine, W. V.: The Problem of Simplifying Truth Functions, *Am. Math. Monthly*, vol. 59, no. 8, pp. 521–531, October, 1952.
4. Caldwell, S. H.: "Switching Circuits and Logical Design," John Wiley & Sons, Inc., New York, 1958.
5. Tanana, E. J.: The Map Method, in E. J. McCluskey, Jr., and T. C. Bartee (eds.), "A Survey of Switching Circuit Theory," chap. 4, pp. 47–65, McGraw-Hill Book Company, 1962.
6. McCluskey, E. J., Jr.: Minimization Theory, in E. J. McCluskey, Jr., and T. C. Bartee (eds.), "A Survey of Switching Circuit Theory," chap. 5, pp. 67–88, McGraw-Hill Book Company, New York, 1962.
7. Lawler, E. L., and G. A. Salton: The Use of Parenthesis-free Notation for the Automatic Design of Switching Circuits, *IRE Trans. on Electronic Computers*, vol. EC-9, no. 3, pp. 342–352, September, 1960.
8. Bartee, T. C.: Computer Design of Multiple-output Logical Networks, *IRE Trans. on Electronic Computers*, vol. EC-10, no. 2, pp 21–30, March, 1961.
9. McCluskey, E. J., and H. Schorr: Essential Multiple-output Prime Implicants, Mathematical Theory of Automata, *Proc. Polytechnic Inst. Brooklyn Symposium*, vol. 12, pp. 437–457, April, 1962.
10. McCluskey, E. J.: Minimization of Boolean Functions, *Bell System Tech. J.*, vol. 35, no. 5, pp. 1417–1444, November, 1956.

11. Beatson, T. J.: Minimization of Components in Electronic Switching Circuits, *Trans. AIEE*, Part I, Communications and Electronics, vol. 77, pp. 283–291, July, 1958.
12. Mueller, R. K., and R. H. Urbano: A Topological Method for the Determination of the Minimal Forms of a Boolean Function, *IRE Trans. on Electronic Computers*, vol. EC-5, no. 3, pp. 126–132, September, 1956.
13. Quine, W. V.: A Way to Simplify Truth Functions, *Am. Math. Monthly*, vol. 62, no. 9, pp. 627–631, November, 1955.
14. Whitesitt, J. E.: "Boolean Algebra and Its Application," Addison-Wesley Publishing Company, Inc., Reading, Mass., 1961.
15. Mott, T. H., Jr.: Determination of the Irredundant Normal Forms of a Truth Function by Iterated Consensus of the Prime Implicants, *IRE Trans. on Electronic Computers*, vol. EC-9, no. 2, pp. 245–252, June, 1960.
16. Gaines, R. S.: Implication Techniques for Boolean Functions, Switching Circuit Theory and Logical Design, *Proc. Fifth Ann. Symposium*, Princeton, N.J., October, 1964, pp. 174–182. Published by the IEEE, New York, *Spec. Publ.* S-164.

5 SEQUENTIAL - CIRCUIT ANALYSIS

The methods developed in the preceding chapters are applicable to *combinational circuits*—circuits whose outputs are determined completely by their present inputs. Many digital circuits satisfy this restriction; however, there are also many circuits which do not. Circuits whose outputs depend not only on the present inputs but also on previous inputs are called *sequential circuits*. This chapter will introduce the various types of sequential circuits and show how they are analyzed. The following chapter will then present techniques for synthesizing sequential circuits.

5.1 INTRODUCTION

The difference between a combinational circuit and a sequential circuit is analogous to the difference between the two types of combination lock shown in Fig. 5.1-1. Whether the lock of Fig. 5.1-1a is open or not depends not only on which number the pointer is selecting but also on which numbers the pointer stopped at previously. Similarly, the output of a sequential circuit depends on previous as well as present inputs. The lock of Fig. 5.1-1b is open or closed depending only on the present setting of its dials; past settings are unimportant, just as, in a combinational circuit, past inputs are unimportant in determining the present circuit outputs.

Illustration

In order to gain some insight into the performance of sequential circuits, the circuit of Fig. 5.1-2 will be analyzed intuitively before any formal methods are developed. The outputs of this circuit (z_1 and z_2) depend not only on the circuit inputs (x_1 and x_2) but also on the outputs of the two flip-flops (y_1 and y_2). The flip-

5.1 introduction

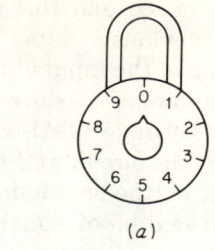

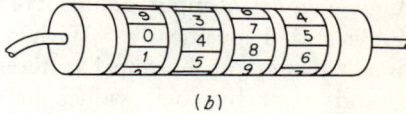

Fig. 5.1-1. Two types of combination lock.

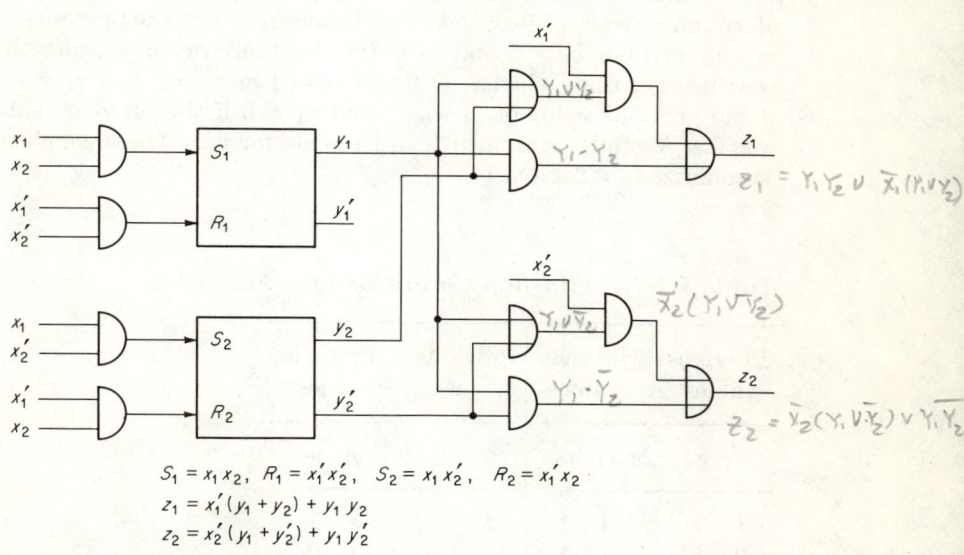

$S_1 = x_1 x_2, \quad R_1 = x_1' x_2', \quad S_2 = x_1 x_2', \quad R_2 = x_1' x_2$
$z_1 = x_1'(y_1 + y_2) + y_1 y_2$
$z_2 = x_2'(y_1 + y_2') + y_1 y_2'$

Fig. 5.1-2. *Illustrative sequential circuit.*

flop outputs in turn depend on the past and present circuit inputs.† If $x_1 = x_2 = 1$, then flip-flop 1 must be set and y_1 must equal 1; and if $x_1 = x_2 = 0$, then flip-flop 1 must be reset and y_1 must equal 0. However, if $x_1 = 0$ and $x_2 = 1$ or $x_1 = 1$ and $x_2 = 0$, neither input to flip-

† The operation of these flip-flops is assumed to be the same as described in Chap. 1: When a 1 signal is present on the S (set) lead ($S = 1$), a 1 signal is present on the y lead ($y = 1$). When a 1 signal is present on the R (reset) lead ($R = 1$), a 0 signal is present on the y lead ($y = 0$). When neither the S lead nor the R lead has a 1 signal ($S = R = 0$), the signal on the y lead is determined by which of the S and R leads last had a 1 signal present. There should never be a 1 signal present on both the S and R leads.

181

Sequential-circuit Analysis

flop 1 is energized and this flip-flop remains in the state (set or reset) caused by previous inputs. It is customary to make the assumption that only one of the inputs to a circuit changes at a time. This is a very reasonable assumption, since physically it is, in effect, impossible to have simultaneous changes. Also, the effects of the input changes will propagate through the circuit at different speeds so that the effects of the input changes will not be simultaneous throughout the circuit. This discrepancy in propagation of signals can lead to unreliable operation, and thus simultaneous changes are usually forbidden. Throughout this book, single changes of input signals will be assumed unless a different assumption is stated specifically. With x_1 and x_2 restricted to changing one at a time, the state of flip-flop 1 is determined either by the present values of x_1 and x_2 or by their values before the last input change. Thus, if $x_1 = 0$ and $x_2 = 1$ or $x_1 = 1$ and $x_2 = 0$, $y_1 = 1$ if the previous values of x_1 and x_2 were $x_1 = 1$ and $x_2 = 1$, and $y_1 = 0$ if the previous values of x_1 and x_2 were $x_1 = 0$ and $x_2 = 0$. Another way of stating this same conclusion is that, when $x_1 = 0$ and $x_2 = 1$ or $x_1 = 1$ and $x_2 = 0$, $y_1 = 1$ if the previous value of y_1 was 1 and $y_1 = 0$ if the previous value of y_1 was 0. A similar statement can be made for y_2. These conclusions are summarized in Table 5.1-1.

Table 5.1-1. Flip-flop Conditions for Fig. 5.1-2

Previous values	Present values	Previous values	Present values
x_1 x_2	x_1 x_2	y_1 y_2	y_1 y_2
1 0 0 1	1 1 1 1	d 1 d 0	1 1 1 0
1 0 0 1	0 0 0 0	d 1 d 0	0 1 0 0
1 1 0 0	1 0 1 0	1 d 0 d	1 1 0 1
1 1 0 0	0 1 0 1	1 d 0 d	1 0 0 0

When $x_1 = x_2$, $y_2 =$ previous value of x_1.
When $x_1 = x_2'$, $y_1 =$ previous value of x_1.

5.1 introduction

In analyzing a sequential circuit, what is desired is some scheme for determining what *sequence* of outputs will be produced by any sequence of inputs. Since the outputs are just switching functions of the circuit inputs and flip-flop outputs, the output sequences can be determined directly once the sequences of flip-flop outputs are known. For any given sequence of inputs, the flip-flop outputs can be determined by using Table 5.1-1. The switching functions can then be used to determine the output sequence. Table 5.1-2 shows the sequences that result from a typical input sequence.

For the sequence in this table, the values of z_1 and z_2 are equal to the values of x_1 and x_2 before the last input change. It can be shown that this relationship is true, not only for the particular sequence of this table, but for *any* input sequence. Thus, this circuit performs like an asynchronous one-unit delay line. In order to establish that this circuit does have this same performance for *any* input sequence, a more general analysis technique is required. Such a technique will be described in the following section. It should be pointed out that the ultimate objective of this investigation is not really an analysis technique but a synthesis technique. The analysis technique to be developed will therefore be one which can be "turned inside out" and used for synthesis.

An assumption about the way in which the circuit inputs are changed is implicit in the preceding discussion. Specifically, it has been assumed that the inputs are never changed unless the circuit is in a stable condition, i.e., unless none of the internal signals are changing. Whenever a sequential circuit's inputs are controlled so that this assumption is valid, the circuit is said to be operating in *fundamental mode* [1]. The following three sections apply specifically to fundamental-mode operation. In Sec. 5.5, the analysis of circuits not operating in fundamental mode is considered. Much of the theory of fundamental-mode operation was developed by Huffman [2].

Table 5.1-2. Input-Output Sequences for Fig. 5.1-2

	a	b	c	d	e	f	g	h
x_1	0	0	1	0	1	1	0	0
x_2	0	1	1	1	1	0	0	1
y_1	0	0	1	1	1	1	0	0
y_2	?	0	0	0	0	1	1	0
z_1	?	0	0	1	0	1	1	0
z_2	?	0	1	1	1	1	0	0

Time ⟶

5.2 FORMAL ANALYSIS OF CIRCUITS CONTAINING S-R FLIP-FLOPS

As was pointed out previously, in a sequential circuit such as that of Fig. 5.1-2, the outputs are switching functions of the circuit inputs and the flip-flop outputs. The determination of the circuit outputs from the circuit inputs and flip-flop outputs is a straightforward combinational-circuit problem, and the techniques of Chaps. 3 and 4 can be used. The novel aspect of sequential circuits is the relationship between the circuit inputs and the flip-flop outputs. The flip-flop inputs are switching functions of the circuit inputs,† but the flip-flop outputs are not switching functions of the circuit inputs, and so the basic problem in analyzing sequential circuits is that of determining the relationship between the circuit inputs and the flip-flop outputs.

Since the relationships between the circuit inputs and the flip-flop outputs are not characterized by switching functions, some new representation of this relationship will have to be developed. This development is one of the major objectives of this section.

In order to investigate the relation between the circuit inputs and flip-flop outputs, let us consider a specific situation in the circuit of Fig. 5.1-2. Let the circuit inputs x_1 and x_2 both be equal to 0 and the flip-flop outputs y_1 and y_2 also both be equal to 0. Now let x_1 change to become equal to 1. Since S_2 is equal to $x_1 x_2'$, S_2 now becomes equal to 1 and flip-flop 2 must change state so that y_2 becomes equal to 1. From this little "experiment" it is clear that the circuit is unstable when $x_1 = 1, x_2 = 0, y_1 = 0, y_2 = 0$ and the y_2 flip-flop must change so that the stable state with $x_1 = 1, x_2 = 0, y_1 = 0, y_2 = 1$ is reached. This transition from an unstable situation to a stable situation by a flip-flop changing is the key to sequential-circuit behavior.

In Table 5.2-1, all possible combinations of values of x_1, x_2, y_1, and y_2 are listed. Each combination of values of the circuit inputs and flip-flop outputs is called a *total state* of the sequential circuit, since all signals in the circuit (including the circuit outputs) can be determined from the circuit inputs and flip-flop outputs. Each *stable* total state in Table 5.2-1 is encircled, and an arrow is drawn from each unstable state to the stable state to which the circuit goes from the unstable state. The stable states and unstable-state behavior can be determined directly from the circuit or from Table 5.1-1. For example, row 1 of the table corresponds to the stable state with $x_1 = x_2 = y_1 = y_2 = 0$. Since the state is stable, the circuit will remain in this state until some input is changed. If x_1 is

† The flip-flop inputs will, in general, depend also on flip-flop outputs. The method to be developed here will be applicable to circuits in which this dependency is present.

5.2 formal analysis of circuits containing S-R flip-flops

Table 5.2-1. Table Showing Stable and Unstable States for the Circuit of Fig. 5.1-2

Row	x_1	x_2	y_1	y_2	z_1	z_2
1	0	0	0	(0)	0	1
2	0	0	0	(1)	1	0
3	0	0	1	1	1	1
4	0	0	1	0	1	1
5	0	1	0	(0)	0	0
6	0	1	0	1	1	0
7	0	1	1	1	1	0
8	0	1	1	(0)	1	1
9	1	1	0	0	0	0
10	1	1	0	1	0	0
11	1	1	1	(1)	1	0
12	1	1	1	(0)	0	1
13	1	0	0	0	0	1
14	1	0	0	(1)	0	0
15	1	0	1	(1)	1	1
16	1	0	1	0	0	1

changed to 1, the circuit is then in the total state represented by row 13. This row represents an unstable state, and the table shows that a flip-flop change must take place so that the circuit enters the stable state represented by row 14.

By means of a table such as Table 5.2-1, the output sequence corresponding to any input sequence can easily be obtained. In a very real sense, the derivation of Table 5.2-1 completes the analysis of the circuit in that it permits the determination of the output sequence corresponding to any input sequence. On the other hand, it could be argued that the analysis is still incomplete because the word statement "The outputs are equal to the previous values of the inputs" has not been obtained. The analysis techniques to be developed here will *not* result in such word statements, for several reasons. First of all, many circuits do not have any simple word statement, since they are part of a larger system and their operation specifications are determined by the system. Second, since English is not a formal language in a mathematical sense, it is not really possible to have a *formal* procedure for obtaining word statements. This does not change the fact that some presentations of the circuit operation are more easily understood than others. The presentation in Table 5.2-1 is not the most acceptable for ease of understanding, and hence a variation of this table is in common use.

Sequential-circuit Analysis

Table 5.2-2. Table Showing Y_1 and Y_2 (Next States of y_1 and y_2) for Each Total State in the Circuit of Fig. 5.1-2

Row	x_1	x_2	y_1	y_2	Y_1	Y_2	z_1	z_2
1	0	0	0	0	(0	0)	0	1
2	0	0	0	1	(0	1)	1	0
3	0	0	1	1	0	1	1	1
4	0	0	1	0	0	0	1	1
5	0	1	0	0	(0	0)	0	0
6	0	1	0	1	0	0	1	0
7	0	1	1	1	1	0	1	0
8	0	1	1	0	(1	0)	1	1
9	1	1	0	0	1	0	0	0
10	1	1	0	1	1	1	0	0
11	1	1	1	1	(1	1)	1	0
12	1	1	1	0	(1	0)	0	1
13	1	0	0	0	0	1	0	1
14	1	0	0	1	(0	1)	0	0
15	1	0	1	1	(1	1)	1	1
16	1	0	1	0	1	1	0	1

Transition Table

In Table 5.2-1, the transitions between unstable and stable states are shown by means of arrows. Another method of describing these transitions is to list, for each total state, the *next* values of the y_1 and y_2 variables. In fundamental-mode operation the inputs do not change until the flip-flops have reached their stable values, so that only the next values of the *flip-flop variables* need be listed. In Table 5.2-2, the *next* values of y_1 and y_2, symbolized by Y_1 and Y_2, are listed for each total state of the circuit of Fig. 5.1-2. The variables y_1, y_2, . . . will be called *internal variables*, or present-state variables, and Y_1, Y_2, . . . will be called *next-state variables*. Values of Y_1 and Y_2 which are the same as the corresponding values of y_1 and y_2 are encircled, since they represent stable states.

It is customary to draw Table 5.2-2 in a slightly different form which distinguishes more strongly between the circuit inputs and the flip-flop outputs. This form, called a *transition table*, is shown in Table 5.2-3a. Each column of this table corresponds to a specific assignment of values to the circuit-input variables, or to an *input state*. Each row of the table corresponds to a specific assignment of values to the flip-flop output variables, or to an *internal state*. Each cell of the table corresponds to an assignment of values to the circuit inputs and the flip-flop outputs, or to a *total*

5.2 formal analysis of circuits containing S-R flip-flops

Table 5.2-3. Tables for the Circuit of Fig. 5.1-1

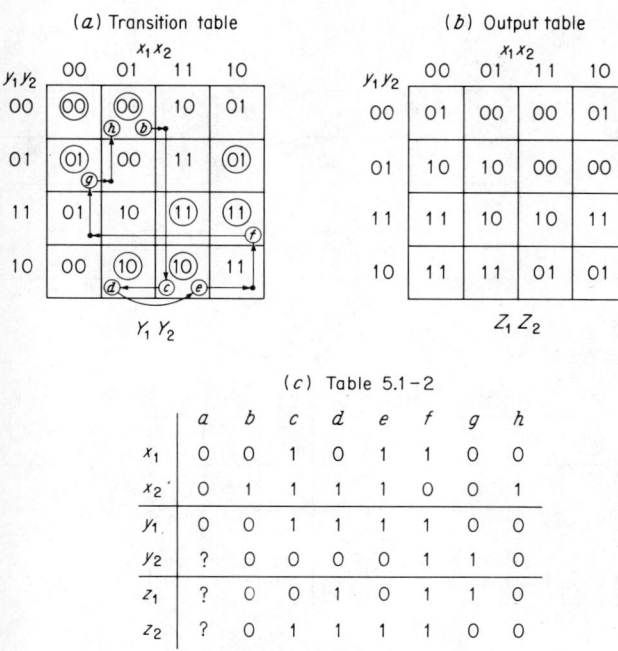

(c) Table 5.1-2

	a	b	c	d	e	f	g	h
x_1	0	0	1	0	1	1	0	0
x_2	0	1	1	1	1	0	0	1
y_1	0	0	1	1	1	1	0	0
y_2	?	0	0	0	0	1	1	0
z_1	?	0	0	1	0	1	1	0
z_2	?	0	1	1	1	1	0	0

state. The entries of the table are the appropriate *next internal states* for each total state. Thus, a change of input variable causes a change from one column of the table to another column without any row change. If, in the new column, the Y_1Y_2 values disagree with the y_1y_2 values for the row, a row change must take place to a new row whose y_1y_2 values are the same as the Y_1Y_2 values of the original row. It is sometimes helpful to think of an "operating point" which represents the total state of the circuit and which moves around on the transition table in accordance with the changes in the total state of the circuit. The operating points which correspond to the sequence of Table 5.1-2 have been plotted on Table 5.2-3a. Table 5.1-2 is repeated as Table 5.2-3c for the sake of convenience. The corresponding output states are shown in the output table of Table 5.2-3b.

The Transition Diagram and State Table

A more pictorial representation of the transition table is sometimes used in which each stable total state is represented by a small circle and each unstable total state is represented by a dot. Arrows are drawn showing the transitions between unstable and stable states. This is called a *transition diagram* and is illustrated in Table 5.2-4c.

Table 5.2-4. Tables for the Circuit of Fig. 5.1-1

(a) Transition table

y_1y_2 \ x_1x_2	00	01	11	10
00	(00)	(00)	10	01
01	(01)	00	11	(01)
11	01	10	(11)	(11)
10	00	(10)	(10)	11

Y_1Y_2

(b) Output table

y_1y_2 \ x_1x_2	00	01	11	10
00	01	00	00	01
01	10	10	00	00
11	11	10	10	11
10	11	11	01	01

Z_1Z_2

(c) Transition diagram

y_1y_2 \ x_1x_2	00	01	11	10
00	0	0	•	•
01	0	•	•	0
11	•	•	0	0
10	•	0	0	•

(d) State table

S \ x_1x_2	00	01	11	10
A	(A)	(A)	D	B
B	(B)	A	C	(B)
C	B	D	(C)	(C)
D	A	(D)	(D)	C

S

(e) State table

S \ x_1x_2	00	01	11	10
1	(1)	(1)	4	2
2	(2)	1	3	(2)
3	2	4	(3)	(3)
4	1	(4)	(4)	3

S

Another form of the transition table is often used in which each internal state is replaced by an arbitrary letter or decimal number. This form of table is called a *state table* and is illustrated in Table 5.2-4d and Table 5.2-4e. By using a state table and an output table it is possible to determine the output sequence produced by any input sequence. It is not possible to determine the sequence of states of the circuit flip-flops, but this information is not important to the *external* performance of the circuit. The state table is important because it will form the starting point for the *synthesis* of sequential circuits.

The Excitation Table

In discussing the transition and state tables and showing that they are reasonable forms in which to present the operation of a sequential circuit, the question of how these tables are obtained from the circuit diagram has not been discussed. In order to complete the formal analysis procedure, a formal technique must be determined for going from a circuit diagram to a transition table. The first step in this technique is to obtain, from the circuit diagram, the switching functions which describe the effect of the circuit inputs and flip-flop outputs on the flip-flop inputs. These switching functions are called *excitation functions*. The excitation functions for the circuit of Fig. 5.1-1 are shown in Table 5.2-5a. The excitation functions

5.2 formal analysis of circuits containing S-R flip-flops

Table 5.2-5. Excitation Table for the Circuit of Fig. 5.1-2

(a) Excitation functions

$S_1 = x_1 x_2 \quad R_1 = x_1' x_2' \quad S_2 = x_1 x_2' \quad R_2 = x_1' x_2$

(b) Excitation table

$y_1 y_2$ \ $x_1 x_2$	00	01	11	10
00	01, 00	00, 01	10, 00	00, 10
01	01, 00	00, 01	10, 00	00, 10
11	01, 00	00, 01	10, 00	00, 10
10	01, 00	00, 01	10, 00	00, 10

$S_1 R_1, S_2 R_2$

are then used to fill in an excitation table. The *excitation table* is the same as the transition table, except that its entries are the values of the flip-flop inputs (S_1, R_1, . . .) rather than the next states (Y_1, Y_2, . . .) of the flip-flop outputs. An excitation table is shown in Table 5.2-5b.

The crucial step in analyzing a sequential circuit is that of going from an excitation table to the corresponding transition table. One method of doing this is to note that, whenever S and R are both equal to 0, the value of Y will be the same as the value of y; whenever $S = 1$ and $R = 0$, Y will be equal to 1; and whenever $S = 0$ and $R = 1$, Y will be equal to 0. The situation where $S = R = 1$ is assumed not to occur. By using these simple rules the transition table can readily be written down from the excitation table. This procedure can also be formalized by writing these rules down in a table of combinations and obtaining a function giving the dependence of Y on S, R, and y. This is done in Table 5.2-6, and the function is shown to be $Y = S + R'y$. This function is called the *characteristic function* for set-reset flip-flops.

It is perhaps well to point out now that the variable Y is quite different from the other variables such as S, R, y, Z in that Y does *not* correspond to any *physical signal* in the circuit. The variable Y is in some sense a fictitious variable representing the *next* condition of the y variable.

A formal technique for analyzing sequential circuits has been presented. The steps in this procedure are illustrated in Fig. 5.2-1. The following sections will show how this technique is applied to sequential circuits using devices other than set-reset flip-flops and to circuits operating in different modes.

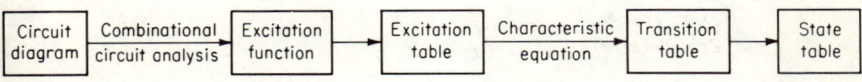

Fig. 5.2-1. *Sequential-circuit analysis.*

Table 5.2-6. Relationship between S, R, y, and Y

(a) Table of combinations

S	R	y	Y
0	0	0	0
0	0	1	1
0	1	0	0
0	1	1	0
1	1	0	d
1	1	1	d
1	0	0	1
1	0	1	1

(b) Map

y \ SR	00	01	11	10
0	0	0	d	1
1	1	0	d	1

Y

(c) Characteristic function for set-reset flip-flops.
$Y = S + R'y$

5.3 VARIOUS "MEMORY" DEVICES FOR SEQUENTIAL CIRCUITS

The preceding discussion has been concerned with sequential circuits containing set-reset flip-flops. There are, of course, many other devices which can be used to construct sequential circuits. The analysis technique which was presented for set-reset flip-flops will still be valid with minor modifications for circuits containing other devices. Specifically, different devices will require changes in the rules used for obtaining the transition table from the excitation table, but the rest of the technique will remain unchanged. Before considering some different devices in detail it seems appropriate to consider the question of what physical properties the devices used to construct sequential circuits must have.

One essential feature of a sequential circuit is that there must be some signal or signals in the circuit whose value is determined not only by the present circuit inputs but by past circuit inputs as well. The devices

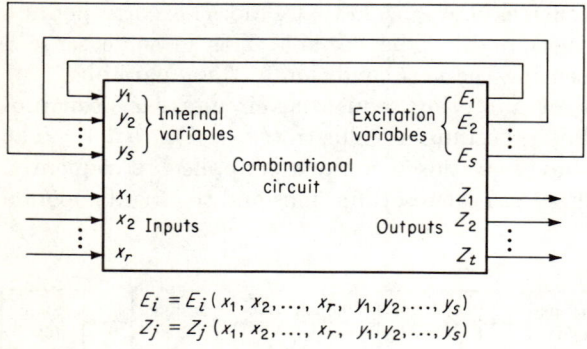

$E_i = E_i(x_1, x_2, \ldots, x_r, y_1, y_2, \ldots, y_s)$
$Z_j = Z_j(x_1, x_2, \ldots, x_r, y_1, y_2, \ldots, y_s)$

Fig. 5.3-1. *General form of a sequential circuit.*

5.3 various "memory" devices for sequential circuits

controlling these signals are usually called *memory* devices, internal devices, or secondary devices. In the circuit analyzed previously, the flip-flops were the internal devices. The mechanism by which these devices operate usually involves some sort of feedback to provide the memory, and it is customary to represent a generalized sequential circuit as a combinational circuit with some feedback loops (Fig. 5.3-1). In a flip-flop, these feedback loops are contained within the flip-flop circuit, but they are present nevertheless.

Physical Requirements

As an introduction to different types of memory devices, the requirements on the electrical properties of the feedback loops will be considered [3]. Suppose that one of the feedback loops has been broken, as in Fig. 5.3-2, and a terminating impedance added to simulate the impedance presented by the rest of the loop. There must be some set of values for the inputs (x_i) and the internal variables (y_j) such that either a 1 or a 0 can be present in this feedback loop. Assume that this set of values is present and that the loop characteristic is measured (e_o versus e_i). When the loop is closed, $e_o = e_i$; therefore, there must be two intersections of the open-loop characteristic with the $e_o = e_i$ line. One of these intersections corresponds to a 1 stored in the loop and therefore occurs with $e_o = e_i = E_H$, and the other intersection occurs with $e_o = e_i = E_L$ and corresponds to a stored 0. These intersections are labeled A and B in Fig. 5.3-3. For these intersections to correspond to stable operating points, the slope of the open-loop characteristic must be less than the slope of the closed-loop characteristic. This is indicated by the heavy portion of the open-loop characteristic. The rest of the open-loop characteristic has to be a continuous curve connecting the two heavy portions. Note that there is

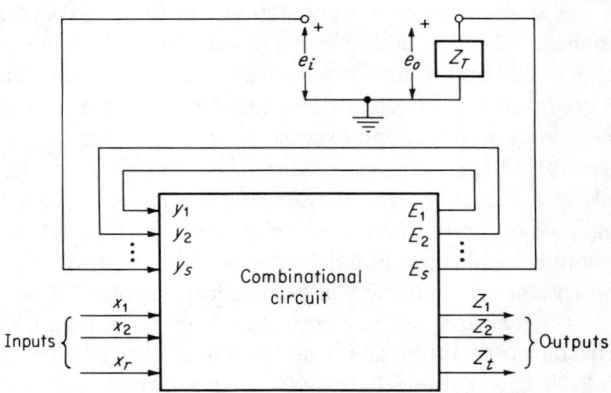

Fig. 5.3-2. *General sequential circuit with feedback loop broken.*

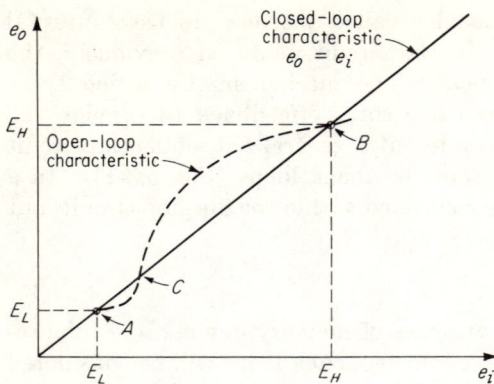

Fig. 5.3-3. Feedback-loop characteristics.

another intersection of the two curves (point C), but this corresponds to an unstable operating point. The important feature of the open-loop characteristic is the region where the slope is greater than 1. In this region a change in e_i will result in a larger change in e_o. Clearly some sort of active device providing amplification must be present in the loop.

There are many ways in which this amplification can be provided. When diode gates are used, it is customary to use vacuum-tube or transistor amplifiers for the gain. When AND-NOT or OR-NOT gates are used, the gain is provided by the transistors contained in the gates themselves. In flip-flops the tubes or transistors provide the gain, and in relay circuits the gain is made possible by the fact that the relay contacts can control more power than is required to operate the relay itself.

Analysis of Sequential Circuits Constructed of Diode Gates

A sequential circuit constructed of diode gates and amplifiers is shown in Fig. 5.3-4. The first step in analyzing such a circuit is to write down the excitation functions, in this case, the switching functions for the signals at the inputs to the feedback loop amplifiers, E_1 and E_2. The next step is to form an excitation table such as that shown in Table 5.3-1a. To obtain the transition table, the effect of E_1 on y_1 must be determined, or, equivalently, the characteristic function must be obtained. In the case of an ordinary amplifier it is clear that the amplifier output y will change to become equal to the amplifier input E. Because of this, the characteristic function is particularly simple, $Y = E$; and *the transition table is identical with the excitation table*. Passing from the transition table to the state table and the output table completes the analysis—Table 5.3-1b and Table 5.3-1c. Since these are the same as Tables 5.2-4e and 5.2-4b, it can be concluded that the circuits of Figs. 5.1-2 and 5.3-4 have the same external performance.

5.3 *various "memory" devices for sequential circuits*

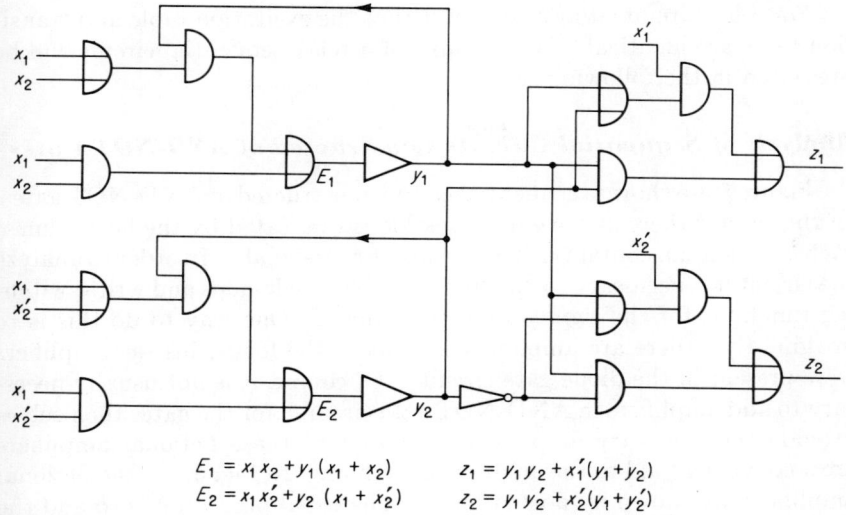

$$E_1 = x_1 x_2 + y_1 (x_1 + x_2) \qquad z_1 = y_1 y_2 + x_1'(y_1 + y_2)$$
$$E_2 = x_1 x_2' + y_2 (x_1 + x_2') \qquad z_2 = y_1 y_2' + x_2'(y_1 + y_2')$$

Fig. 5.3-4. *A sequential circuit using diode gates and amplifiers.*

Table 5.3-1. Tables for the Circuit of Fig. 5.3-4

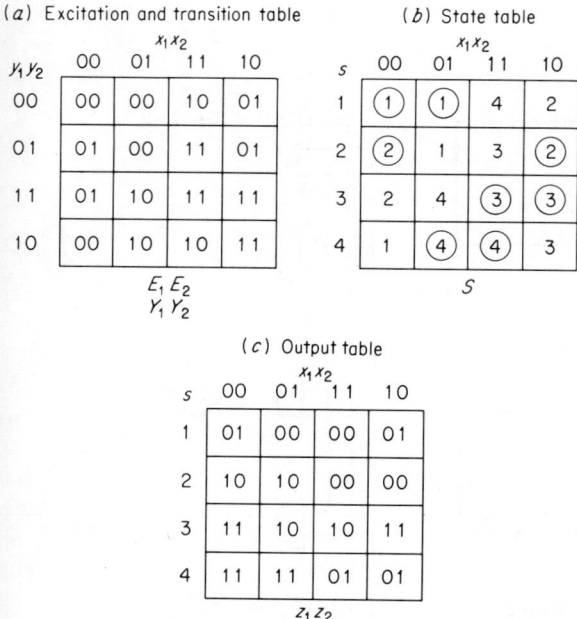

(a) Excitation and transition table

$y_1 y_2$ \ $x_1 x_2$	00	01	11	10
00	00	00	10	01
01	01	00	11	01
11	01	10	11	11
10	00	10	10	11

$E_1 E_2$
$Y_1 Y_2$

(b) State table

s \ $x_1 x_2$	00	01	11	10
1	①	①	4	2
2	②	1	3	②
3	2	4	③	③
4	1	④	④	3

S

(c) Output table

s \ $x_1 x_2$	00	01	11	10
1	01	00	00	01
2	10	10	00	00
3	11	10	10	11
4	11	11	01	01

$z_1 z_2$

Sequential-circuit Analysis

For relay circuits it also turns out that the excitation table and transition table are identical. An example of a relay sequential circuit will be presented in the following section.

Analysis of Sequential Circuits Constructed of AND-NOT Gates

Figure 5.3-5 shows a sequential circuit constructed of AND-NOT gates. In this circuit there are two feedback loops (indicated by the heavy lines) each of which can contain either a 1 signal or a 0 signal. In order to analyze the circuit it is necessary to "break" these feedback loops and write switching functions for the signals present in them. One way to do this is to imagine that there are amplifiers present in the loops, just as amplifiers were present in the diode gate circuit. Of course, it is not usually necessary to add amplifiers in AND-NOT gate circuits, for the gates themselves provide the necessary amplification; however, these fictional amplifiers are a convenient aid to the analysis procedure. In Fig. 5.3-5, the fictional amplifiers are shown by dotted lines. The usual procedure is to add the minimum number of fictional amplifiers which is sufficient to break all the feedback loops. Once the amplifiers have been added, the analysis is identical to the analysis of a diode gate circuit. In Fig. 5.3-5, the inputs to the amplifiers have been labeled Y_1 and Y_2 because the excitation variables E_1, E_2 have been shown to be equal to the next-state variables Y_1, Y_2.

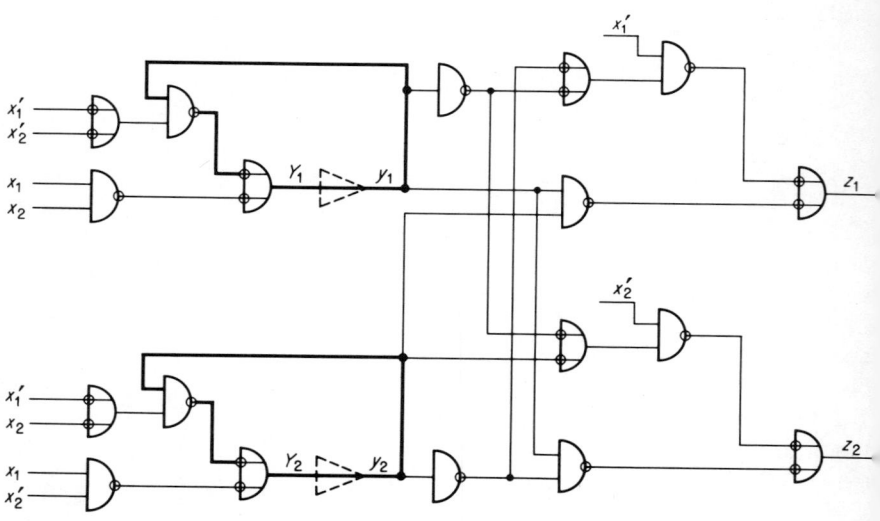

Fig. 5.3-5. *A sequential circuit constructed of AND-NOT gates.*

The next-state functions and output functions are

$Y_1 = x_1 x_2 + y_1(x_1 + x_2)$
$Y_2 = x_1 x_2' + y_2(x_1 + x_2')$
$z_1 = y_1 y_2 + x_1'(y_1 + y_2)$
$z_2 = y_1 y_2' + x_2'(y_1 + y_2')$

Since these are identical with the functions for Fig. 5.3-4, the external performance of the circuits of Figs. 5.3-4 and 5.3-5 is the same.

It is an interesting thing to compare the results of this analysis with the analysis obtained by "breaking" the feedback loops in different places. Since this operation of inserting fictional amplifiers is merely a convenience for analysis and does not affect the circuit performance, the same result should be obtained for different locations of the fictional amplifiers. This question will be studied further in the problems at the end of the chapter.

5.4 RACES IN SEQUENTIAL CIRCUITS

In Chap. 3 a mathematical model for digital circuits was developed, and a switching algebra was formulated. In order to obtain a simple model and an algebra which could be easily used, several idealizing assumptions were made. One of the most serious of these was the assumption that no delay is involved in changing a signal from one of its values to its other value. While this assumption was not stated explicitly, it is implicit in the convention to represent all make contacts on the same relay by the same variable (any "stagger" in the opening or closing of the contacts being thus neglected) and in the theorems $x + x' = 1$ and $xx' = 0$ (which neglect the possibility of x changing before x' changes, etc.). The techniques of Chap. 3 are satisfactory for combinational circuits in which only the steady-state performance is important. As soon as the behavior of the circuit during input changes must be controlled, the Chap. 3 techniques are no longer sufficient and additional techniques must be developed. This situation will be treated further in Chap. 7; however, a very important aspect of sequential circuits involves the delay in the feedback loops, and it seems appropriate to consider some of the effects of this delay now.

A relay sequential circuit is shown in Fig. 5.4-1, and the tables for the analysis of this circuit are shown in Table 5.4-1. When the inputs are both equal to 1 and $y_1 = 0$, $y_2 = 1$, the transition table shows that $Y_1 = 1$ and $Y_2 = 0$ (indicated by the * in Table 5.4-1). This means that the Y_1 relay is released and energized and the Y_2 relay is operated and not energized, so that the Y_1 relay must operate and the Y_2 relay must release. If these two events occur simultaneously the transition specified in the transition table will actually take place. However, it is extremely unlikely

Sequential-circuit Analysis

Table 5.4-1

(a) Excitation and transition table

$y_1 y_2$ \ $x_1 x_2$	00	01	11	10
00	⓪⓪	01	10	11
01	00	⓪1	10*	11
11	01	⑪	⑪	⑪
10	⑩	⑩	⑩	11

$Y_1 Y_2$

(b) Output table

$y_1 y_2$ \ $x_1 x_2$	00	01	11	10
00	0	0	0	0
01	0	0	0	0
11	0	0	0	0
10	1	0	0	0

Z

(c) Transition diagram

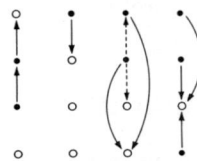

that all the y_1 and y_2' contacts will close at exactly the same time as all the y_1' and y_2 contacts open. In practice, either Y_1 will operate before Y_2 releases, or vice versa. Thus, instead of going directly to the $y_1 = 1$, $y_2 = 0$ state, the circuit will go either to the $y_1 = 1$, $y_2 = 1$ state or to the $y_1 = 0$, $y_2 = 0$ state. If the circuit goes to the $y_1 = 0$, $y_2 = 0$ state, the Y_1 relay will still be unstable (since in this state $Y_1 = 1$ and $Y_2 = 0$) and the circuit will now go to the $y_1 = 1$, $y_2 = 0$ state. In this case the desired operation is achieved. On the other hand, if the circuit goes to the $y_1 = 1$, $y_2 = 1$ state, it will remain there, since this state is stable and the circuit operation will be incorrect. This situation where more than one of the internal variables are unstable is called a *race*. If the final stable state to which the circuit goes depends on the order in which the internal variables change, then a *critical race* is said to be present.

In order to obtain reliable circuit operation, critical races must be avoided. One approach to eliminating these is to "fix" the races, that is, choose the relays so that the order in which they change is controlled by their electrical characteristics. For the circuit of Fig. 5.4-1, this would mean using a Y_2 relay whose operate time was definitely longer than the release time of the Y_1 relay. There are two difficulties with this technique for handling critical races. First of all, the electrical characteristics of the relays will change. What was a reliable circuit when first constructed may no longer be a reliable circuit after it has been in operation for some time. Also, there are tables in which it is impossible to fix the critical races. For example, if the entry in the $y_1 = 0$, $y_2 = 0$ row of the $x_1 = 1$, $x_2 = 1$

5.4 races in sequential circuits

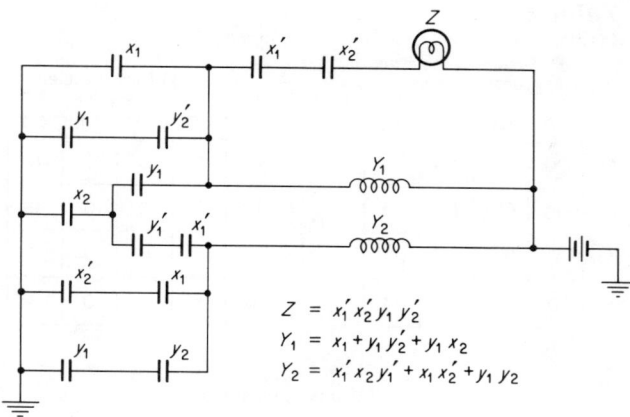

Fig. 5.4-1. *A relay sequential circuit illustrating races.*

column of Table 5.4-1a were 00 instead of 10, the critical race could not be avoided by controlling the electrical properties of the internal relays. In gate circuits, it is frequently possible to resolve critical races by using additional "slow-up" capacitors or gates with slower switching times.

A more generally satisfactory technique for avoiding critical races is to arrange the transition table so that they never occur. A technique for doing this will be presented in Chap. 6. Avoiding the situation where more than one of the internal variables are unstable at one time is one method

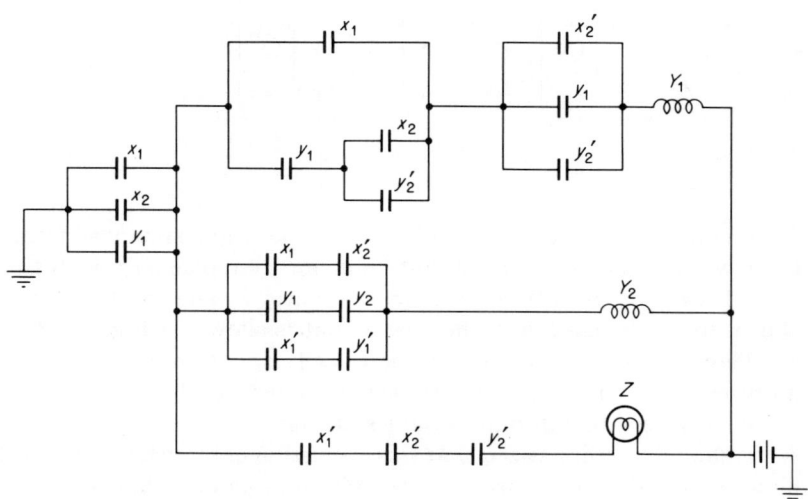

Fig. 5.4-2. *Circuit for Table 5.4-2.*

197

Table 5.4-2

(a) Excitation table and transition table

y_1y_2 \ x_1x_2	00	01	11	10
00	⓪⓪	01	10	11
01	00	⓪1	00*	11
11	01	⑪	⑪	⑪
10	⑩	⑩	⑩	11

Y_1Y_2

(b) Output table

y_1y_2 \ x_1x_2	00	01	11	10
00	0	0	0	0
01	0	0	0	0
11	0	0	0	0
10	1	0	0	0

Z

(c) Transition diagram

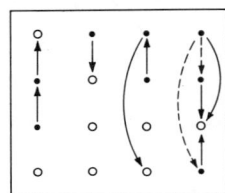

(d) State table

s \ x_1x_2	00	01	11	10
1	①	2	4	3
2	1	②	1	3
3	2	③	③	③
4	④	④	④	3

S

(e) Flow table

s \ x_1x_2	00	01	11	10
1	①	2	4	3
2	1	②	4	3
3	1	③	③	③
4	④	④	④	3

S

for eliminating critical races. The critical race can be removed from Table 5.4-1 without changing the circuit behavior by replacing the 10 entry for $x_1 = 1$, $x_2 = 1$, $y_1 = 0$, $y_2 = 1$ by a 00 entry. The resulting tables are shown in Table 5.4-2, and the new circuit is shown in Fig. 5.4-2.

There is still a race present in the $x_1 = 1$, $x_2 = 0$ column, but this is not a critical race, since the final stable state must be 11, independent of the order in which the internal variables change.

It should be emphasized that this discussion of races holds for electronic circuits as well as relay circuits, since flip-flops and amplifiers have a delay in changing state just as relays do. The circuit corresponding to Table 5.4-2a and b is shown in Fig. 5.4-2.

The Flow Table

A state table such as Table 5.4-2d describes some of the internal behavior of the corresponding circuit, since for $x_1 = x_2 = 1$ it specifies the multiple transition from state 2 to state 4 via state 1. As far as the external circuit performance is concerned this multiple transition is usually not important.† It is customary to describe the circuit performance by means of a *flow table* which is identical with the state table except that multiple transitions are not shown; when the circuit goes from one unstable state to another unstable state, the entry specifying the first unstable state is replaced by an entry specifying the final stable state which is reached. Table 5.4-2e is the flow table corresponding to the state table of Table 5.4-2d. The 1 entry corresponding to $x_1 = x_2 = 1$, $s_j = 2$ has been replaced by a 4 entry, since the final stable state reached is 4.

The major importance of the flow table is in synthesis, since in specifying the desired performance of a circuit no information is available about internal multiple transitions. It may be necessary to introduce multiple transitions in the design procedure for economy or to avoid critical races. In summary, a flow table specifies the final stable state which is reached, and a state table specifies the next state, whether this state is stable or unstable.

5.5 PULSE-MODE OPERATION

Very often sequential circuits are designed to operate with inputs which are in the form of pulses. In its present form this statement does not have too much meaning, for it is not clear exactly what is meant by a pulse in connection with sequential circuits. A precise interpretation of the expression *pulse input* will be developed in the course of the analysis of the circuits to be presented in this section. For the present, it will be sufficient to regard a pulse as a signal that is in the 1 state for a much shorter period of time than it is in the 0 state. Certainly there is a large intuitive appeal associated with the concept of a pulse.

The circuit of Fig. 5.5-1 is designed to operate with a series of pulses occurring on the input lead. The tables for the analysis of this circuit are shown in Table 5.5-1. It is clear from this analysis that there will be an output pulse on the z lead with every fourth pulse on the x lead and that the (level) signals on the Z_1 and Z_2 leads will be a binary representation of the number of input pulses which have been received modulo four. Thus this circuit can be considered both a modulo-four counter with respect to the Z_1, Z_2 leads and a frequency divider with respect to the z lead.

Under the proper conditions, the circuit shown in Fig. 5.5-2 will also

† Unless a transient output is developed in the intermediate state.

Sequential-circuit Analysis

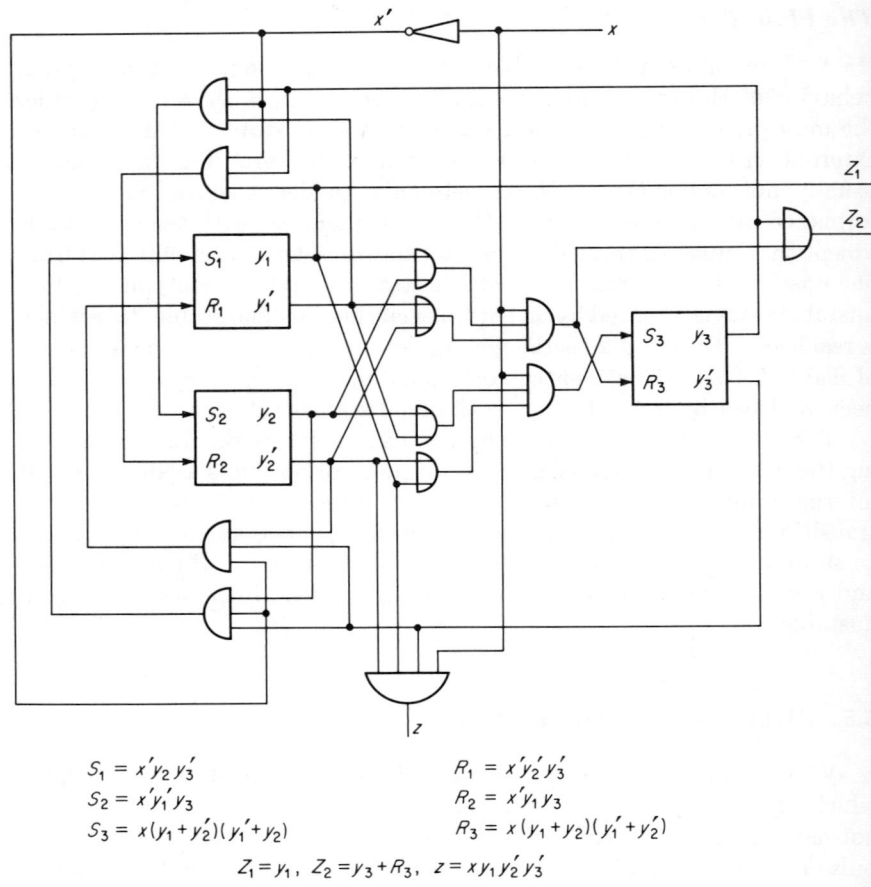

$S_1 = x'y_2 y_3'$ $R_1 = x'y_2' y_3'$
$S_2 = x'y_1' y_3$ $R_2 = x'y_1 y_3$
$S_3 = x(y_1 + y_2')(y_1' + y_2)$ $R_3 = x(y_1 + y_2)(y_1' + y_2')$

$Z_1 = y_1$, $Z_2 = y_3 + R_3$, $z = xy_1 y_2' y_3'$

Fig. 5.5-1. *Fundamental-mode counter circuit.*

perform as a modulo-four counter and frequency divider. The analysis for this circuit is shown in Table 5.5-2. According to this analysis, when $x = 1$ the circuit should just cycle through its four internal states. The final state reached when x again becomes equal to 0 should be determined on a random basis, depending on which state the circuit is "caught in" when the transition to 0 takes place. This reasoning is valid for pulses whose duration is very long compared with the time required for a flip-flop to change state. For very short pulses the situation is quite different. In order to see this, the detailed circuit actions will be traced for short input pulses.

The operation of the circuit will be traced starting with the condition where $x = 0$ and $y_1 = y_2 = 0$. This initial state is indicated in Fig. 5.5-2a by the labels on the leads. (For those leads with two labels, the

5.5 pulse-mode operation

Table 5.5-1. *Tables for the Circuit of Fig. 5.5-1*

(a) Excitation table

$y_1 y_2 y_3$	0	1
000	01 00 00	00 00 10
001	00 10 00	00 00 10
011	00 10 00	00 00 01
010	10 00 00	00 00 01
110	10 00 00	00 00 10
111	00 01 00	00 00 10
101	00 01 00	00 00 01
100	01 00 00	00 00 01

$S_1 R_1$ $S_2 R_2$ $S_3 R_3$

(b) Combined state – output table

s	0	1
1	①, 000	2, 000
2	3, 010	②, 010
3	③, 010	4, 010
4	5, 000	④, 010
5	⑤, 100	6, 100
6	7, 110	⑥, 110
7	⑦, 110	8, 110
8	1, 100	⑧, 111

$S, Z_1 Z_2$ z

(c) Word statement

z pulse out with every fourth x pulse in
$Z_1 Z_2$ counts number of pulses in modulo four

Table 5.5-2. *Analysis for Fig. 5.5-2*

(a) Excitation functions

$$S_1 = xy_2 \quad R_1 = xy_2' \quad S_2 = xy_1' \quad R_2 = xy_1$$
$$z = xy_1 y_2' \quad Z_1 = y_1 \quad Z_2 = y_1 y_2' + y_1' y_2$$

(b) Excitation table

$y_1 y_2$	0	1
00	00 00	01 10
01	00 00	10 10
11	00 00	10 01
10	00 00	01 01

$S_1 R_1$ $S_2 R_2$

(c) Output table

$y_1 y_2$	0	1
00	0, 00	0, 00
01	0, 01	0, 01
11	0, 10	0, 10
10	0, 11	1, 11

$z, Z_1 Z_2$

(d) Transition table

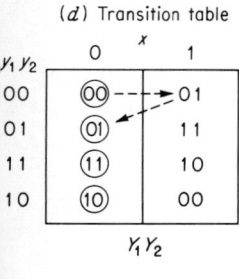

$y_1 y_2$

(e) State table

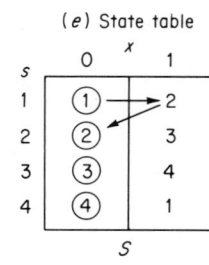

s

Sequential-circuit Analysis

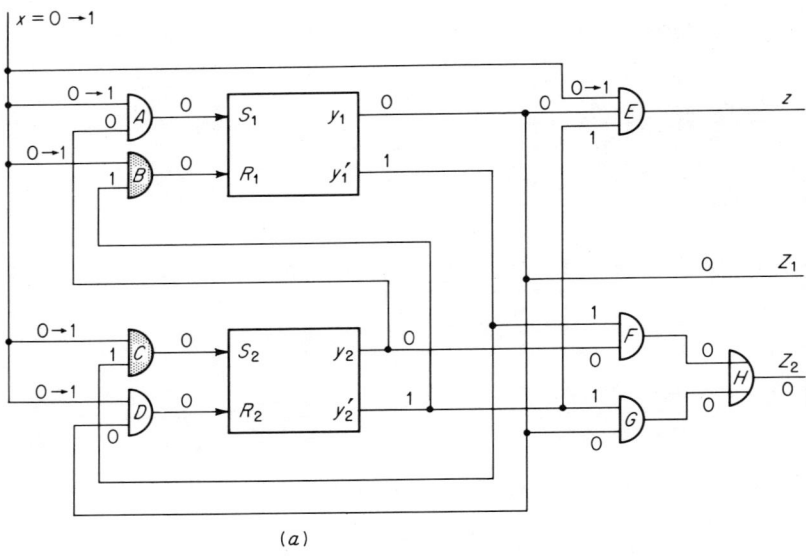

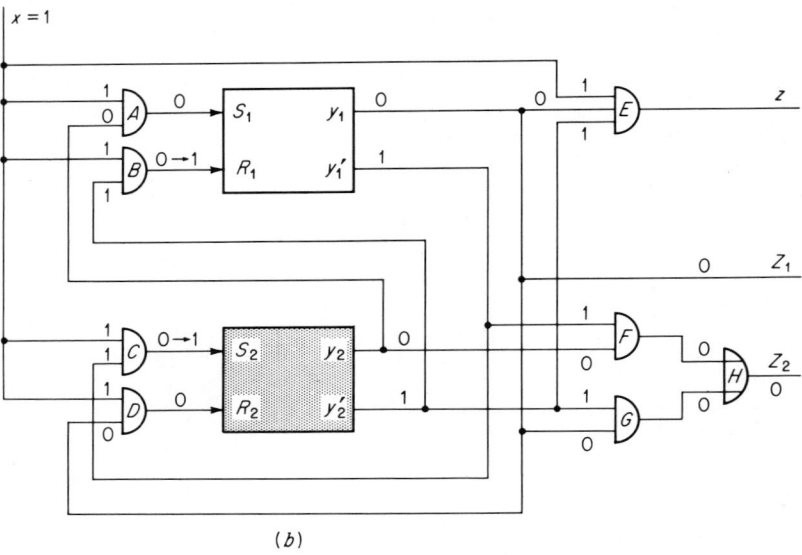

Fig. 5.5-2

5.5 pulse-mode operation

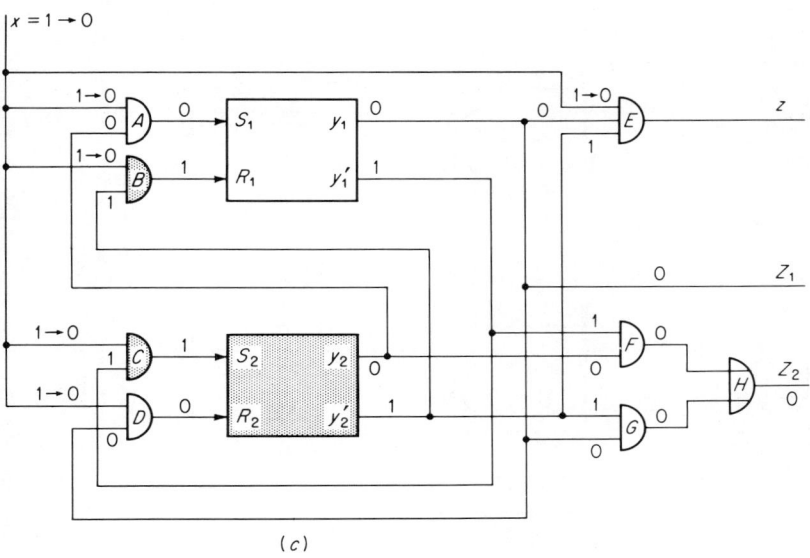

(c)

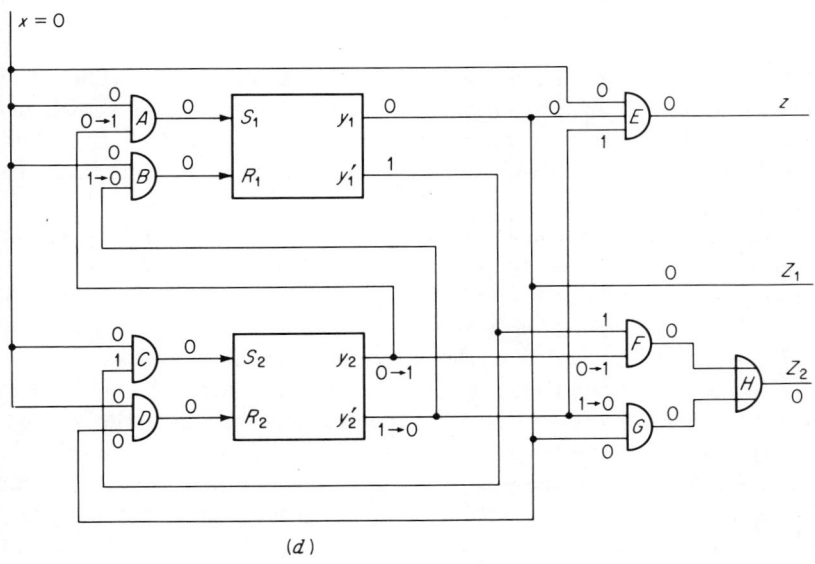

(d)

Fig. 5.5-2 *(Continued)*

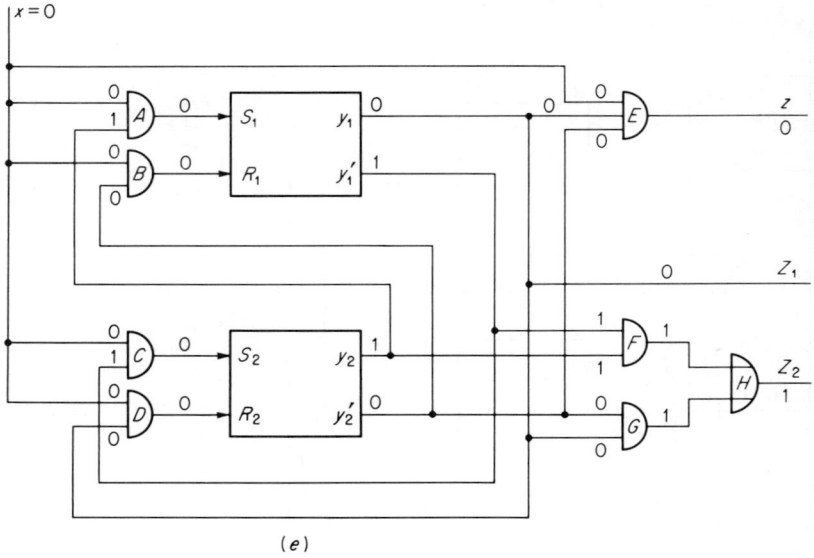

Fig. 5.5-2 *(Continued)*

label preceding the arrow pertains to the initial state.) The situation immediately following a change of x from 0 to 1 is also shown in Fig 5.5-2a. This change is represented symbolically as $0 \to 1$. Thus the labels following the arrows represent the circuit conditions immediately after x changes.

In assigning the labels on Fig. 5.5-2a, the assumption has been made that the delay between the time when x changes and the time when this change appears at the inputs to the gates driven directly by x is the same for all such gates. This is not a crucial assumption for the situation being studied here and is made just to simplify the discussion.

At the end of the time required to propagate the x change to the gate inputs, gates B and C of Fig. 5.5-2a will be unstable, since their inputs are all equal to 1 and the outputs equal 0. The next change in the circuit will therefore occur when these gate outputs become equal to 1. This change is shown in Fig. 5.5-2b. (Here the assumption has been made that the delays of these two gates in changing are equal. Again this is solely a matter of convenience.) At this time flip-flop 2 becomes unstable, since there is a 1 signal on its set lead, but it is in the reset state ($y_2 = 0$). Let us assume that, while flip-flop 2 is in the process of changing state, the input x changes back to 0. This situation is shown in Fig. 5.5-2c. Following this, flip-flop 2 will change, producing the situation in Fig. 5.5-2d. (Actually, it is possible that gates B and C may change their outputs before

5.5 pulse-mode operation

y_2 changes. Again this will not affect the circuit operation.) Finally, the unstable gates, B, C, and F, will have their outputs change, and the final stable state shown in Fig. 5.5-2e will result. In order to ensure the possibility of returning the x input of gate A to 0 before the signal from y_2 changes to 1, it may be necessary to insert a delay in the lead from the y_2 output of flip-flop 2 to the input of gate A. The circuit can always be operated in the manner described if such a delay is provided.

The result of the circuit actions just described is that a single input pulse has caused the circuit to change state from $y_1 = 0$, $y_2 = 0$ to $y_1 = 0$, $y_2 = 1$. The pulse count indicated by the Z_1, Z_2 leads has been caused to increase by 1 (from 00 to 01). A similar detailed analysis for additional pulses would show that each pulse causes the circuit to change from one state to another well-determined state and the count indicated on the Z_1, Z_2 leads to increase by 1 (modulo four). With every fourth input pulse an output pulse will occur on the z lead. This is a circuit with pulses as inputs and with one output lead on which pulse signals occur and two output leads on which level signals are present.†

In more general terms, the foregoing analysis has involved two basic assumptions about the length of the input pulses. First, the pulses were assumed to last long enough to cause the appropriate flip-flops to change state. This assumption must be satisfied for any sequential circuit operation, since the input signals must always be of sufficient duration to cause the circuit to respond. Second, the pulse is assumed to be short enough so that it is no longer present at the circuits which generate the flip-flop input signals when the change in flip-flop outputs has propagated to the input circuitry. If this assumption is valid, each input pulse will cause only one change of internal state and a succession of internal-state changes in response to a single input pulse will be impossible.

A circuit which is operated in this fashion is said to be operated in *pulse mode*. It is clear that a circuit can be operated in pulse mode only if more time is required to propagate changes in flip-flops to the flip-flop input circuitry than is required to cause a flip-flop to change state. This time difference can always be guaranteed by placing delays at the flip-flop outputs. The general form of a sequential circuit constructed with set-reset flip-flops is shown in Fig. 5.5-3 with the delays necessary for pulse-mode operation explicitly indicated. Only minor modifications in this figure are necessary if some other type of flip-flop is used. Since it is customary to employ circuits with flip-flops for pulse-mode operation, operation of circuits with feedback loops in pulse mode will not be discussed here. A discussion of this question can be found in [1]. Another possibility is to make use of flip-flops which change state only when the input

† Sequential circuits with pulse inputs and pulse outputs are sometimes called *Mealy model circuits* [4], and sequential circuits with pulse inputs and level outputs are sometimes called *Moore model circuits* [5].

Sequential-circuit Analysis

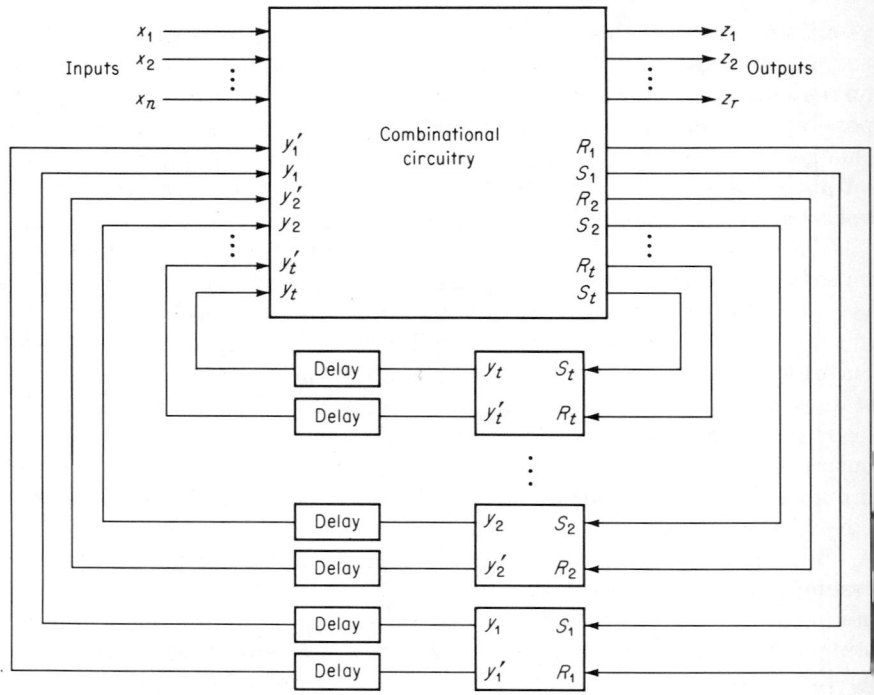

Fig. 5.5-3. *General form of sequential circuit with set-reset flip-flops showing delays for pulse-mode operation.*

signal changes from 1 to 0 (flip-flops which trigger on the falling edges of pulses) [6, p. 99]. This guarantees that the input pulse is terminated before the flip-flop outputs change. The delays at the flip-flop outputs are not required when this type of flip-flop is used, but the flip-flop itself is more complex than the ordinary type of flip-flop which responds to the presence of signals.

The circuits of Figs. 5.5-1 and 5.5-2 differ in the following respects. The Fig. 5.5-2 circuit requires one fewer flip-flop but must have delays at the flip-flop outputs and must have the maximum length of the input pulses controlled. Thus, designing a circuit for pulse-mode operation can result in a saving in the number of flip-flops required; however, a requirement on the maximum length of the input pulses which is not present for fundamental-mode operation is introduced.

The circuit of Fig. 5.5-2 cannot possibly be operated in fundamental mode, since it never stabilizes internally when x is equal to 1. However, it is possible to operate this circuit with the input signal held at 1 long enough for several changes of internal state to take place. In this case the circuit is not being operated in either pulse mode or fundamental mode. This

5.5 pulse-mode operation

illustrates the fact that other modes of operation besides fundamental mode and pulse mode are possible. Emphasis is being placed on pulse- and fundamental-mode operation because these are by far the most common modes in which sequential circuits are operated.

Operating Points [7]

In Sec. 5.2 the concept of an operating point on a state table or transition table was introduced in connection with fundamental-mode operation. The motion of the operating point is different for pulse-mode operation from that for fundamental-mode operation. In pulse-mode operation, a change of input produces a change in the excitation variables and a corresponding change in the next-state variables Y_i, but the present-state variables y_i do not change to become equal to the next-state variables until after the input has changed back to its original value. Thus, for the operation which was considered in detail in connection with Fig. 5.5-2d, the sequence of circuit conditions is that shown in Table 5.5-3. Note the third row of this table: the internal variables y_1 and y_2 are both equal to 0, the excitation variables are all equal to 0, but the next-state variable Y_2 is equal to 1. There is no cell of the transition table which corresponds to this circuit condition. Because of this it is not possible to show the exact sequence of circuit conditions on the transition table. Rather than introduce a more complicated table in which all the circuit conditions are explicit, it is customary for pulse-mode operation to represent the operating-point path as a horizontal line and a diagonal line, as shown in Table 5.5-2d. The horizontal line represents the change of the input variable when the input pulse starts, and the diagonal line represents the termination of the input pulse *and* the change of the internal variables.

Thus for pulse-mode operation the operating point on the transition table follows a path composed of horizontal and slant line segments, and for fundamental-mode operation the operating-point path consists of horizontal and vertical line segments.

Table 5.5-3. Sequence of Circuit Conditions Shown in Table 5.5-2d

	x	y_1y_2	S_1R_1	S_2R_2	Y_1Y_2
1	0	00	00	00	00
2	1	00	01	10	01
3	0	00	00	00	01
4	0	01	00	00	01

Sequential-circuit Analysis

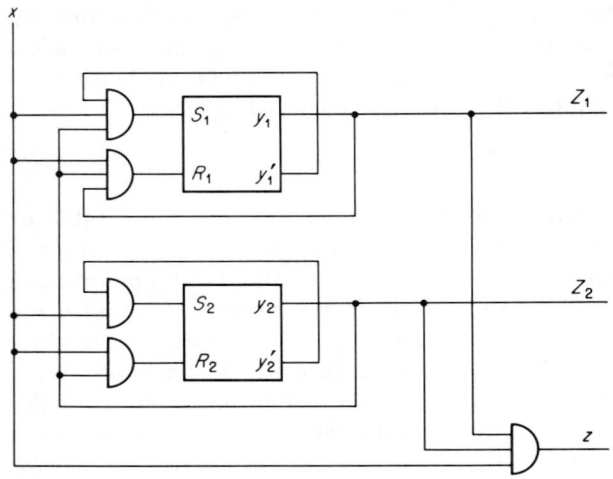

Fig. 5.5-4. Sequential circuit in which several internal variables can be simultaneously unstable.

Races in Pulse-mode Operation

In Sec. 5.4 the problem of races in sequential circuits was discussed, and it was shown that a race is present whenever more than one internal device is unstable. It was pointed out that critical races must be avoided in order to ensure proper circuit operation. These remarks apply only to fundamental-mode operation, since *races cannot result in improper operation of a circuit operating in pulse mode.*†

The circuit of Fig. 5.5-4 is an example of a circuit designed to operate in pulse mode with several internal variables unstable simultaneously. The transition table for this circuit, Table 5.5-4d, shows that when $y_1 = 0$, $y_2 = 1$ or $y_1 = 1$, $y_2 = 1$, and $x = 1$, both the internal variables are unstable. This is a race situation, but it has no effect on the circuit operation. Because of the restriction on the pulse width, multiple transitions cannot occur. It is only because of the possibility of an incorrect multiple transition that races have to receive special attention.

The circuit of Fig. 5.5-4 has the same behavior as the circuit of Fig. 5.5-1 but requires three fewer gates. Thus use of the fact that races can be present without affecting the pulse-mode operation of a circuit has resulted

---

† In order to be completely accurate, this statement needs to be qualified. It is true provided that no changes of internal variables occur when the input pulses are not present. This means that the circuit must always go directly to a stable state when the input pulse terminates. Common practice is to design pulse-mode circuits so that this criterion is met; however, there is nothing to prevent a designer from violating this condition.

5.5 pulse-mode operation

Table 5.5-4. Tables for the Circuit of Fig. 5.5-4

(a) Excitation functions

$$S_1 = xy_1'y_2 \quad R_1 = xy_1y_2 \quad S_2 = xy_2' \quad R_2 = xy_2$$
$$z = xy_1y_2 \quad Z_1 = y_1 \quad Z_2 = y_2$$

(b) Excitation table

y_1y_2	x = 0	x = 1
00	00 00	00 10
01	00 00	10 01
10	00 00	00 10
11	00 00	01 01
	$S_1R_1\ S_2R_2$	

(c) Output table

y_1y_2	x = 0	x = 1
00	0, 00	0, 00
01	0, 01	0, 01
10	0, 10	0, 10
11	0, 11	1, 11
	z, Z_1Z_2	

(d) Transition table

y_1y_2	x = 0	x = 1
00	(00)	01
01	(01)	10
10	(10)	11
11	(11)	00
	Y_1Y_2	

(e) State table

S	x = 0	x = 1
1	(1)	2
2	(2)	3
3	(3)	4
4	(4)	1
	S	

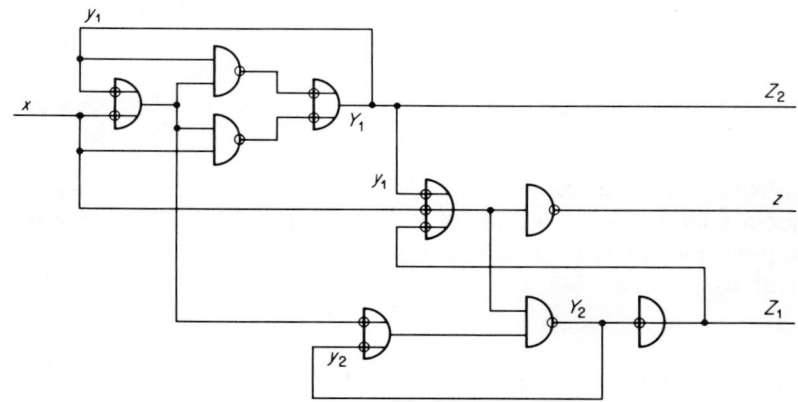

$$Y_1 = y_1(y_1' + x') + x(y_1' + x')$$
$$Y_2 = \{(y_1' + x' + y_2)\ [(x' + y_1') + y_2']\}'$$
$$Z_1 = y_2' \quad Z_2 = y_1 \quad z = (y_1' + x' + y_2)'$$

Fig. 5.5-5. OR-NOT gate circuit having the same behavior as the circuit of Fig. 5.5-4.

Table 5.5-5. Tables for the Circuit of Fig. 5.5-5

(a) Excitation functions

$$Y_1 = y_1(y_1' + x') + x(y_1' + x') = x'y_1 + xy_1'$$

$$Y_2 = \{(y_1' + x' + y_2)([x' + y_1']' + y_2')\}'$$

$$Y_2 = \{(y_1' + x' + y_2)(xy_1 + y_2')\}'$$

$$Y_2 = xy_1 y_2' + y_2(x' + y_1')$$

$$Z_1 = y_2' \quad Z_2 = y_1 \quad z = (y_1' + x' + y_2)' = xy_1 y_2'$$

(b) Excitation table, transition table

$y_1 y_2$	$x=0$	$x=1$
01	(01)	11
11	(11)	00
00	(00)	10
10	(10)	01

$Y_1 Y_2$

(c) Output table

$y_1 y_2$	$x=0$	$x=1$
01	0, 00	0, 00
11	0, 01	0, 01
00	0, 10	0, 10
10	0, 11	1, 11

$z, Z_1 Z_2$

(d) State table

s	$x=0$	$x=1$
1	(1)	2
2	(2)	3
3	(3)	4
4	(4)	1

S

in a more economical circuit. Another circuit to have the same pulse-mode operation is shown in Fig. 5.5-5. This circuit is included to illustrate a pulse-mode circuit which does not use flip-flops. The analysis is presented in Table 5.5-5.

5.6 CLOCKED SEQUENTIAL CIRCUITS

One common technique for designing digital systems makes use of a *master-clock circuit* which produces a periodic train of pulses [8]. The operation of each of the circuits composing the system is synchronized with these *clock pulses*. It is the purpose of this section to investigate sequential circuits which have such a train of pulses as one of their inputs and have the remaining inputs as level-type signals (such as the outputs of

5.6 clocked sequential circuits

Table 5.6-1. Tables for the Circuit of Fig. 5.6-1

(a) Functions

$$S = c\,x_1 x_2 \qquad R = c\,x_1' x_2' \qquad z = c(y' + x_1' x_2' + x_1 x_2)(y + x_1' x_2 + x_1 x_2')$$

(b) Excitation table

		$c=0$				$c=1$		
			$x_1 x_2$					
y	00	01	11	10	00	01	11	10
0	00	00	00	00	01	00	10	00
1	00	00	00	00	01	00	10	00

SR

(c) Condensed form of excitation table

	$c=0$		$c=1$		
			$x_1 x_2$		
y		00	01	11	10
0	00	01	00	10	00
1	00	01	00	10	00

SR

(d) Transition table

	$c=0$		$c=1$		
			$x_1 x_2$		
y		00	01	11	10
0	0	0	0	1	0
1	1	1	0	1	1

Y

(e) Output table

	$c=0$		$c=1$		
			$x_1 x_2$		
y		00	01	11	10
0	0	0	1	0	1
1	0	1	0	1	0

z

(f) Combined state and output table

	$c=0$		$c=1$		
			$x_1 x_2$		
S		00	01	11	10
A	(A), 0	(A), 0	(A), 1	B, 0	(A), 1
B	(B), 0	A, 1	(B), 0	(B), 1	(B), 0

S, z

flip-flops). A sequential circuit which has this type of input will be called a *clocked*† *sequential circuit,* and the pulse input will be referred to as the clock pulse, symbolized by c. As far as the circuit is concerned, whether or not the pulses are equally spaced is of no importance. Furthermore, in a digital system there may be several clocks which are *not* synchronized with one another. Again, this will be of no importance in the design procedure for the individual sequential circuits. (Of course, the design *specifications* will have to take into account the presence of more than one clock.) The assumption will be made that the level inputs $x_1, x_2, \ldots$ do not change while the clock pulse is present. Clocked sequential circuits are most commonly designed for pulse-mode operation, for this typically leads to more economical circuits than those which operate in fundamental mode. There is nothing to prevent the design of fundamental-mode

† These circuits are sometimes also called *synchronous sequential circuits* because of the synchronizing effect of the clock pulses.

Sequential-circuit Analysis

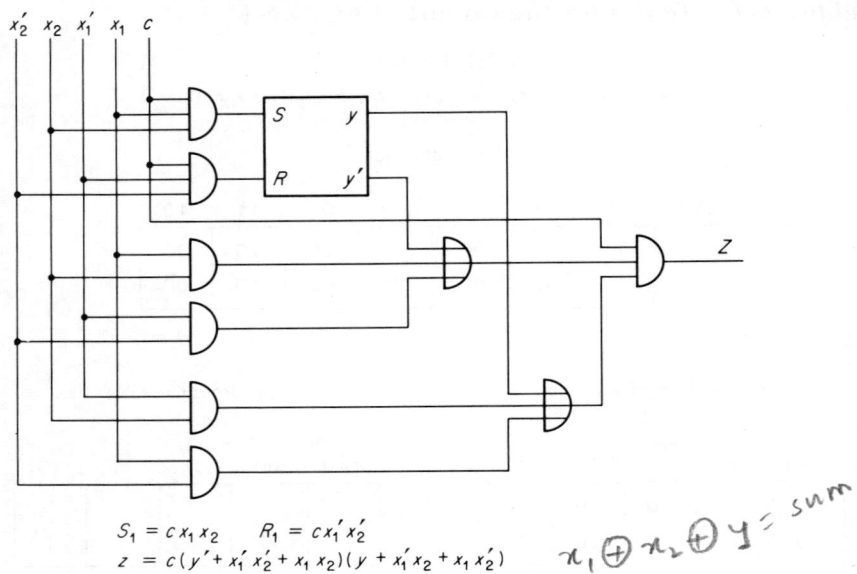

$$S_1 = c x_1 x_2 \quad R_1 = c x_1' x_2'$$
$$z = c(y' + x_1' x_2' + x_1 x_2)(y + x_1' x_2 + x_1 x_2')$$

$x_1 \oplus x_2 \oplus y = sum$

Fig. 5.6-1. A serial adder circuit (*pulse mode*). can be operated fundamental mode

clocked circuits, and this may be desirable where strict margins on the clock-pulse width are difficult to maintain, or reliability is of special importance, or the technology being used (e.g., relays, cryotrons) does not favor pulse-mode operation.

A clocked pulse-mode sequential circuit is shown in Fig. 5.6-1, and the analysis of this circuit is presented in Table 5.6-1. The fact that this is a clocked circuit does not necessitate any change in the analysis method presented in the preceding sections. The *nature* of the inputs becomes important only when the results of the analysis are being interpreted, i.e., when some description or word statement of the circuit behavior is being sought.

Table 5.6-1*b* illustrates a typical characteristic of clocked sequential circuits designed for pulse-mode operations: all the states for which $c = 0$ are stable. Since this is typical of this kind of circuit, it is customary to include only one column in the transition table for *all* the input conditions for which $c = 0$ (Table 5.6-1*c*). Indeed, tables are often written for clocked pulse-mode circuits in which only the input conditions for which $c = 1$ are shown explicitly. No information is lost by omitting the other columns, since all the entries in these columns correspond to stable conditions.

Some sample waveforms for the circuit of Fig. 5.6-1 are shown in Fig. 5.6-2. This circuit is commonly called a *serial adder*. If the values

5.6 clocked sequential circuits

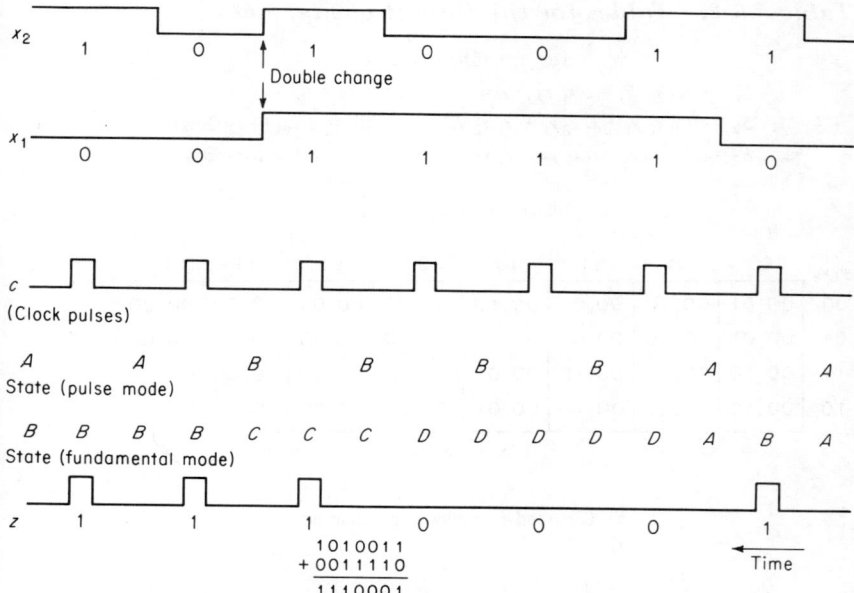

Fig. 5.6-2. *Sample waveforms for the circuits of Figs. 5.6-1 and 5.6-3.*

of the two inputs at each clock-pulse time are interpreted as the bits of two binary numbers, then the values of the output at each clock-pulse time correspond to the bits of the arithmetic sum of the binary numbers represented by the inputs. Notice that in Fig. 5.6-2 time increases from right to left. For the circuit to function as an adder, it is necessary to have the first values of the inputs correspond to the low-order bits of the numbers. This corresponds to performing "hand" addition from right to left.

There are two internal states in the circuit: state A corresponds to a 0 value for the arithmetic carry, and state B corresponds to a carry of 1. Of course, it is necessary to have the circuit in state A before starting an addition. An extra input would normally be required to perform this function of placing the circuit in the proper initial state. This input could be added to the circuit of Fig. 5.6-1 by changing R_1 to $R_1 = cx_1'x_2' + I$, where I is the *initializing input*.

In Fig. 5.6-2 there is a time (marked by an arrow) when both the inputs x_1 and x_2 change simultaneously. It is possible to allow such double input changes in clocked circuits because input changes produce no change in the internal state of the circuit when the clock pulse is not present. The restriction that must be placed on clocked circuits in order to ensure proper operation is that *the circuit inputs must not change while the clock pulse is present*.

Sequential-circuit Analysis

Table 5.6-2. Tables for the Circuit of Fig. 5.6-3

(a) Excitation functions

$$S_1 = c x_1 x_2, \quad R_1 = c x_1' x_2', \quad z = c y_2$$
$$S_2 = c'(x_1 x_2' y_1' + x_1' x_2 y_1' + x_1' x_2' y_1 + x_1 x_2 y_1) = c'(x_1 \oplus x_2 \oplus y_1)$$
$$R_2 = c'(x_1' x_2 y_1 + x_1 x_2' y_1 + x_1 x_2 y_1' + x_1' x_2' y_1') = c'(x_1 \oplus x_2 \oplus y_1')$$

(b) Excitation table

| $y_1 y_2$ | \multicolumn{4}{c}{$c=0$} | \multicolumn{4}{c}{$c=1$} |
	$x_1 x_2$ 00	01	11	10	00	01	11	10
00	00,01	00,10	00,01	00,10	01,00	00,00	10,00	00,00
01	00,01	00,10	00,01	00,10	01,00	00,00	10,00	00,00
11	00,10	00,01	00,10	00,01	01,00	00,00	10,00	00,00
10	00,10	00,01	00,10	00,01	01,00	00,00	10,00	00,00

$S_1 R_1, \; S_2 R_2$

(c) Combined transition and output table

| $y_1 y_2$ | \multicolumn{4}{c}{$c=0$ \quad $x_1 x_2$} | \multicolumn{4}{c}{$c=1$} |
	00	01	11	10	00	01	11	10
00	⓪⓪, 0	01, 0	⓪⓪, 0	01, 0	⓪⓪, 0	⓪⓪, 0	10, 0	⓪⓪, 0
01	00, 0	⓪①, 0	00, 0	⓪①, 0	⓪①, 1	⓪①, 1	11, 1	⓪①, 1
11	⑪, 0	10, 0	⑪, 0	10, 0	01, 1	⑪, 1	⑪, 1	⑪, 1
10	11, 0	⑩, 0	11, 0	⑩, 0	00, 0	⑩, 0	⑩, 0	⑩, 0

$Y_1 Y_2, \; z$

(d) Combined state and output table

| s | \multicolumn{4}{c}{$c=0$ \quad $x_1 x_2$} | \multicolumn{4}{c}{$c=1$} |
	00	01	11	10	00	01	11	10
A	Ⓐ, 0	B, 0	Ⓐ, 0	B, 0	Ⓐ, 0	Ⓐ, 0	D, 0	Ⓐ, 0
B	A, 0	Ⓑ, 0	A, 0	Ⓑ, 0	Ⓑ, 1	Ⓑ, 1	C, 1	Ⓑ, 1
C	Ⓒ, 0	D, 0	Ⓒ, 0	D, 0	B, 1	Ⓒ, 1	Ⓒ, 1	Ⓒ, 1
D	C, 0	Ⓓ, 0	C, 0	Ⓓ, 0	A, 0	Ⓓ, 0	Ⓓ, 0	Ⓓ, 0

S, z

It is also possible to design clocked circuits in which the *output is a level signal* (like the inputs) rather than a pulse. This will be illustrated by a problem. Formal definitions of pulse- and fundamental-mode operation can now be given.

Definition. A sequential circuit is said to be operating in *pulse mode* if the following conditions are satisfied:

These two defn's are not disjoint

5.6 clocked sequential circuits

1. At least one of the inputs is a pulse signal.
2. Changes in internal state occur only in response to the occurrence of a pulse at one of the pulse inputs.
3. Each input causes only one change in internal state.

Definition. A sequential circuit is said to be operating in *fundamental mode* if and only if the inputs are never changed unless the circuit is stable internally.

It is clear that other modes of operation are possible. These are not explicitly defined, for normally circuits are operated in either fundamental or pulse mode.

Fundamental-mode Clocked Circuits

Although it is not commonly done, it is possible to design *clocked circuits to operate in fundamental mode*. A fundamental-mode serial adder is shown in Fig. 5.6-3, and the analysis for this circuit is given in Table 5.6-2. The waveforms of Fig. 5.6-2 apply also to this circuit, and the internal states which this circuit assumes are also shown explicitly on Fig. 5.6-2. The circuits of Figs. 5.6-1 and 5.6-3 illustrate why pulse-mode circuits are preferred: the pulse-mode circuit typically requires less equipment than the corresponding fundamental-mode circuit. Fundamental-mode cir-

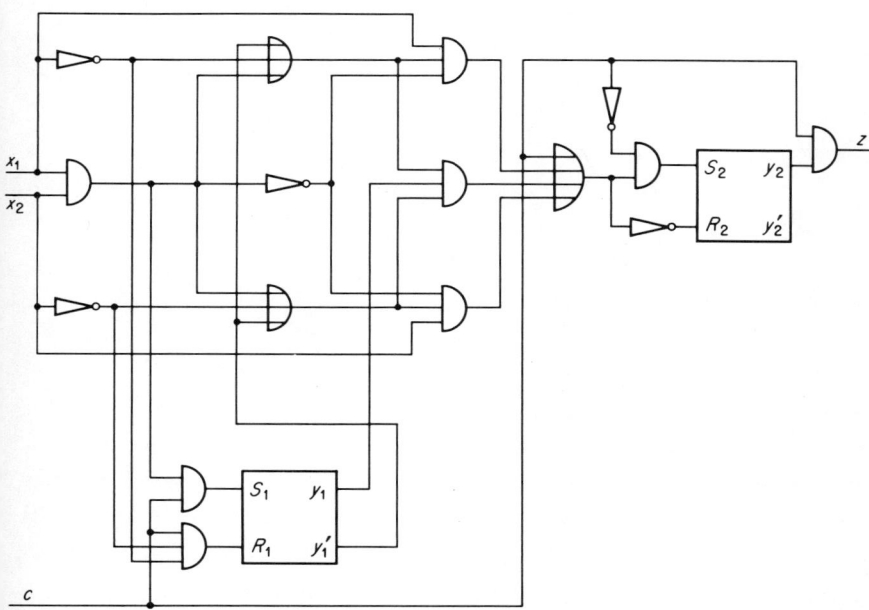

Fig. 5.6-3. *A serial adder circuit (fundamental mode).*

cuits are ordinarily used when reliability is critical or when it is difficult to control the pulse widths accurately.

5.7 STATE DIAGRAMS

In addition to the flow table, there are several other formal descriptions of sequential-circuit operation in common use. The most widely used of these is the *state diagram* [5]. The state diagram for the fundamental-mode circuit of Fig. 5.1-2 is shown in Fig. 5.7-1b. Each of the internal states of the circuit is represented by a node of the state diagram, and each of the transitions between internal states is represented by a directed line connecting nodes. The lines are labeled with the corresponding values of the input variables, a solidus (/), and the corresponding values of the

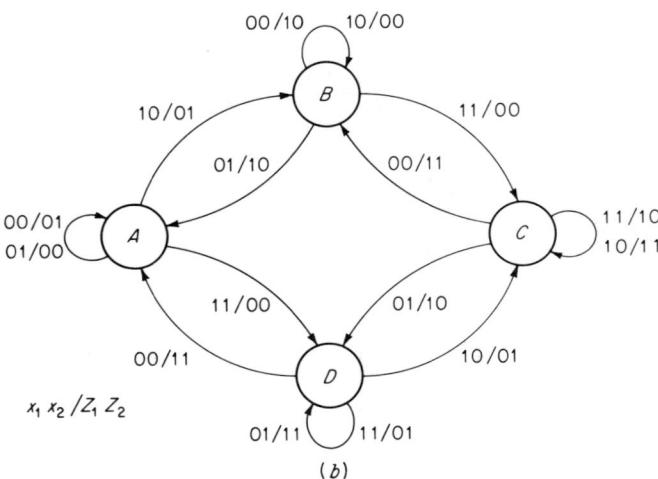

Fig. 5.7-1. Representations of the performance of the fundamental-mode circuit of Fig. 5.1-1. (a) Flow-output table; (b) state diagram.

5.7 state diagrams

output variables. A line which begins and ends at the same node corresponds to a stable state of the circuit. Just as it is possible to determine the output sequence resulting from a given input sequence by means of a flow table, it is possible to determine the resulting output sequence from a state diagram.

Figure 5.7-2 shows the state diagram for the pulse-mode circuit of Fig. 5.5-2. This state diagram differs from that of Fig. 5.7-1 in that some of the outputs Z_1 and Z_2 are included in the nodes, along with the states. The reason for doing this is that the outputs Z_1 and Z_2 depend only on the internal state and are independent of the value of the input. The output z cannot be included in the nodes, since its value depends both on the internal state and on the output state (that is, on the total state). Generally, in circuits with pulse inputs it is possible to have either pulse outputs or level outputs. The pulse outputs z are determined by the total state, while the level outputs Z_1 and Z_2 are determined by the internal state only.

A state diagram for the clocked pulse-mode circuit of Fig. 5.6-1 is

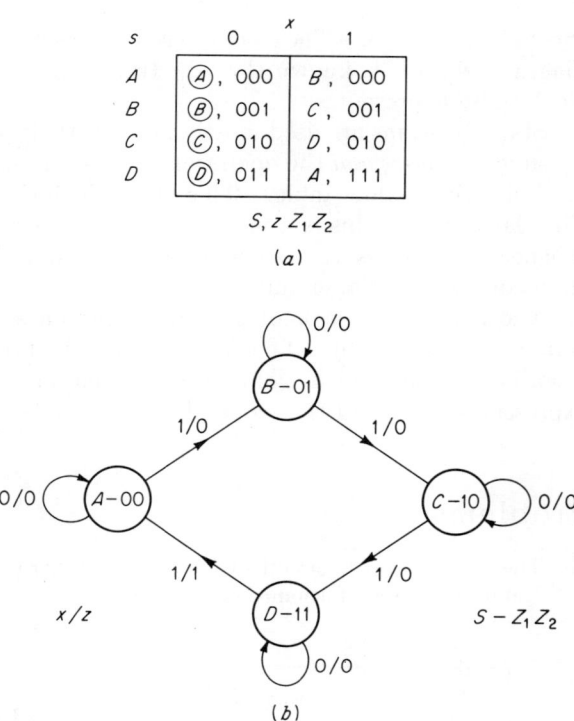

Fig. 5.7-2. Representations of the performance of the pulse-mode circuit of Fig. 5.5-2. (a) Flow-output table; (b) state diagram.

Sequential-circuit Analysis

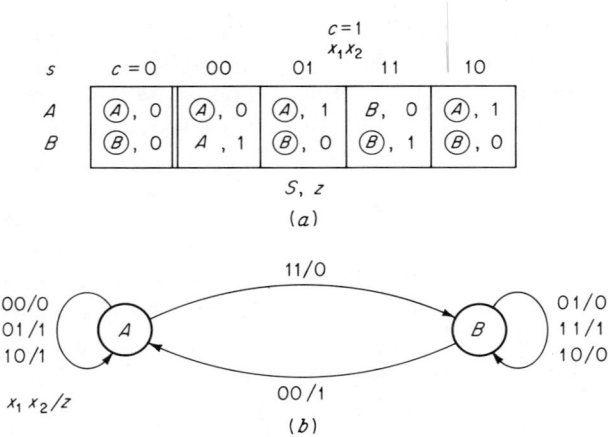

Fig. 5.7-3. Representations of the performance of the clocked circuit of Fig. 5.6-1. (a) Flow-output table; (b) state diagram.

shown in Fig. 5.7-3. The clock-pulse input is not shown explicitly on this diagram, since it is known that the transitions take place only when the clock pulse is present.

State diagrams are used mostly because their pictorial nature makes it possible to *understand* the operation of some circuits more easily than can be done with a flow table. This is particularly true for pulse circuits. To a large extent this is a matter of personal preference. In any event, it is necessary to resort to some type of table in order to carry out the synthesis of a sequential circuit.

Another type of sequential-circuit representation which is used is the *regular expression* [9]. This is an algebraic type of expression which describes the performance of a sequential circuit. A discussion of regular expressions is beyond the scope of this text.

Problems

1. The input x to the circuit of Fig. P5-1 is either grounded or open-circuited repeatedly. Determine the performance of the circuit.

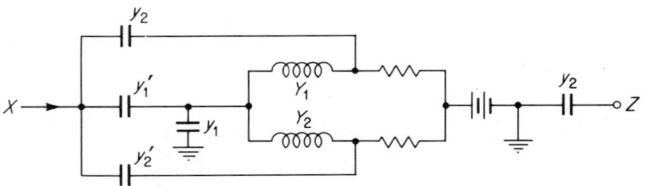

Fig. P5-1

problems

2. Analyze the circuit shown in Fig. P5-2.

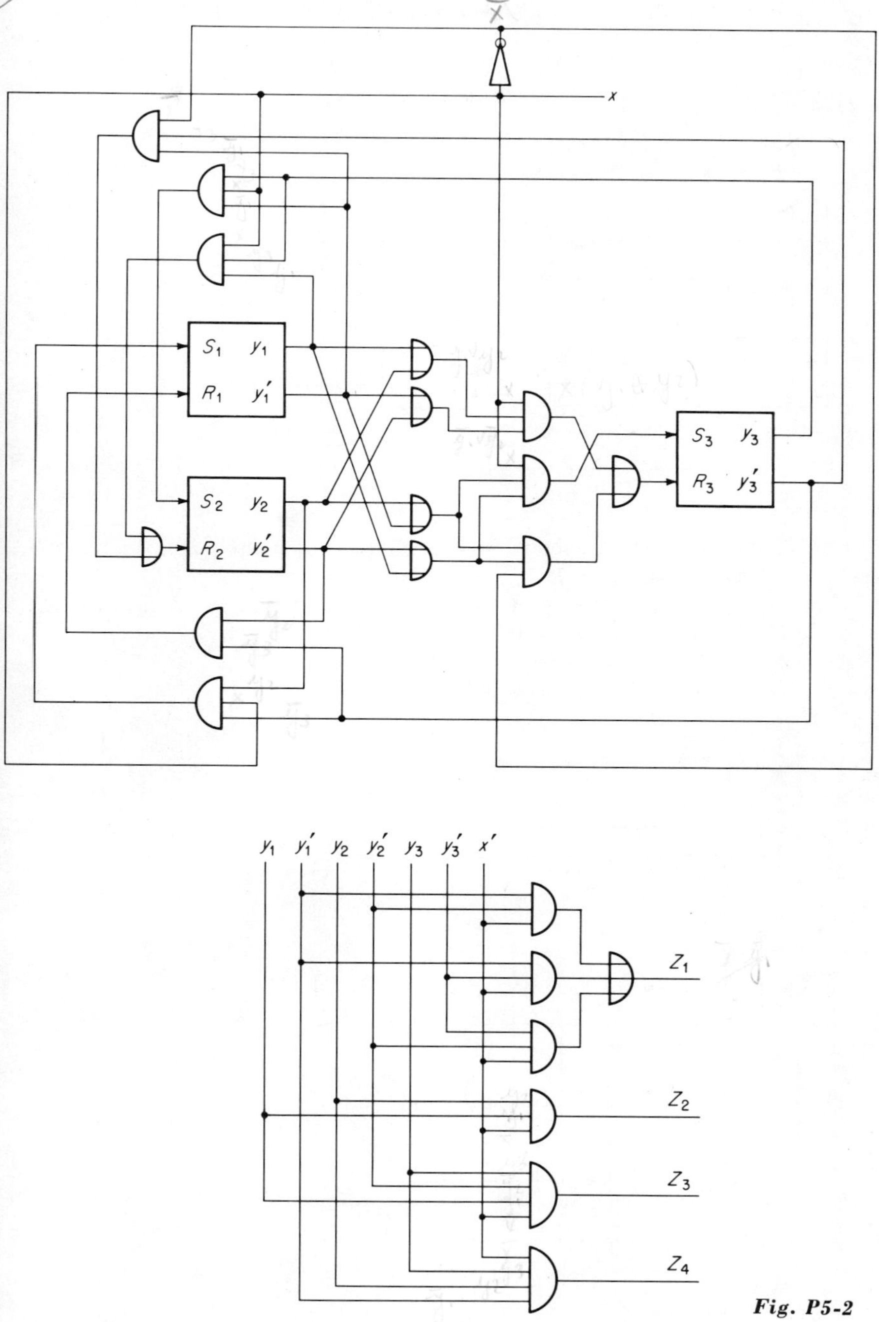

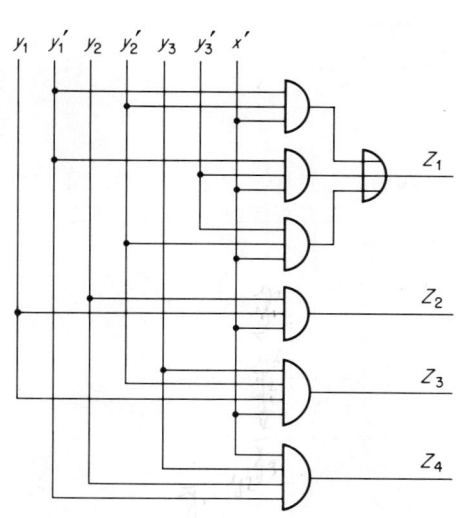

Fig. P5-2

Sequential-circuit Analysis

3. Analyze the circuit shown in Fig. P5-3. The flip-flops have the property that, when both the set and reset leads are pulsed, the flip-flop goes to the set state.

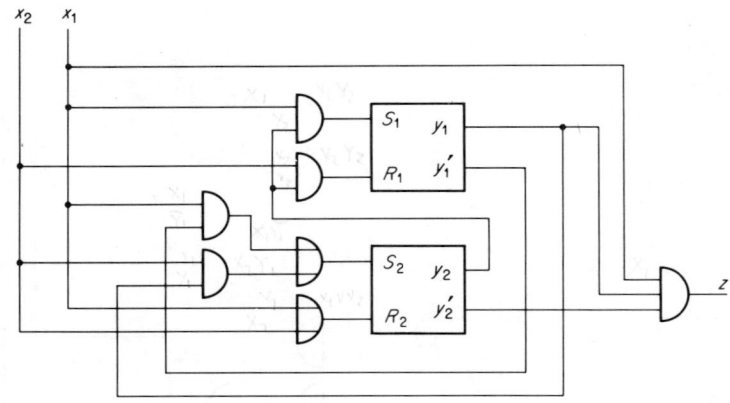

Fig. P5-3

4. Analyze the circuit shown in Fig. P5-4. The input x is a level, and c is a (clock) pulse.
   (a) Write down the excitation and output functions S_1, R_1, S_2, R_2, Z.
   (b) Form the excitation table.
   (c) Form the state table.
   (d) What does the circuit do?

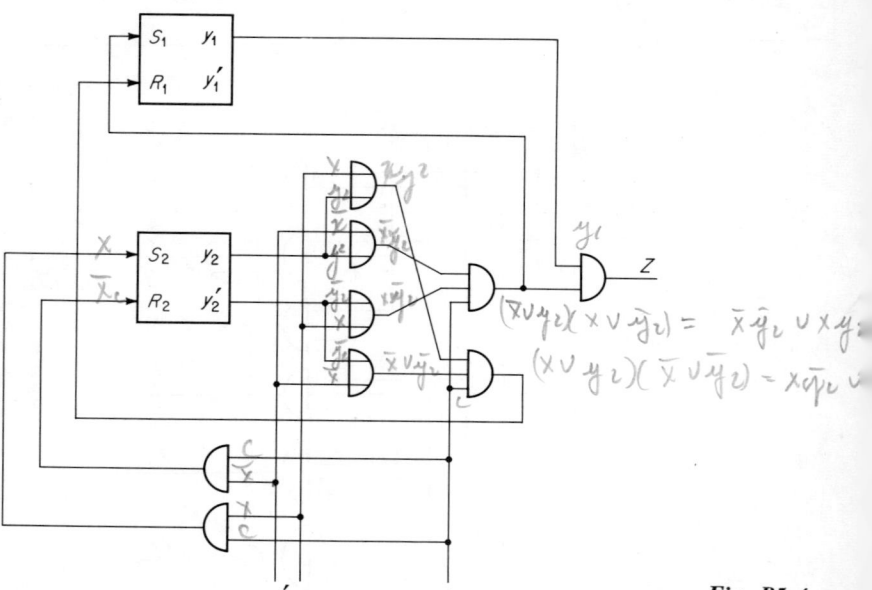

Fig. P5-4

problems

5. In the circuit of Fig. 5.3-5, the feedback loops are to be broken (fictional amplifiers inserted) in a place other than that shown in the figure. Choose another location to break the loops, and analyze the circuit, forming the excitation table, transition table, output table, and state table. Explain any discrepancies between this analysis and the analysis given in Sec. 5.3.

6. (a) Analyze the circuit of Fig. P5-6. (Form the excitation, transition, output, state, and flow tables.)
   (b) Derive the sequence of internal states and output states which correspond to the input sequence x_1x_2 = 00, 10, 11, 10, 11, 01, 00, 10, 11, 01, 00. (Assume $Y_1 = Y_2 = 0$ initially.)

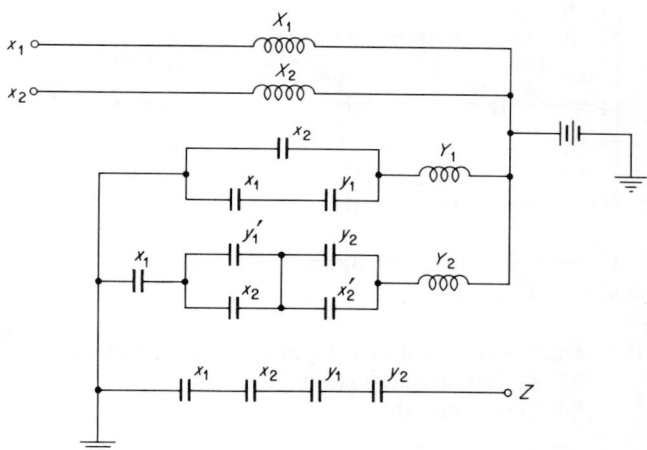

Fig. P5-6

7. Redesign the circuit of Prob. 6:
   (a) Using AND gates, OR gates, and inverters.
   (b) Using OR-NOT gates.
   (c) Using AND-NOT gates.

8. Analyze the circuit of Fig. P5-8, forming the:
   (a) Excitation table.
   (b) Transition table.
   (c) State table.
   (d) Output table.

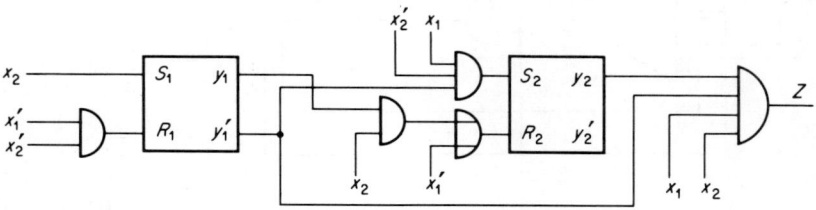

Fig. P5-8

Sequential-circuit Analysis

9. Analyze the circuit of Fig. P5-9:
   (a) Form the appropriate tables.
   (b) Develop a precise word statement of the circuit performance.

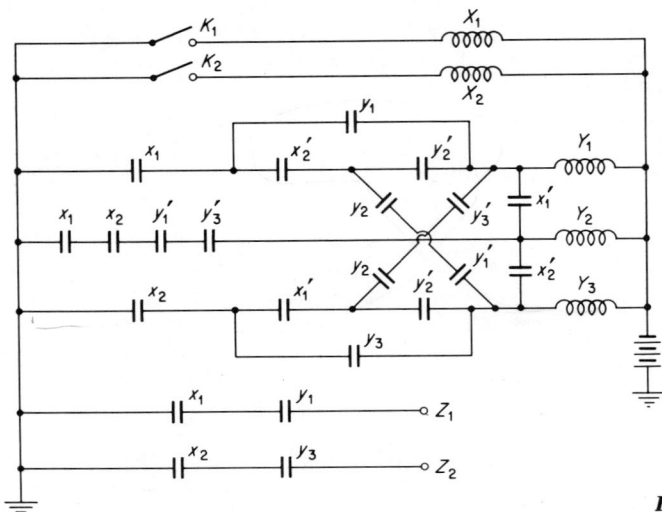

Fig. P5-9

10. Analyze the circuit shown in Fig. P5-10, forming the:
    (a) Excitation functions and table.
    (b) Transition table.

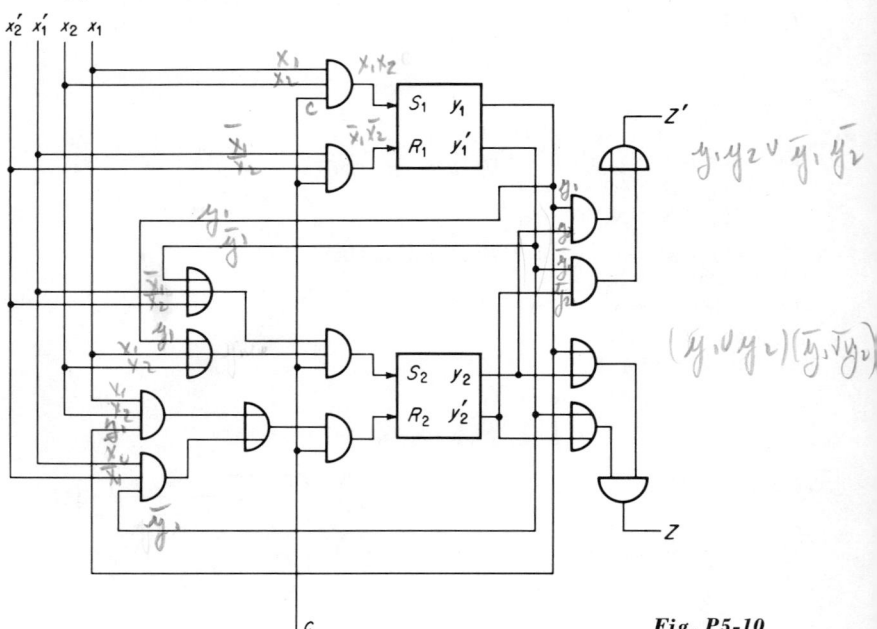

Fig. P5-10

(c) Flow table and output table.
(d) Illustrate the performance with waveforms for the input sequences,

X_1: 1 0 1 1 0 1 1 0 1
X_2: 0 1 0 1 1 1 0 1 0

(e) Describe in words the circuit performance.
(f) Draw a state diagram for the circuit.

REFERENCES

1. McCluskey, E. J.: Fundamental Mode and Pulse Mode Sequential Circuits, *Proc. IFIP Congress* 1962, *Intern. Conf. Information Processing*, Munich, Aug. 27–Sept. 1, 1962, pp. 725–730, Cicely M. Popplewell (ed.), North Holland Publishing Company, Amsterdam, 1963.
2. Huffman, D. A.: The Synthesis of Sequential Switching Circuits, *J. Franklin Inst.*, vol. 257, no. 3, pp. 161–190, no. 4, pp. 275–303, March, April, 1954.
3. Unger, S. H.: A Study of Asynchronous Logical Feedback Networks, *Research Lab. Electronics, Mass. Inst. Technol., Tech. Rept.* 320, Apr. 26, 1957.
4. Mealy, G. H.: A Method for Synthesizing Sequential Circuits, *Bell System Tech. J.*, vol. 34, no. 5, pp. 1045–1080, September, 1955.
5. Moore, E. F.: Gedanken-experiments on Sequential Machines, in C. E. Shannon and J. McCarthy (eds.), "Automata Studies," Princeton University Press, Princeton, N.J., 1956.
6. Bartee, T. C.: "Digital Computer Fundamentals," McGraw-Hill Book Company, New York, 1960.
7. Cadden, W. J.: Equivalent Sequential Circuits, *IRE Trans. on Circuit Theory*, vol. CT-6, no. 1, pp. 30–34, March, 1959.
8. Phister, M., Jr.: "Logical Design of Digital Computers," John Wiley & Sons, Inc., New York, 1958.
9. Brzozowski, J. A.: A Survey of Regular Expressions and Their Applications, *IRE Trans. on Electronic Computers*, vol. EC-11, no. 3, pp. 324–335, June, 1962.

6 SEQUENTIAL - CIRCUIT SYNTHESIS

In the preceding chapter a method for analyzing sequential circuits was presented, and various modes of operation of sequential circuits were discussed. The purpose of this chapter is to develop techniques for designing sequential circuits. The approach to be followed is to reverse the steps which were used in the analysis procedure. Thus, the first step in the design process is to form a flow output table describing the desired circuit performance. This table must then be checked to determine whether or not it is more complicated than is actually necessary. The transition table is then obtained from the simplified flow table by assigning combinations of internal variables to the internal states. An excitation table is derived from the transition table, and finally the circuit diagram is drawn. Each of these steps in the synthesis procedure will be considered in order.

6.1 FORMATION OF THE FLOW TABLE [1,2]

Just as it was not possible to give a procedure for going from a flow table to a word statement of the circuit performance, it is impossible to develop a formal procedure for going from a word statement to a flow table. The basic reason for this difficulty is that the word statement is not a formal description of the circuit action. The best that can be done is to illustrate the formation of flow tables by means of some examples. Once the flow table is obtained, formal procedures can be used to complete the synthesis of the final circuit.

At first glance, it might seem that forming *exactly* the correct flow table for the desired circuit performance would be very difficult. In a sense this is true; but, in actual fact, there are *very many* flow tables which correspond to the same circuit performance. All

6.1 formation of the flow table

that is really necessary is to form one of the many flow tables which are suitable. Formal procedures exist for removing superfluous internal states or for transforming one flow table into another flow table without changing the corresponding circuit performance. Thus, in forming a flow table, it is not necessary to be concerned about adding internal states which may turn out to be superfluous. All that is required is that a sufficient number of internal states are included so that all possible contingencies are accounted for. Examples will be given first for pulse-mode specifications since these are usually easier to handle.

Example 6.1-1: Pulse Inputs, Level Outputs, Pulse Mode. A *Moore* (pulse-mode) circuit having two inputs x_1 and x_2 and one output Z is to be designed. The desired performance is as follows:

1. Pulses occur on the input leads. These pulses never overlap and never occur simultaneously.
2. The output is to equal 0 as long as the input pulses alternate between the two input leads.
3. Whenever two or more successive pulses occur on the same input lead, the output is to become equal to 1. The output is to remain equal to 1 until the input pulses again alternate between the input leads.
4. The output is to be a level (d-c) signal.

The flow table can be formed as follows:

1. Since the output is a level signal, while the inputs are pulses, the output must depend only on the internal state of the circuit. There are two possible values for the circuit output, and thus there must be at least two internal states.

2. Assume that there are only two internal states—A and B—and let the output equal 0 ($Z = 0$) when the circuit is in state A and the output equal 1 ($Z = 1$) when the circuit is in state B.

3. The next-state entries corresponding to each total state must now be determined. Assume that the circuit is in state A and that no input pulses are present ($x_1 = x_2 = 0$). It is possible for the circuit to remain in this condition indefinitely so that this must be a stable state and the next state must be A.†

4. Now consider the next state for the condition when the circuit is in state A and a pulse occurs on the x_1 lead. If state A ($Z = 0$) is chosen for this next-state entry, it will be possible successively to apply pulses on the x_1 lead without changing the internal state A of the circuit and without making the output equal to 1. Since this violates the condition that

† It is typical of pulse-mode circuits that the circuit is stable whenever there are no input pulses present.

successive pulses on the same lead should cause Z to equal 1, this choice for the next state must be incorrect.

5. The only remaining possibility is to choose B for the next state when the present state is A and a pulse appears on the x_1 lead. Now suppose that pulses have been alternating on the two input leads and the last pulse to occur was an x_2 pulse. The circuit must be in state A because the output is to be 0 for alternating pulses. If the next pulse is an x_1 pulse, the output should still be 0, since the input pulses are still alternating. However, the chosen next-state entry of B will cause a 1 output to appear. Thus, the choice of B also leads to incorrect operation.

6. Since both possible next-state choices result in incorrect operation, there are only two alternatives: either the specified performance is impossible,† or more internal states are needed.

7. An attempt will be made to form a suitable flow table by adding more states. The difficulty with the 2-state table is caused by the fact that the next state is dependent only on the present input pulse and the present output condition. Actually, the next state should be determined by the present input pulse and the last previous input pulse. Since the output can be either 0 or 1 and the last input pulse can be x_1 or x_2, four states are needed corresponding to the four combinations of present output and last previous input. The accompanying table shows the correspondence between states and circuit conditions which will be used.

State	Output Z	Last previous input
A	0	x_1
B	1	x_1
C	0	x_2
D	1	x_2

8. The flow table is filled in as follows: In the column corresponding to a pulse on the x_2 input the next-state entries must be either C or D (see the table). If the present state is A or B, the next state should be C, since the pulses are alternating and a 0 output is desired. If the present state is C or D, the next state should be D, since the x_2 pulse has repeated and a 1 output is required. Similar reasoning holds for the x_1 column. The entries in the column for $x_1 = x_2 = 0$ are all stable, as discussed previously. The flow table is shown in Table 6.1-1.

† Word specifications can sometimes correspond to circuits which are not physically realizable. This will be illustrated by a later example.

6.1 formation of the flow table

Table 6.1-1. Flow Table for Example 6.1-1

		$x_1 x_2$			
Last input	S	00 (x_2)	01	11	10 (x_1)	Z
x_1	A	Ⓐ	C	–	B	0
x_1	B	Ⓑ	C	–	Ⓑ	1
x_2	C	Ⓒ	D	–	A	0
x_2	D	Ⓓ	Ⓓ	–	A	1

S

9. The column for $x_1 = x_2 = 1$ has dashes entered, since this input combination does not occur. The convention which will be used in flow tables will be to enter a dash in any cell for which the corresponding total state *cannot* occur in the circuit. In this case no total state can occur with $x_1 = x_2 = 1$.

10. The state diagram corresponding to this flow table is shown in Fig. 6.1-1. When pulses are alternating on the two input leads, the circuit goes between states A and C via the heavy lines. Successive pulses on the same lead will cause the circuit to go to either state B or state D, where it will remain until a pulse on the other lead arrives.

Output Specifications—Pulse Mode

Actually, the flow table alone does not specify the circuit performance; in order to determine outputs, it is necessary to have the output table as well as the flow table. Rather than actually writing down two tables, it is more convenient to include the information of both the flow and output tables in one single table. From now on, when a flow table is written down, the outputs as well as the next states will be included.

Normally, the outputs must be specified for each total state of the cir-

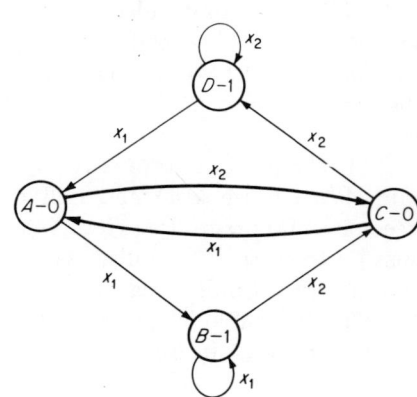

Fig. 6.1-1. State diagram for Table 6.1-1.

cuit. An exception occurs for a pulse-mode circuit with level (d-c) outputs. In this case, the outputs depend only on the internal state and consequently must be specified not for each *total* state (cell of the flow table) but only for each *internal* state (row of the flow table). Thus, the level (d-c) outputs in a pulse-mode circuit can be specified by means of a single column added to the flow table, for example, the column labeled Z in Table 6.1-1. Use of this Z column is merely a matter of convenience to avoid repeating the same output specification in each column of a row, as was done previously (Z_1Z_2 in Table 5.5-4c). By convention, capital Z's will be used for level (d-c) outputs, and lower-case z's will be used for pulse outputs. This is not really necessary but is a convenience in interpreting tables and circuits.

Usually, for each pulse-input circuit with level outputs there is an analogous circuit with pulse outputs. The pulse output circuit has an output pulse whenever (1) there is an input pulse present and (2) the corresponding output of the level-output circuit is equal to 1. The formation of a flow table for a pulse-mode circuit with a pulse output is illustrated in the following example:

Example 6.1-2: Pulse Inputs, Pulse Output, Pulse Mode. A pulse-mode circuit having two inputs x_1 and x_2 and one output z is to be designed. The desired performance is as follows:

1. Pulses occur on the input leads. These pulses never overlap or occur simultaneously.

2. A pulse is to occur on the output (z) lead whenever a pulse occurs on an input lead *and* the last previous input pulse occurred on the same input lead. (There are no output pulses as long as the input pulses alternate between the two input leads.)

The flow table can be formed as follows:

1. When an input pulse occurs, whether or not an output pulse occurs depends upon which lead the last previous input pulse occurred. There must be at least two internal states, A and B. The circuit is in state A when an x_1 pulse occurred last and is in state B when an x_2 pulse occurred last.

2. If the present state is A (x_1 pulse last), the circuit must remain in state A (x_1 pulse last) and deliver an output pulse (two x_1 pulses in succession) if an x_1 pulse is received. If an x_2 pulse is received, the circuit must go to state B (x_2 pulse last) and the output must remain 0 (inputs alternating). Similar reasoning applies when the circuit present state is B. The flow table is shown in Table 6.1-2, and the corresponding state diagram is shown in Fig. 6.1-2.

6.1 formation of the flow table

Table 6.1-2. Flow Table for Example 6.1-2

Last input s		$x_1 x_2$		
		(x_2)		(x_1)
		00	01	11	10
x_1	A	$\widehat{A}$, 0	B, 0	—	$\widehat{A}$, 1
x_2	B	$\widehat{B}$, 0	$\widehat{B}$, 1	—	A, 0

S, z

Fig. 6.1-2. State diagram for Table 6.1-2.

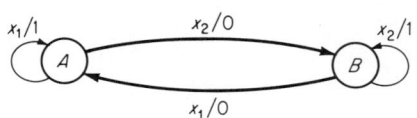

Example 6.1-2 is the analogous pulse-output circuit for the level-output circuit of Example 6.1-1. It is typical that the pulse-output circuit requires fewer internal states than the level-output circuit. Two additional states are required in Example 6.1-1 to hold the output value between input pulses. Of course, it is possible to have a pulse-circuit which has both pulse and level outputs, for example, the circuit of Fig. 5.5-4.

The process of writing a flow table for clocked pulse-mode circuits does not differ significantly from that for nonclocked circuits. The following example illustrates the formation of a flow table for a clocked pulse-mode circuit:

Example 6.1-3: Clocked, Level-Output, Pulse Mode. A circuit having one level (d-c) input x, one clock-pulse input c, and one level output Z is to be designed. The desired performance is as follows:

1. The output changes only when the clock pulse arrives.
2. The output becomes (or remains) equal to 0 when a clock pulse arrives only if the present value of x is opposite to the value of x when the last previous clock pulse arrived.
3. The output becomes (or remains) equal to 1 when a clock pulse arrives only if the present value of x is the same as the value of x when the last previous clock pulse arrived.

Four internal states are required, since all combinations of two independent pairs of possibilities—the value of x at the last clock pulse and the value of Z—must be "stored" or "remembered." The formation of this flow table, shown in Table 6.1-3, is similar to the formation of Table 6.1-1. The state diagram corresponding to Table 6.1-3 is shown in Fig. 6.1-3.

Table 6.1-3. *Flow Table for Example 6.1-3*

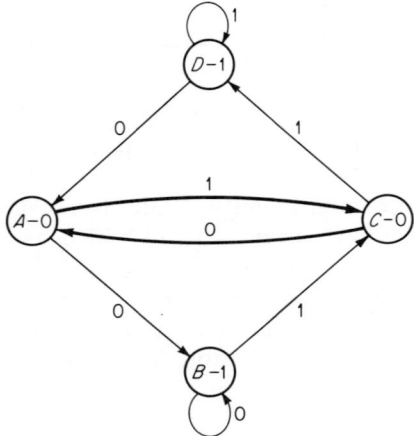

Fig. 6.1-3. State diagram for Table 6.1-3.

Fundamental Mode—Primitive-form Flow Tables

These three examples illustrate the considerations which are involved in forming pulse-mode flow tables. The formation of fundamental-mode tables is more complicated, as will be seen in the following example. In order to simplify the writing of fundamental-mode tables, it is customary to write them in a special format. As has been mentioned previously, there are many flow tables which correspond to the same circuit performance. In writing pulse-mode tables, an attempt is usually made to arrive at a table which has no unnecessary internal states. Fundamental-mode tables are written so that there is *only one stable (total) state in each row of the table*. This is called the *primitive form* for a fundamental-mode flow table. It is possible to write these tables in other forms and with fewer states than the primitive form, but it is usually *easier* to write a table in primitive form.

In the following example, a fundamental-mode table will be written for a circuit which is to have the same performance as the pulse-mode circuit of Example 6.1-2:

6.1 formation of the flow table

Example 6.1-4: Fundamental Mode. A (fundamental-mode) circuit with two pulse inputs and one pulse output is to be designed:

1. The input pulses do not overlap or occur simultaneously.
2. An output pulse occurs with an input pulse whenever the last previous input pulse occurred on the same input lead as the present input pulse. Otherwise there are no output pulses.

The table is formed as follows:

1. When no pulses are present ($x_1 = x_2 = 0$), there must be two stable states—one for the condition where the last input was x_1 and one for the condition where the last input was x_2. This is necessary because the output when an input pulse arrives is determined by the previous input pulse. The output for each of these states is 0 (see Table 6.1-4a).
2. The next-state entries in the 01 column of rows A and B will be considered next. Since a primitive-form flow table (one stable state per row) is being written, neither of these entries can be A or B.† Without loss of generality the entry in row A can be chosen to be C, an arbitrary new state. This means that there must be a stable C entry in the C row of column 01 (since this is a flow table). The output associated with this stable C entry must be 0 since the present input is x_2 (01 column) and the previous input was x_1 (state A). The entry in row B can only be C or a new state D. If the C entry is chosen, a 0 output will be developed. This is incorrect, for the present input is x_2 (01 column) and the previous input was x_2 (state B). Thus the row B entry must be D, and there must be a stable D entry in the 01 column. The output for this stable D entry must be 1 because this state will occur when two successive x_1 pulses have been received (Table 6.1-4b).
3. The same reasoning applied to the 10 column shows that two additional internal states must be introduced. The outputs associated with these states are determined in the same fashion as the C and D outputs were determined (Table 6.1-4c).
4. A dash (–) is placed in each of the cells of the 11 column, since it is assumed that the pulses do not overlap. A dash can also be placed in the 10 column of row C: it is impossible to enter this state, for this would require a double change of input (from 01 to 10). The stable entry in row C represents the presence of an x_2 pulse; this must be followed by a transition to the 00 column (no pulse present). For similar reasons, dashes can be placed in the 10 column of row D and in the 01 column of rows E and F.
5. Four entries, in the 00 column and rows C, D, E, and F, remain to be specified. The entries in rows C and D must be B, since they are

† The entry in row A cannot be A, for this would be stable. The entry in row B cannot be A, for in this *flow* table each unstable state must go directly to a stable state.

Table 6.1-4. Flow Table for Example 6.1-4

(a)

		00	01 x_1x_2	11	10
Last input – x_1	A	(A), 0		–	
Last input – x_2	B	(B), 0		–	

(b)

		00	01 x_1x_2	11	10
Last input – x_1	A	(A), 0	C		–
Last input – x_2	B	(B), 0	D		–
	C		(C), 0		
	D		(D), 1		

(c)

		00	01 x_1x_2	11	10
Last input – x_1	A	(A), 0	C	–	E
Last input – x_2	B	(B), 0	D	–	F
	C		(C), 0	–	–
	D		(D), 1	–	–
	E	–		–	(E), 1
	F	–		–	(F), 0

(d)

		00	01 x_1x_2	11	10
Last input – x_1	A	(A), 0	C ,	–	E ,
Last input – x_2	B	(B), 0	D ,	–	F ,
	C	B ,	(C), 0	–	–
	D	B ,	(D), 1	–	–
	E	A ,	–	–	(E), 1
	F	A ,	–	–	(F), 0

reached from the 01 column and therefore must lead to the state corresponding to a previous x_2 input. For analogous reasons, the entries in rows E and F must be A. All the next states have now been entered (Table 6.1-4d), and only the outputs corresponding to the unstable states remain unspecified.

6. One possibility is to assign to each unstable state the output value which is associated with the stable state to which the unstable state goes. Thus, the output entry in row A of the 01 column would be 0, since the

6.1 formation of the flow table

Table 6.1-5. Different Output Specifications for Table 6.1-4

(a) Unstable entries same as stable entries

s	00	01	11	10
A	Ⓐ, 0	C, 0	—	E, 1
B	Ⓑ, 0	D, 1	—	F, 0
C	B, 0	Ⓒ, 0	—	—
D	B, 0	Ⓓ, 1	—	—
E	A, 0	—	—	Ⓔ, 1
F	A, 0	—	—	Ⓕ, 0

$x_1 x_2$

S, Z

(b) Unstable entries same as stable entries only when necessary

s	00	01	11	10
A	Ⓐ, 0	C, 0	—	E, d
B	Ⓑ, 0	D, d	—	F, 0
C	B, 0	Ⓒ, 0	—	—
D	B, d	Ⓓ, 1	—	—
E	A, d	—	—	Ⓔ, 1
F	A, 0	—	—	Ⓕ, 0

$x_1 x_2$

S, Z

next-state entry is C and the output associated with the stable C entry in this column is 0 (see Table 6.1-5). This output specification is satisfactory, for the output associated with a stable state will occur when the stable state is entered. Associating the same output with the preceding unstable state can only speed the appearance of output changes.

Output Specifications–Fundamental Mode

There are other output specifications which are also possible. In order to determine the precise constraints on these outputs, one entry will be considered in detail. If the output for an unstable state is made the same as the output for the next stable state, a 0 output must be specified in the A row of the 01 column. The question which naturally arises is whether or not the circuit will still perform correctly if a 1 output is specified instead.

There is only one way in which the circuit can enter the total state $s-x_1 x_2 = A$–01: the circuit must initially be in the state A–00, and the x_2 input must change from 0 to 1. If this happens, the circuit will start in state A–00, change to state A–01 because of the x_2 change, and finally change to state C–01 because of the unstable next state associated with state A–01.

In a primitive flow table, each unstable total state can be entered from only one stable state (there is only one stable state per row), and each unstable total state is followed directly by a unique stable state (by definition of flow table). It is thus possible to associate with each unstable total state of a flow table a *predecessor state* and a *successor state*.[†] For state A–01, the predecessor is A–00, and the successor is C–01.

The output is 0 for total states A–00 and C–01. If the output is 1 for state A–01, an output sequence 0–1–0 will result when x_2 changes from 0 to 1 with the circuit initially in state A. In this output sequence the 0 outputs will be levels which remain indefinitely until an input is changed. On the other hand, the 1 output will be a transient "spike" whose duration is determined by the circuit parameters. Moreover, this spike is a false output since the inputs are alternating. Thus, a 1 entry for the output in the A–01 state is incorrect because it produces a false output.[‡]

In general, *the output associated with an unstable state must be the same as the output associated with the successor state whenever the successor state and the predecessor state have the same output.* The entries of Table 6.1-5 in rows C and F of the 00 column, row A of the 01 column, and row B of the 10 column fall into this category.

The other type of unstable entry is one in which the predecessor and successor outputs differ, for example, the B–01 state of Table 6.1-5. In this case the output associated with the unstable state is unimportant. If the unstable-state output is the same as the predecessor output, no change in the output will take place until the stable state is reached. At worst, this situation can introduce a slight delay in the formation of the new output value. No spurious transient outputs are developed. Thus, *the output associated with an unstable state is unspecified (d) whenever the output of the successor state and the output of the predecessor state differ.* Table 6.1-5b shows the output specification with d entries for Table 6.1-4d. Usually the d entries will be used rather than having the unstable-state outputs always agree with the successor-state outputs. The small amount of output delay is usually compensated for by the economy that results from using the d entries.

Because Table 6.1-5 was written in primitive form it contains more internal states than are actually necessary. A technique for reducing the

---

[†] These are stable total states, but since there is only one stable total state for each internal state, it is sufficient to specify the internal state only.

[‡] If the output is being used to light an incandescent lamp or energize some other device which does not respond to short pulses, the false spike output may be unimportant. This should be stated in the specifications of the circuit and will not be assumed unless explicitly mentioned.

6.1 formation of the flow table

number of internal states without changing the circuit performance will be given in the following section.

Initial States

Thus far nothing has been said about the performance of a circuit when it is first put into operation, i.e., when the "power is turned on." For all the circuits which have been presented previously, the stable state in which the circuit will be when power is applied, the *initial state*, will depend on the specific electrical properties of the circuit elements. Nothing is included in these circuits to constrain the initial state.

In connection with the serial adder circuit of Fig. 5.6-1 it was pointed out that the initial state must be controlled somehow if the circuit is to function properly as a serial adder. On the other hand, it would not be at all unreasonable to omit any control over the initial state from the circuits specified in the present section. If the initial state is not controlled in these circuits, there will possibly be a few incorrect outputs when the circuit is first put into operation. However, after this initial transient disappears, the steady-state circuit outputs will henceforth be correct, independent of what happened initially. This is not true for the serial adder circuit. If this circuit starts in the wrong initial state, all its outputs will be incorrect until either a 00 or 11 input is received. The difference between these two types of sequential circuits is that in the circuit of Example 6.1-1 the output is determined only by the two most recent previous inputs, while in the serial adder circuit (Fig. 5.6-1) the output can be influenced by inputs in the very remote past. When a circuit output can depend on an indefinite number of previous inputs, the initial state of the circuit is commonly controlled by means of an extra "initialize" or "reset" input. Since it is a simple task to add this reset input after the main portion of the circuit has been designed, it usually will not be shown explicitly.

Impossible Specifications

It is not always possible to write down a flow table corresponding to a sequential-circuit specification, because the corresponding circuit may not be physically realizable. The following is an example of such a specification:

> **Example 6.1-5: A Specification for a Nonrealizable Sequential Circuit.** A circuit is to be designed to operate in pulse mode. There are two inputs: a clock pulse and a level input which represents the bits of a binary number $(b_n b_{n-1} \cdots b_1 b_0)$. The order of appearance of signals on the level input is such that less significant bits precede more significant bits in time. Thus, at the first clock pulse a signal representing b_0 will occur, at the next clock-pulse

Sequential-circuit Synthesis

time a signal representing b_1 will occur, etc. There is to be one output lead on which pulses representing the bits of the corresponding Gray-code number $(g_n g_{n-1} \cdots g_1 g_0)$ are to occur. Specifically, at the clock-pulse time when a signal representing b_i occurs at the input, a pulse representing g_i is to occur at the output.

It is impossible to design a circuit to satisfy these specifications. The reason is that the rule for translating from binary to Gray code is given by $g_i = b_i \oplus b_{i+1}$. Since the ith Gray bit depends on both the ith and the $(i + 1)$st binary bits, the output of the specified circuit at one clock time depends on the input at the same clock time and on the input at the *next* clock time.

In general, any circuit whose output at a given time depends on some future input cannot be realized physically. In order to realize a circuit for the specifications of Example 6.1-5, it would be necessary to modify the specifications so that either the ith Gray bit appears at the output when the $(i + 1)$st binary bit appears on the input or the binary bits appear at the input in reverse order, with the more significant bits preceding less significant bits.

Even if the circuit specifications do not require the present output to depend on future inputs, it may not be possible to design a realizable circuit.

Example 6.1-6: A Specification for a Nonrealizable Sequential Circuit Which Does Not Involve Future Dependence. The circuit is to operate in pulse mode with two inputs x_1 and x_2 and with a reset input r. There is one level output, z, which is set equal to 0 by the occurrence of a pulse on the reset lead. The output becomes (and remains) equal to 1 when a pulse arrives on the x_1 input lead provided that (1) there has been exactly one x_2 pulse since the last reset pulse and (2) there have been an equal number of x_1 pulses occurring after the x_2 pulse and before the x_2 pulse (but after the reset pulse). Thus, the output will equal 1 after the sequences $r x_1 x_2 x_1$, $r x_1 x_1 x_1 x_2 x_1 x_1 x_1$, whereas the output will be 0 after the sequences $r x_1$, $r x_1 x_1 x_2 x_2 x_1 x_1$, $r x_1 x_1 x_2 x_1$. In order to satisfy these specifications, the circuit would have to be capable of counting the number of x_1 pulses which occur after the last r pulse and before the first x_2 pulse. Since there is no limit on the number of such x_1 pulses which can occur, the circuit must be able to store an unbounded count. The circuit can contain only a finite number of states, say, m, and can count at most up to 2^m. Any circuit built to satisfy these specifications could always be forced to operate incorrectly by applying a sequence of x_1 pulses containing more than 2^m pulses.

6.2 simplification of completely specified flow tables

A sequential-circuit specification can be nonrealizable either because the output depends on a future input or because an unbounded number of internal states are required by the specification. A more formal discussion of realizability is given in the papers by Rabin and Scott [3] and Kleene [4].

6.2 SIMPLIFICATION OF COMPLETELY SPECIFIED FLOW TABLES

For each performance specification for a sequential circuit there will be many suitable flow tables. Corresponding to each particular flow table there will be several different circuit realizations. Although all the circuits formed from flow tables corresponding to the same specification will have the same performance, the actual circuits will differ greatly. Any one of these circuits is acceptable if the only objective is to meet the performance specifications; however, the cost and complexity of the circuit must also be considered. No simple relationship between the form of a flow table and the cost of the corresponding circuits is known. There is thus no technique possible for choosing the flow table that will result in a minimum-cost circuit. A very simple relation does exist between one factor of the circuit cost and the flow table: the number of internal states (rows) of the flow table is directly related to the number of internal variables. If there are r rows in the flow table, there must be at least $\langle \log_2 r \rangle$† internal variables. There are 2^m different values for m internal variables, and each internal state must be represented by an assignment of values to the internal variables. In a relay circuit each internal variable corresponds to an internal relay, and the circuit cost depends significantly on the number of such relays. In electronic circuits, the internal variables correspond to flip-flops or feedback loops. Their influence on circuit cost is not so significant as in the relay case, but it is still quite important. By using a flow table with the fewest possible rows it is possible to reduce one significant factor in the final circuit cost: the number of internal variables. In order to make this possible, it is necessary to develop a technique for transforming a given flow table into another flow table which corresponds to the same performance specification and also contains the fewest possible rows for such a table. The development of such a technique is the object of this section.

Inaccessible States

Before discussing the transformation of flow tables it is appropriate to consider the relationships among different sequential circuits which exhibit

† The symbol $\langle p \rangle$ is equal to p if p is an integer and is equal to the next highest integer if p is a fraction. Thus $\langle 11 \rangle = 11$ and $\langle 12.7 \rangle = 13$.

Table 6.2-1. Inaccessible States

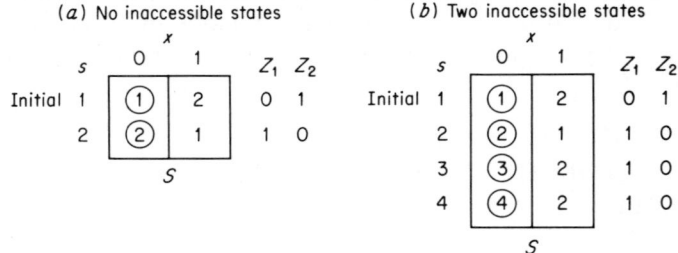

(a) No inaccessible states

(b) Two inaccessible states

(c) An inaccessible set of states (3,4)

identical performance. Table 6.2-1a is a flow table for a pulse-mode circuit in which the output changes whenever an input pulse is received. The circuit corresponding to Table 6.2-1b will also have its output change with each input pulse. States 3 and 4 of Table 6.2-1b have the property that the circuit will never enter these states unless it is placed in one of them by some external mechanism. Such states are called *inaccessible*.

Definition. An internal state is called an *inaccessible state* if it does not occur as a next-state entry in any row of the flow table other than its own row.

It will be assumed that there is no external mechanism for placing a sequential circuit in a given state other than that associated with placing the circuit in its specified initial state (if any such state exists). It is clear that any inaccessible state which is not an initial state can be removed from a flow table without changing the performance of the corresponding circuit. (The circuit never is in such a state, and therefore the state cannot affect the circuit performance.)

It is also possible to have a set of internal states such that no state of the set is entered from any state *not* in the set. Such a set is called an *inaccessible set of states*. States 3 and 4 of Table 6.2-1c form such a set. For this table, a corresponding circuit will never be in either state 3 or state 4 since the initial state is state 1 and there is no way for the circuit to enter states 3 or 4 from state 1. Clearly such an inaccessible set of states can be removed from the circuit if no one of the states is an initial state.

6.2 simplification of completely specified flow tables

Table 6.2-2. Indistinguishable Flow Tables

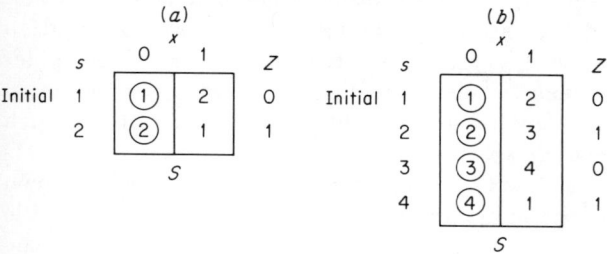

It has been shown that all the flow tables of Table 6.2-1 correspond to the same circuit performance. In the following it will be assumed that any inaccessible states or inaccessible sets of states which can be removed have already been removed. More precisely, the assumptions will be made that:

1. For flow tables in which an initial state is specified, there are no states in the table which cannot be "reached" from the initial state by the application of an appropriate input sequence.
2. For flow tables with no initial-state specification, it is assumed that a corresponding circuit can be started in any one of the states and therefore that for any ordered pair of states there is an input sequence which will cause the circuit to reach the second state of the pair if started initially in the first state of the pair.

Indistinguishable Circuits

Even after the removals described in the previous paragraph have been carried out, it is still possible to have different flow tables which correspond to the same circuit behavior [5]. Table 6.2-2a is identical with Table 6.2-1a, and Table 6.2-2b is a flow table with no inaccessible states which corresponds to exactly the same circuit behavior as does Table 6.2-2a. If circuits corresponding to Tables 6.2-2a and 6.2-2b were constructed and both circuits were placed initially in state 1, then identical input sequences applied to both circuits would produce identical output sequences, from both circuits. Two such tables which represent the same external behavior should be equally acceptable in satisfying a set of performance specifications. A pair of tables or circuits having this property will be said to be *indistinguishable*. The problem to be considered here is that of finding, for any given flow table, a flow table which is indistinguishable from the original table and which also has the fewest internal states of any such table.

The technique for simplifying flow tables results from a formalization of the relationship illustrated by these tables. Before proceeding with this formalism, it should be pointed out that this is basically a discussion of relationships among sequential circuits rather than among flow tables. All the tables discussed here are *fully specified;* i.e., a next state and an output are specified for each total state. Because of this, these tables have a direct correspondence to circuit realizations. It is very common to have some entries of a flow table unspecified, e.g., Table 6.1-5. Such tables are called *incompletely specified.*

In circuit synthesis, a flow table is a formal means for writing down the specifications which the final circuit must satisfy. Thus a flow table can correspond to a large class of circuits—all those circuits satisfying the design specifications. It is possible that some of the circuits may differ in their performance, for the specifications may not specify the behavior in all situations. For example, if the specifications state that input pulses never occur simultaneously at more than one input, the b.navior of the various circuits if simultaneous inputs do occur may be different for different circuits. Yet all these circuits can still satisfy the original specifications which assumed that multiple inputs would never occur. The relationships among incompletely specified flow tables are similar to, but more complicated than, those among completely specified tables. For ease of understanding, completely specified tables (circuits) will be considered first and the results then extended to incompletely specified tables.

> ***Definition: Initial States Specified.*** Two sequential circuits P and Q are *indistinguishable*, written $P \sim Q$, if and only if, with P and Q in their respective initial states, any input sequence applied to both circuits results in identical output sequences from both circuits.

For sequential circuits without specified initial states it has been assumed that any state can be the starting state. In this case the definition of indistinguishability must specify a correspondence for each state of P or Q rather than only for the initial states.

> ***Definition: No Initial-state Specification.*** Two sequential circuits P and Q are *indistinguishable* ($P \sim Q$) if and only if for each state p_i of P there exists at least one state q_j of Q such that, with P initially in p_i and Q initially in q_j, identical input sequences applied to P and Q result in identical output sequences from P and Q. For each state of Q there must be at least one state of P satisfying the same criterion.

Although these definitions are given in terms of sequential circuits, they apply equally well to flow tables. It will be assumed that they serve also to define what is meant by indistinguishable flow tables,

6.2 simplification of completely specified flow tables

and the terms sequential circuit and flow table will be used somewhat interchangeably.

Separate definitions of indistinguishability have been given for circuits with initial states and circuits without initial states. While it might seem that the definition for the initial-state circuits is weaker than the other, this turns out to be untrue. With the assumption that no inaccessible states are present, the two definitions are completely equivalent. Both these definitions specify a relationship which must be satisfied by pairs of states. The difference between the two definitions is that, for circuits with initial states, only the two initial states are required to satisfy this relation, while, for circuits without initial states, this relation must be satisfied by all the states. A formal definition will be given for this relation between states, and it will be shown that the two definitions of indistinguishability of circuits are equivalent.

Indistinguishable States

Definition. An internal state p_i of a sequential circuit P is *indistinguishable* from a state q_j of a circuit Q (P and Q can be the same circuit) if and only if, when circuit P is placed in state p_i, circuit Q is placed in state q_j, and the same input sequence is applied to both circuits; the same output sequence results from both circuits. This must be true for all possible input sequences. The fact that states p_i and q_j are indistinguishable will be written symbolically as $p_i \sim q_j$. This definition is illustrated in Fig. 6.2-1. It is now possible to restate the definitions of circuit indistinguishability in terms of indistinguishability of states.

Definition: Initial States Specified. Two sequential circuits P and Q are indistinguishable if and only if the initial states of P and Q are indistinguishable.

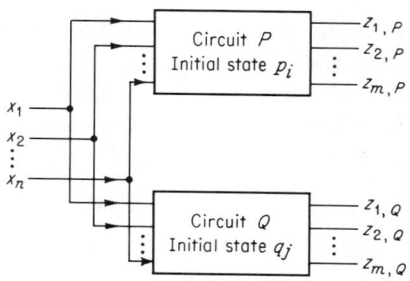

Fig. 6.2-1. *Indistinguishable states.*

If $z_{k,P} = z_{k,Q}$ for all k and for all input sequences, then state p_i of P is indistinguishable from state q_j of Q.

Sequential-circuit Synthesis

Definition: No Initial-state Specification. Two sequential circuits P and Q are indistinguishable if and only if for each state p_i of P there exists at least one state of Q which is indistinguishable from p_i and for each state q_j of Q there exists at least one state of P which is indistinguishable from q_j.

If two states are indistinguishable, they must both have the same output for each possible input. If this were not true, there would be some input which would produce different outputs and which would thus violate the indistinguishability condition. In order to state this condition succinctly, and for convenience in the discussion to follow, some additional symbols will now be introduced. The symbol $\mathbf{x}^i$ will represent the ith input condition, where i is a decimal number whose binary equivalent gives the values assumed by the circuit inputs. Thus, for a circuit with two inputs, the correspondence between the $\mathbf{x}^i$ and the input conditions is as shown in the accompanying table. The symbol $Z(s_i,\mathbf{x}^k)$ will represent

Symbol	Input Condition	
	x_1	x_2
$\mathbf{x}^0$	0	0
$\mathbf{x}^1$	0	1
$\mathbf{x}^2$	1	0
$\mathbf{x}^3$	1	1

the output produced by a circuit in internal state s_i when the input condition is $\mathbf{x}^k$. Thus, for Table 6.2-2b, $Z(3,\mathbf{x}^0) = 0$. The symbol $S(s_i,\mathbf{x}^k)$ will represent the next state for present state s_i and input condition $\mathbf{x}^k$. For Table 6.2-2b, $S(3,\mathbf{x}^1) = 4$, and $S(1,\mathbf{x}^0) = 1$. In terms of these symbols, the condition stated at the beginning of this paragraph is

If $p_i \sim q_j$ then $Z(p_i,\mathbf{x}^k) = Z(q_j,\mathbf{x}^k)$ for all $\mathbf{x}^k$

For a pair of indistinguishable states, not only must the present outputs be identical for all possible inputs, but all future outputs must also be identical. For this to be true, it is necessary that the next states be indistinguishable. (The requirement that future outputs be identical is precisely the requirement for the next states to be indistinguishable.)

If $p_i \sim q_j$ then $S(p_i,\mathbf{x}^k) \sim S(q_j,\mathbf{x}^k)$ for all $\mathbf{x}^k$

Theorem 6.2-1. Two states p_i and q_j are indistinguishable if and only if (1) $Z(p_i,\mathbf{x}^k) = Z(q_j,\mathbf{x}^k)$, for all $\mathbf{x}^k$, and (2) $S(p_i,\mathbf{x}^k) \sim S(q_j,\mathbf{x}^k)$ for all $\mathbf{x}^k$.

The previous discussion demonstrated that, if a pair of states are indistinguishable, they must satisfy these two conditions.

6.2 simplification of completely specified flow tables

It follows directly from the definition of indistinguishability that, if these two conditions are satisfied, the states must be indistinguishable.

When the definition of indistinguishability of circuits with initial states is reexamined, it is evident that the requirement that the initial states of P and Q be indistinguishable induces a requirement that each of the other states of P must be indistinguishable from some state of Q, and vice versa. This results directly from the requirement that the next states corresponding to a pair of indistinguishable states must be indistinguishable and that all states of a circuit must occur as next states either of the initial state or of some state which can be reached from the initial state by means of a suitable input sequence.

Indistinguishability Classes

For any given sequential circuit there will be a family or class of other sequential circuits which are indistinguishable from the original circuit. Each of these circuits will exhibit the same internal behavior: it would be impossible to determine by external measurements which particular member of the class was being tested. Thus, for any given *performance* specification, all circuits or flow tables in a class of indistinguishable circuits or flow tables should be equally acceptable. The costs of the circuits or flow tables will vary widely, and a minimal-cost circuit is usually desired. As discussed previously, the assumption that a flow table with the fewest states is a good approximation to a minimal-cost circuit will be made. It will be shown that in each family of indistinguishable flow tables there is a unique table having the fewest number of states.† The objective of the present discussion is the development of a procedure for obtaining, for any given, fully specified flow table, the minimum-state table which is indistinguishable from it. The properties of two indistinguishable flow tables having an unequal number of states will be considered next.

Assume that P and Q are two indistinguishable flow tables and that P has fewer states than Q. Each state of Q must be indistinguishable from some state of P. Since Q has more states than P, there must be at least two states of Q, say, q_i and q_j, which are both indistinguishable from the *same* state of P, say, p_k. The relationship of two states which are both indistinguishable from the same state is basic to the procedure being developed and must be examined in detail. It will be shown that if $q_i \sim p_k$ and $q_j \sim p_k$ then it must be true that $q_i \sim q_j$. A theorem to this effect can be quite easily proved by means of the general properties of the indistinguishability relation to be presented next.

† This flow table is unique in the sense that any other flow table in the class which has the same number of states will be identical with the original table except perhaps for a relabeling of the states. This discussion applies only to fully specified flow tables.

It is clear that any state is always indistinguishable from itself (two circuits started in the same state and sent identical inputs have identical outputs). Also, if state p_i is indistinguishable from state q_j, then state q_j is indistinguishable from p_i. These properties are called the *reflexive property* and the *symmetric* (or *commutative*) *property*, respectively:

equivalence Relation

(P1) $p_i \sim p_i$ (Reflexive)
(P2)   If $p_i \sim q_j$ then $q_j \sim p_i$ (Symmetric)

There is a third general property which this relation satisfies.

(P3)   If $p_i \sim q_j$ and $q_j \sim r_k$ then $p_i \sim r_k$ (Transitive)

The transitive property states that, if p_i is indistinguishable from q_j and q_j is indistinguishable from r_k, then p_i is indistinguishable from r_k. The proof that this property must hold is quite simple. Suppose that three circuits P, Q, and R are placed in states p_i, q_j, and r_k and are all sent identical inputs. Since $p_i \sim q_j$, the outputs of P and Q must be identical, and since $q_j \sim r_k$, the outputs of Q and R must be identical. Therefore the outputs of P and R must be identical, and $p_i \sim r_k$.

Any relation which satisfies the three properties P1, P2, and P3 is said to be an *equivalence relation*. The important characteristic of an equivalence relation is that it divides the set of objects on which it is defined (the states of sequential circuits) into disjoint (nonoverlapping) *equivalence classes* [6]. In any sequential circuit, the internal states can be grouped into sets of states, called *indistinguishability classes*, so that all states in each set are indistinguishable from each other and no state in one set is indistinguishable from any state in any other set. In Table 6.2-2b, states 1 and 3 form one indistinguishability class, and states 2 and 4 form another indistinguishability class.

The following theorem can now be easily demonstrated:

> **Theorem 6.2-2.** If flow tables P and Q are indistinguishable, then any two states of Q are indistinguishable from each other if and only if they are indistinguishable from the same state of P.
>
> PROOF. First assume that states q_i and q_j of Q are both indistinguishable from state p_k of P. Then, $q_i \sim p_k$, and $q_j \sim p_k$. By means of the symmetric property P2 this can be written as $q_i \sim p_k$ and $p_k \sim q_j$. The transitive property P3 shows that it follows that $q_i \sim q_j$. To prove the other half of the theorem, assume that $q_i \sim q_j$. Since $P \sim Q$, there must be some state of P, say, p_k, such that $q_j \sim p_k$. Again, by using P3, it can be concluded that $q_i \sim p_k$.

Minimum-state Flow Tables

It was pointed out previously that the states of any flow table can be divided into disjoint sets or classes of states, called indistinguishability

6.2 simplification of completely specified flow tables

classes, such that all states in the same class are indistinguishable and no pair of states from different classes are indistinguishable. The preceding theorem shows that flow table Q must have at least as many states as there are indistinguishability classes in P.

It is possible to form a flow table having exactly this number of states. This is done by forming a new flow table P^* in which each state corresponds to one of the indistinguishability classes of P.

Since all the states in one indistinguishability class of P are indistinguishable, they must have the same output entries (Theorem 6.2-1). These outputs are used for the outputs of the corresponding state of P^*. The next-state entries for two indistinguishable states need not be identical, but they must correspond to indistinguishable states (Theorem 6.2-1). Thus the next-state entries in P^* can be filled in with the state of P^* corresponding to the appropriate indistinguishability class of P. This process is illustrated in Table 6.2-3, in which it is assumed that the indistinguishability classes of indistinguishable states are known. A procedure for determining these will be presented below.

Several remarks can be made about the table P^* formed by this process. First of all, P^* is indistinguishable from table P: if table P is initially in state p_k and table P^* is initially in state q_j, which corresponds to p_k's equivalence class, then the first outputs produced must be identical. Furthermore, both tables must go to next states, which again have identical outputs, etc. It can also be shown that P^* is unique and that no pair of states in P^* are indistinguishable. With the exception of a technique for obtaining the indistinguishability classes, the procedure for obtaining

Table 6.2-3. Formation of Minimum-state Flow Table

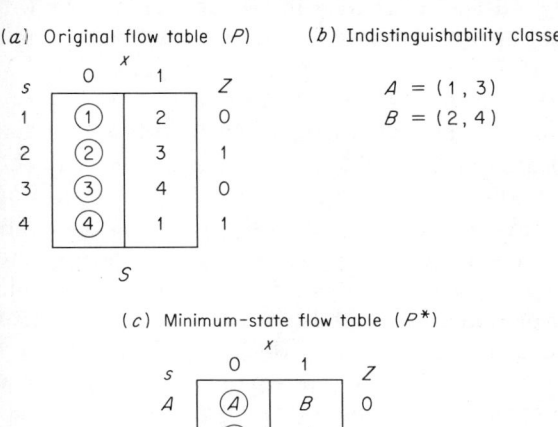

P^* has now been completely specified. A method for determining the indistinguishability classes will be presented next.

Distinguishable States

Two states which are not indistinguishable are said to be *distinguishable*. It follows from the definition of indistinguishable states that two states p_i and q_j are distinguishable if and only if there exists an input sequence which results in different output sequences for a circuit initially in state p_i and for a circuit initially in state q_j. This concept of distinguishability is important because it is sometimes easier to show that two states do not satisfy the requirement for being distinguishable than to show that two states do satisfy the indistinguishability criterion.

Determination of Indistinguishability Sets

In order to determine the indistinguishability classes for a flow table, it is necessary to consider each pair of states individually. A convenient tabular form for recording the results of these considerations is illustrated by Table 6.2-4b. Each cell of this table corresponds to two states—one defined by the row in which the cell appears and one by the column. As explained previously, any pair of states which do not have identical output entries are distinguishable. In Table 6.2-4a, the output entries of states 5 and 6 differ from the output entries of states 1, 2, 3 and 4. Thus the state pairs (1,5), (1,6), (2,5), (2,6), (3,5), (3,6), (4,5), and (4,6) are distinguishable. This is indicated on Table 6.2-4b by placing ×'s in the cells corresponding to these state pairs. Examination of the entries for states 2 and 4 leads directly to the conclusion that these states are indistinguishable. All their output entries agree, and their next-state entries are either identical (for x^1 and x^2) or are stable. Two circuits started in states 2 and 4, respectively, either will remain in these two states or will both be in the same state. If both circuits are in the same state, their outputs must obviously be identical, and if they remain in states 2 and 4, their outputs are identical because the output entries for 2 and 4 are identical. From this it must be concluded that states 2 and 4 are indistinguishable. This is indicated in Table 6.2-4b by placing a check in the 2–4 cell. It should be pointed out that the same reasoning would still be valid if, instead of having stable entries in some of the columns, the next-state entry for state 2 were state 4 and the next state entry for state 4 were state 2.† The remaining state pairs do not satisfy either of these simple tests for distinguishability or indistinguishability and must be examined further. For states 1 and 2, $S(1,x^0) = 3$ and $S(2,x^0) = 2$, $S(1,x^1) = 1$ and $S(2,x^1) = 6$. Thus states 1 and 2 are indistinguishable if and only if states 2 and 3 are indistinguishable and states 1 and 6 are indistinguisha-

† This situation is appropriate for pulse-mode rather than fundamental-mode circuits.

6.2 simplification of completely specified flow tables

Table 6.2-4. Determination of Indistinguishability Sets

(a) Flow table

S	x^0 00	x^1 01	x^3 11	x^2 10
1	3, 1	①, 0	①, 1	①, 0
2	②, 1	6, 0	②, 1	1, 0
3	③, 1	5, 0	③, 1	1, 0
4	④, 1	6, 0	④, 1	1, 0
5	2, 1	⑤, 1	4, 0	⑤, 0
6	2, 1	⑥, 1	3, 0	⑥ 0

S, Z

(b) First step in determining sets

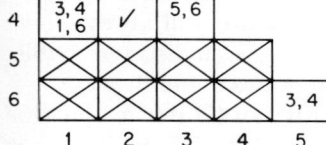

(c) Final step in determining sets

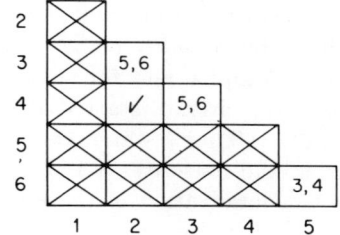

(d) Indistinguishability sets

A : (1)
B : (2,3,4)
C : (5,6)

(e) Minimum-state flow table

S	00	01	11	10
A	B, 1	Ⓐ, 0	Ⓐ, 1	Ⓐ, 0
B	Ⓑ, 1	C, 0	Ⓑ, 1	A, 0
C	B, 1	Ⓒ, 1	B, 0	Ⓒ, 0

S, Z

ble. This is indicated in Table 6.2-4b by entering the state pairs (2,3) and (1,6) in the cell for the (1,2) state pair. Similar entries are placed in the other appropriate cells of Table 6.2-4b. This table must now be examined to determine whether or not any of the state pairs listed *within* cells of the table have already been discovered to be distinguishable or indistinguishable. For example, the 1–2 cell contains the entry 1,6. The 1–6 cell contains an × since the (1,6) state pair has already been determined distinguishable. This means that the (1,2) state pair is distinguishable, and an × can be placed in the 1–2 cell of the table. Similar reasoning applies to the 1–3 cell and the 1–4 cell. This process must be continued until there are no entries within cells which correspond to cells containing ×'s. A table such as Table 6.2-4c will result. *All the cells of this table which do not contain ×'s correspond to indistinguishable state pairs.* In order to see why this is true, consider states 5 and 6. The flow table of Table 6.2-4a shows that two circuits started in states 5 and 6 will, after one input has been applied, both be in the same state (state 2), or still be in states 5 and 6, or be in states 4 and 3, respectively. As discussed previously, when both circuits are in the same state or when the circuits are in states 5 and 6, the outputs will always be identical. The only possibility for states 5 and 6 to be distinguishable is that different outputs will result after the circuits have entered states 4 and 3. It is therefore appropriate to examine the possibilities for the circuit outputs after the internal states have changed from 5 and 6 to 4 and 3. Examination of the flow table shows that the application of an input can cause both circuits to now enter the same state (state 1), to remain in states 3 and 4, or to return to states 5 and 6. To summarize these considerations, two circuits started in states 5 and 6 will always enter (and remain) in the same states, or be in states 5 and 6, or be in states 4 and 3. Since the output entries for states 5 and 6 and for states 4 and 3 are identical, it can be concluded that two circuits started in states 5 and 6 will always have identical outputs. Thus, states 5 and 6 must be indistinguishable. Similar remarks hold for states 3 and 4 and for states 2 and 3.

By a direct generalization of these remarks it can be shown that the procedure described in connection with the flow table of Table 6.2-4a is valid for any flow table. In the table formed by this procedure (such as Table 6.2-4c) any cells not containing ×'s correspond to indistinguishable state pairs. The indistinguishability classes can be determined directly from the pairs of indistinguishable states. The minimum-state flow table is then formed as described previously. These steps are illustrated in Table 6.2-4d and e. A complete procedure or algorithm for forming the minimum-state flow table which is indistinguishable from any given fully specified flow table has now been described.

6.3 SIMPLIFICATION OF INCOMPLETELY SPECIFIED FLOW TABLES

The first step in the sequential-circuit-design procedure described in this chapter consists in writing down a flow table describing the desired circuit performance. This flow table is a formal specification of the requirements which the circuit must satisfy. Often certain circuit input combinations or total states will not occur when the circuit is in operation. Since the action of the circuit in situations which do not arise is unimportant, it is common practice to leave this action unspecified. As illustrated by the flow tables of Sec. 6.1, a dash is entered into all cells of the flow

6.3 simplification of incompletely specified flow tables

table for which the circuit action is unspecified. The discussion of Sec. 6.2 concerning flow-table simplification is not valid for incompletely specified tables. This section presents a discussion of simplification techniques for incompletely specified tables [7,8,9].

Unspecified Outputs

Table 6.3-1a shows a flow table in which one of the output entries is unspecified. It usually happens that, for each total state, the next state and output are either both specified or both unspecified. Situations do arise, however, in which only one is unspecified, and thus it is convenient to consider next states and outputs separately.

It will usually be quite difficult to determine any word statement or reasonable explanation for the behavior specified by the flow tables of this section. This is because these tables have been designed to illustrate certain features of the simplification discussion rather than to appear "practical." More palatable tables could be formulated only with a considerable increase in complexity. The important point to emphasize is that all the entries of the flow table must be accepted as requirements to be satisfied; it is not permissible to *change* any of the entries even though the change may seem perfectly sensible.

The flow table which results if the unspecified entry of Table 6.3-1a is replaced by a 0 is shown in Table 6.3-1b. There are no indistinguishable states in Table 6.3-1b, and therefore no reduction in the number of states is possible. The effect of using a 1 rather than a 0 to replace the dash is shown in Table 6.3-1c. Again no reduction is possible. From

Table 6.3-1. A Flow Table with an Unspecified Output

(a) Original flow table

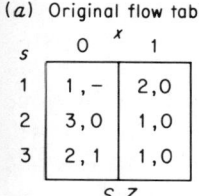

(b) Unspecified entry replaced by 0

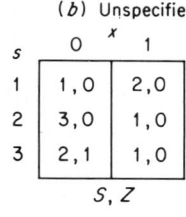

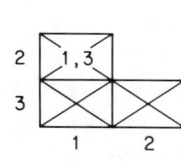

(c) Unspecified entry replaced by 1

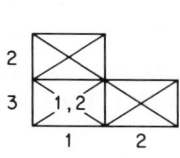

Table 6.3-2. A Flow Table Satisfying Table 6.3-1a

(a) Flow table

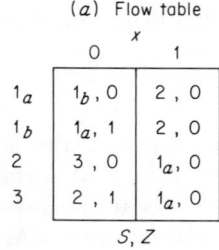

(b) Test for indistinguishability

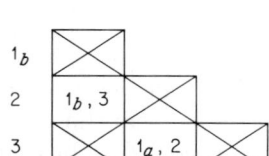

(c) Indistinguishability sets

A : (1_a, 2)
B : (1_b, 3)

(d) Minimum-state flow tables

	0	1
A	B, 0	A, 0
B	A, 1	A, 0

this it would seem that there is no circuit having fewer than three internal states which can satisfy the requirements of the flow table of Table 6.3-1a. However, this conclusion is false, and such a circuit having only two states does exist. This circuit will be presented, and the fallacy in the reasoning illustrated by Table 6.3-1 will be pointed out.

Table 6.3-2a shows a flow table for a circuit which satisfies Table 6.3-1a and contains four internal states. This table has been obtained from Table 6.3-1a by replacing state 1 by two states, 1_a and 1_b. The fact that a circuit constructed from Table 6.3-2a will satisfy Table 6.3-1a can easily be verified. The new table (Table 6.3-2a) has been formed from the original table (Table 6.3-1a) by "splitting" state 1 into two states, 1_a and 1_b, both of which have the same entries as state 1 wherever they are specified. One of the new states has a 0 output where the state 1 output is unspecified, and the other new state has a 1 output in place of the unspecified output. A circuit constructed from Table 6.3-2a will have the same performance as that specified by Table 6.3-1a, but its performance will differ from that specified by Table 6.3-1b or c. Table 6.3-1a has one unspecified output: in circuits constructed from Table 6.3-1b and c either a 0 or a 1 output will occur consistently whenever the conditions correspond to the unspecified output of Table 6.3-1a. However, in a circuit constructed from Table 6.3-2a, sometimes a 0 output will occur, and sometimes a 1 output will occur. This performance should be just as acceptable as the performance where the dash is in effect replaced by either a 0 or a 1: placing a dash in the table is equivalent to stating that the corresponding output is of no interest. The fact that this output sometimes takes on one value and at other times takes on another should cause no concern. If the flow table is thought of as a means of specifying the desired behavior,

6.3 simplification of incompletely specified flow tables

then the dash can be thought of as an agreement to refrain from observing the output in particular situations. In practice the dash usually occurs because a particular total state is known not to occur in the expected use of the circuit. If the particular situation is assumed not to arise, it is surely reasonable not to object to anything that the circuit would do in this nonoccurring situation. In the remainder of this section it will be assumed that an unspecified entry in a flow table means that any behavior is acceptable in place of the dash and that this replacement need not be consistent—the dash can in effect be replaced by one specification at one time and by another specification at another time. In fact with this interpretation there would be nothing wrong with using a probabilistic element to determine the behavior for unspecified entries.

The flow table of Table 6.3-2a can be simplified, and the minimum-state table which is indistinguishable from it is shown in Table 6.3-2d. Since this table with two states is indistinguishable from Table 6.3-2a, this two-state table satisfies the requirements given by Table 6.3-1a in the same sense that the four-state Table 6.3-2a does. Thus a circuit built from Table 6.3-2d should be an acceptable realization for the specification of Table 6.3-1a. The objective of this section is to determine a technique for deriving simplified tables such as Table 6.3-2d from incompletely specified flow tables such as Table 6.3-1a. Table 6.3-2a has been presented solely for purposes of exposition. The technique to be developed will be one for forming the reduced-state table directly from the original table. Before proceeding it is necessary to consider the effect of having next states unspecified.

Unspecified Next States

Table 6.3-3a shows a flow table in which one of the next-state entries is unspecified. If the next state is unspecified, then all future circuit outputs must be unimportant. The only other possible interpretation would be that various different behaviors would be acceptable but that each acceptable behavior would have to be one which could be obtained by replacing the unspecified next state by one of the states of the table. It is a little difficult to imagine a situation for which this interpretation would be valid, and it is customary to assume that all future outputs are unimportant. Clearly, if the unspecified entry is present because the circuit can never enter a total state, then any future outputs need not be constrained.

With this interpretation, it is always possible to replace a table in which some of the next states are unspecified by a flow table in which all the next states are specified, but some of the outputs are unspecified [10]. Thus, Table 6.3-3b requires the same circuit behavior as does Table 6.3-3a but does not contain any next states which are unspecified. Table 6.3-3b has been formed by adding a new state (designated T for terminal) to Table 6.3-3a and replacing the unspecified next state by T. The state T

Sequential-circuit Synthesis

Table 6.3-3. A Flow Table with a Next State Unspecified

(a) Flow table

s	x_1x_2 00	01	11	10
1	1,0	2,0	1,0	–,1
2	2,0	2,0	2,0	2,0

S,z

(b) Flow table with unspecified next state replaced by unspecified outputs

s	x_1x_2 00	01	11	10
1	1,0	2,0	1,0	T,1
2	2,0	2,0	2,0	2,0
T	T,–	T,–	T,–	T,–

S,z

contains only stable next-state entries and has all its outputs unspecified. Thus, once a circuit enters state T, it will always remain in state T, and since all the outputs are unspecified, any output sequence will be acceptable. This same technique will work for any table with unspecified next states. One terminal state T is added with all its outputs unspecified, and all unspecified next-state entries are replaced by a T. The formal development which follows will be stated in terms of flow tables with only outputs unspecified. This does not restrict the generality of the treatment because of the technique just discussed for converting any table with unspecified next states into a corresponding table with only outputs unspecified. In fact the purpose for introducing this conversion technique is to simplify the formalism to be presented.

Covering of Flow Tables

Definition: Initial State Specified. A flow table P is said to cover another table Q, written $P \supset Q$, if and only if, for any input sequence applied to both tables started in their respective initial states, the output sequences are identical *whenever the output of Q is specified.*

The assumption will be made that any table which covers a given table can be substituted for the original table in the sense that a circuit designed

6.3 simplification of incompletely specified flow tables

for the covering table will automatically satisfy the requirements of the covered table. Notice that this relation among flow tables is not symmetric: it is not necessarily true that a covered table will satisfy the requirements of the covering table. If $P \supset Q$, it is possible for an output of P to be specified when the corresponding output of Q is unspecified. Thus P satisfies Q's requirements, but the converse is not true. The preceding definition does not apply to circuits without initial states, and another definition must be formulated for this class of circuits.

Definition: No Initial State. A flow table P is said to cover another table Q if and only if, for each state q_i of Q, there is a state p_j of P such that for any input sequence applied to both tables initially in states q_i and p_j respectively, the output sequences are identical *whenever the output of Q is specified*.

This definition shows that when one flow table covers another there must be a correspondence between the states of the two tables and the corresponding states must have certain properties. This relation between states can be stated formally as follows:

Definition. A state p_i of a flow table P is said to cover a state q_j of a flow table Q, written $p_i \supset q_j$, if and only if for any input sequence applied to P and Q initially in states p_i and q_j, respectively, the outputs are identical whenever the output of Q is specified.

This relation of one state covering another state is analogous to the indistinguishability relation for completely specified flow tables. If a flow table P covers a flow table Q, each state of Q must be covered by at least one state of P. In order for P to have fewer states than Q, it is necessary that at least one state of P cover more than one state of Q. It is not possible for any arbitrary pair of states of a flow table to be covered by a single state of another flow table. In particular, only pairs of states which never require different outputs can be covered by a single state. This relation between states will be called *compatibility* and is defined precisely as follows:

Definition. Two internal states q_i and q_j of a flow table Q are *compatible* if and only if, for all input sequences, the output sequence which results when Q is initially in q_i is the same as the output sequence which results when Q is initially in q_j whenever *both outputs are specified*.

The compatibility relation between states is important because of the following theorem:

Theorem 6.3-1. If internal state p_i of P covers both internal states q_j and q_k of Q, then states q_j and q_k must be compatible.

Sequential-circuit Synthesis

PROOF. By the definition of covering, the output obtained for P initially in p_j must be the same as the output for Q initially in q_j whenever the Q output is specified, and the P output must also be the same as the output for Q initially in q_k whenever the Q output is specified. From this it follows that the outputs for Q initially in q_j and for Q initially in q_k must be identical whenever both are specified.

The compatibility relation is not, in general, an equivalence relation. The reflexive and symmetric properties are satisfied by the compatibility relation, but the transitive property need not be satisfied. To see why this is true, consider three states q_i, q_j, and q_k of a flow table Q. Suppose that the outputs for Q initially in each of these states always agree except for one particular output. Let this particular output be unspecified for Q initially in q_i, be 0 for Q initially in q_j, and be 1 for Q initially in q_k. From this it follows that q_i and q_j are compatible, q_i and q_k are compatible, but q_j and q_k are not compatible.

It is still possible to collect the states of a flow table into compatibility classes where a set of states is a compatibility class if and only if each pair of states in the set are compatible. Since compatibility is not an equivalence relation, these compatibility classes will not necessarily be disjoint. However, they are important because of the following corollary:

Corollary. If internal state p_i of P covers internal states q_{j_1}, q_{j_2}, . . . , q_{j_k} of Q, then states q_{j_1}, q_{j_2}, . . . , q_{j_k} must form a *compatibility class* of Q; that is, each pair of the q_{j_i} must be compatible.

A restatement of this corollary is that, if $P \supset Q$, then each internal state p_i of P must cover a compatibility class of the states of Q. Although this relation between the states of P and the states of Q is a necessary condition if $P \supset Q$, it is not sufficient and there is an additional property that these states must satisfy. This is the closure property, which is defined as follows:

Definition. A collection of compatibility classes is *closed* if and only if, for each compatibility class $\{s_1, s_2, \ldots, s_m\}$, all the states $S(s_1, \mathbf{x}^\alpha)$, $S(s_2, \mathbf{x}^\alpha)$, . . . , $S(s_m, \mathbf{x}^\alpha)$ are included in a single compatibility class in the collection. This must be true for all choices of $\mathbf{x}^\alpha$.

Example 6.3-1. For the flow table of Table 6.3-4a the compatibility classes are $\{1,2\}$, $\{1,3\}$, $\{2,4\}$, $\{3,4\}$. (The procedure for deriving these will be described below.) The collection of classes $\{1,2\}$ and $\{3,4\}$ is not closed because $S(1, \mathbf{x}^0) = 1$, $S(2, \mathbf{x}^0) = 3$, and class $\{1,3\}$ is not included in the collection. For this particular table, the only closed collection of compatibility classes must include all the four compatibility classes.

Table 6.3-4. Closure Example

(a) Flow table R

s	00	01	11	10
1	①,0	2,–	①,0	2,–
2	3,–	②,0	②,0	②,0
3	③,–	4,1	2,0	4,1
4	④,1	④,–	1,0	④,–

S, z

(b) Flow table T

	s	00	01	11	10
{1,2}	A	?,0	Ⓐ,0	Ⓐ,0	Ⓐ,0
{3,4}	B	Ⓑ,1	Ⓑ,1	A,0	Ⓑ,1

S, z

Necessary and sufficient conditions for one flow table to cover another can now be stated in terms of this closure property.

Theorem 6.3-2. A flow table P covers a flow table Q if and only if:

1. Each internal state of Q is included in at least one compatibility class which is covered by an internal state of P.
2. The compatibility classes of Q which are covered by internal states of P form a closed collection.

PROOF. The sufficiency of this theorem will be proved by giving a procedure for forming the flow table P from a closed collection of compatibility classes of states of Q. Let $\{q_{j_1}, q_{j_2}, \ldots, q_{j_h}, \ldots, q_{j_m}\}$ be one of the compatibility classes of Q, and let p_j be the state of table P which is to cover this compatibility class. The outputs $Z(q_{j_1}, \mathbf{x}^\alpha), Z(q_{j_2}, \mathbf{x}^\alpha), \ldots, Z(q_{j_m}, \mathbf{x}^\alpha)$ which are specified must be identical for all α. By the definition of compatibility the specified outputs of Q must be the same, independent of which of the q_{j_h} that Q is started in, for all input sequences and in particular for the input sequence consisting only of $\mathbf{x}^\alpha$. Therefore set $Z(p_j, \mathbf{x}^\alpha)$ equal to this common output if any of the $Z(q_{j_h}, \mathbf{x}^\alpha)$ are specified, and let $Z(p_j, \mathbf{x}^\alpha)$ be unspecified otherwise.

Because of closure, the next states $S(q_{j_1}, \mathbf{x}^\alpha), S(q_{j_2}, \mathbf{x}^\alpha), \ldots, S(q_{j_m}, \mathbf{x}^\alpha)$ must all be included in one of the compatibility classes of the collection. Let p_k be the state of P which corresponds to the compatibility class including the $S(q_{j_h}, \mathbf{x}^\alpha)$, and set $S(p_j, \mathbf{x}^\alpha) = p_k$. It follows from the construction technique that $Z(p_k,\mathbf{x}^\beta) = Z(S(q_{j_h},\mathbf{x}^\alpha),\mathbf{x}^\beta)$. A continuation of this reasoning shows that state p_j of P covers states q_{j_h} of Q and therefore that $P \supset Q$.

The necessity of condition 1 of this theorem follows directly from the corollary to Theorem 6.3-1 and the definition of the covering relation. Suppose now that closure is not satisfied. Specifically, let p_j cover all the states of the compatibility class $\{q_{j_1}, q_{j_2}, \ldots, q_{j_h}, \ldots, q_{j_m}\}$, and let $S(p_j,\mathbf{x}^\alpha) = p_i$. If closure is not satisfied, there must be some state $S(q_{j_k},\mathbf{x}^\beta)$ which is not covered by p_i. This means that there must be some sequence of inputs, $\vec{\mathbf{x}}$, which will produce a different output sequence for table P started in state p_i than it produces for table Q started in state $S(q_{j_k},\mathbf{x}^\beta)$. This in turn means that different output sequences will result when the input sequence $\mathbf{x}^\alpha\vec{\mathbf{x}}$ is applied to table P started in state p_j and to table Q started in state q_{j_k}. Thus state p_j cannot cover state q_{j_k}, which contradicts the original assumption and thereby completes the proof.

In the proof it was shown that, if states q_i and q_j are compatible, then $Z(q_i, \mathbf{x}^\alpha)$ and $Z(q_j,\mathbf{x}^\alpha)$ must be identical for all α for which they are both specified. By an argument similar to that used in connection with Theorem 6.2-1 it can also be shown that, if q_i and q_j are compatible, then $S(q_i,\mathbf{x}^\alpha)$ and $S(q_j,\mathbf{x}^\alpha)$ must be compatible for all α. Thus the following corollary can be proved:

Corollary. Two states q_i and q_j are compatible if and only if:

1. $Z(q_i,\mathbf{x}^\alpha)$ and $Z(q_j,\mathbf{x}^\alpha)$ are identical for all $\mathbf{x}^\alpha$ for which they are both specified.
2. $S(q_i,\mathbf{x}^\alpha)$ and $S(q_j,\mathbf{x}^\alpha)$ are compatible for all $\mathbf{x}^\alpha$.

Theorem 6.3-2 shows that the problem of finding a minimum-state flow table which covers a given table Q is equivalent to that of finding a closed collection of compatibility classes of Q which contains a minimum number of classes. The technique for finding the compatibility classes and forming the new flow table is illustrated in the following example:

Example 6.3-2. A flow table is shown in Table 6.3-5a, and the *compatibility table* for determining the compatibility classes is shown in Table 6.3-5b. The method of forming a compatibility table is identical with the method discussed in connection with Table 6.2-4. (An × is entered into the table only when two outputs disagree, not when one is unspecified.) The compatibility classes determined from Table 6.3-5b are listed in Table 6.3-5c. Any collection of these classes which satisfies the conditions of Theorem 6.3-2 must include either the {1,2} class or the {1,3} class since state 1 must be included in at least one class in the collection. In order to satisfy the closure requirement, *both* {1,2} and {1,3} must be included, since $S(\mathbf{x}^0,1) = 1$, $S(\mathbf{x}^0,2) = 3$, requiring {1,3} if {1,2}

6.3 simplification of incompletely specified flow tables

is included, and $S(\mathbf{x}^3,1) = 1$, $S(\mathbf{x}^3,3) = 2$, requiring $\{1,2\}$ if $\{1,3\}$ is included. Similar reasoning shows that the inclusion of $\{1,3\}$ requires the inclusion of $\{2,4\}$ and that the inclusion of $\{2,4\}$ requires the inclusion of $\{3,4\}$. Thus the only closed collection of these compatibility classes must include all four of the classes. Since there are four states in the original table, no reduction in the number of states is possible. A flow table formed from the compatibility classes is shown in Table 6.3-5d, merely to illustrate the formation of such a table.

In order to give a specific example of the closure property, an attempt will be made to form a table from compatibility classes which do not satisfy the closure requirement. Specifically, Table 6.3-6 shows a table which has two states A and B which correspond to classes $\{1,2\}$ and $\{3,4\}$ of Table 6.3-5. The same procedure can be used for filling in all the next-state entries of this table, except for the $S(A,\mathbf{x}^0)$ entry, since $S(1,\mathbf{x}^0) = 1$ and $S(2,\mathbf{x}^0) = 3$. There are only two choices for this entry, because there are only two states in the table. Table 6.3-6a shows the flow table which results if state A is chosen for the $S(A,\mathbf{x}^0)$ entry. The fact that Table 6.3-6a does not cover Table 6.3-5a is illustrated in Table 6.3-6b. If

Table 6.3-5

(a) Flow table

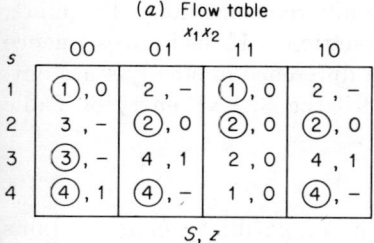

(b) Compatibility table

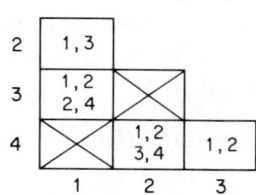

(c) Compatibility classes
$\{1,2\}, \{1,3\}, \{2,4\}, \{3,4\}$

(d) Table formed from the classes of c

Table 6.3-6

(a) Flow table with $S(A, \mathbf{x}^0) = A$

s	$x_1 x_2$ 00	01	11	10
$\{1,2\} \subset A$	Ⓐ, 0	Ⓐ, 0	Ⓐ, 0	Ⓐ, 0
$\{3,4\} \subset B$	Ⓑ, 1	Ⓑ, 1	A, 0	Ⓑ, 1

S, z

(b)

	$x_1 x_2 =$	00	01	01	00	00	01	01	00	00
Table 6.3-5	state	1		2		3		4		4
	output	0	–	0	–	–	①	–	1	1
Table 6.3-6a	state	A		A		A		A		A
	output	0	0	0	0	0	⓪	0	0	0

Tables 6.3-6b and 6.3-5a are placed initially in states A and 1, respectively, and the input sequence 00, 01, 00, 01 is applied, the resulting output sequences are those shown in Table 6.3-6b. The outputs for unstable states are shown explicitly in this table since they are specified in the original table. The unstable outputs which occur between the second 00-01 inputs of the test sequence produce different outputs. Therefore, the two tables do not satisfy the covering relation. If the input sequence is continued and the input 00 is applied, a difference in steady-state outputs results. Similar results are obtained if the $S(A, \mathbf{x}^0)$ entry of Table 6.3-6a is set equal to B.

Maximum Compatibility Classes

Table 6.3-5a is not typical, since each compatibility class contains only two states. Table 6.3-7a shows a flow table which illustrates some more typical features of the compatibility relation. This table has some of the next-state entries unspecified. The theory of this section is directly applicable if the table is transformed by the technique of Table 6.3-3 into an equivalent table having only the outputs unspecified. However, this transformation is unnecessary, for the new state T which would be introduced will always be compatible with all the other states and will appear in all the maximum compatibility classes. The first step in finding a minimum-state flow table which covers the given table is to find the compatibility classes. Actually, only the largest, or maximum, compatibility classes need be found; all other compatibility classes are subsets of these. There are many procedures possible for obtaining the maximum compatibility classes from the compatibility table. One of these involves con-

6.3 simplification of incompletely specified flow tables

Table 6.3-7

(a) Flow table

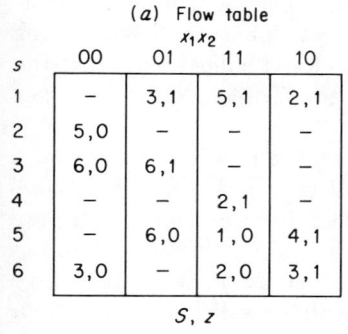

S	00	01	11	10
1	—	3,1	5,1	2,1
2	5,0	—	—	—
3	6,0	6,1	—	—
4	—	—	2,1	—
5	—	6,0	1,0	4,1
6	3,0	—	2,0	3,1

S, z

(b) Compatibility table

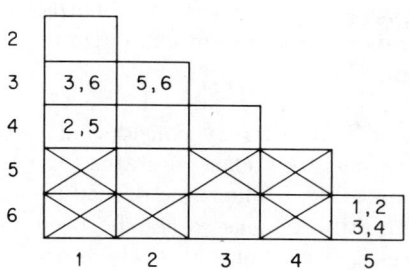

(c) Maximum compatibility classes

{1,2,3,4}, {2,5}, {3,6}, {5,6}

(d) Closed collections

(i) {1,2}, {3,4}, {5,6}
(ii) {1,4}, {2,5}, {3,6}

(e) Minimum-state flow table

S	00	01	11	10
1,2 ⊂ A	C,0	B,1	C,1	A,1
3,4 ⊂ B	C,0	C,1	A,1	—
5,6 ⊂ C	B,0	C,0	A,0	B,1

S, z

sidering each of the columns separately, starting at the rightmost column. This procedure is as follows:

1. List those pairs of states which are shown to be compatible in the rightmost column of the table for which any such pairs exist. For Table 6.3-7 this means that the pair 5, 6 is listed.

2. Proceed to the next column to the left. If the state to which this column corresponds is compatible with all members of a previously determined compatibility class, add this state to the class. If the state is not compatible with all members of a class but is compatible with a subset of the class, form a new class including the current state and the subclass. Finally list any compatible pairs which are not included in any already determined class. Do not retain any classes which are subsets of other classes. Repeat this step until all columns of the compatibility table have been considered. The classes remaining are the maximum compatibility classes.

For Table 6.3-7, this process will result in the following sequence of classes:

{5,6}
{3,4}, {3,6}, {5,6}
{2,3,4}, {3,6} {5,6}, {2,5}
{1,2,3,4}, {3,6}, {5,6}, {2,5}

Bounds on Minimum Number of States

The collection of all maximum compatibility classes will always constitute a closed collection because of the corollary to Theorem 6.3-2. Thus an upper bound on the number of states in a closed collection is just the number of maximum compatibility classes. For Table 6.3-7, this upper bound is 4.

It is also possible to derive a lower bound on the number of states in a closed collection by considering incompatibility classes. An *incompatibility class* is a class of states for which no pair of states in the class are compatible. The same procedure as was described for finding maximum compatibility classes can be used for finding maximum incompatibility classes if pairs of states which are not compatible are listed, rather than pairs of compatible states. For Table 6.3-7, the maximum incompatibility classes are

$\{1,5\}, \{1,6\}, \{2,6\}, \{3,5\}, \{4,5\}, \{4,6\}$

Since no two states from the same incompatibility class can occur in the same compatibility class, there must be at least as many compatibility classes in a closed collection as there are states in the largest incompatibility class. Thus a lower bound on the number of states in a closed collection is the maximum number of states in any incompatibility class. For Table 6.3-7, this lower bound is 2.

Forming Closed Collections of Compatibility Classes

For Table 6.3-7, there is only one collection of compatibility classes which includes all the states and has only two members, $\{1,2,3,4\}$ and $\{5,6\}$. This is not a closed collection, for the inclusion of states 1 and 3 in the same class requires states 3 and 6 also to be included in the same class. Thus, the lower bound cannot be met, and it is necessary to consider a collection containing three classes. If class $\{1,2,3,4\}$ is included in the collection, then closure can be satisfied only if classes $\{2,5\}$, $\{3,6\}$, and $\{5,6\}$ are also included. This shows that there is no closed collection which includes class $\{1,2,3,4\}$ and has only three classes in it. On the other hand, $\{1,2,3,4\}$ is the only maximum compatibility class which includes state 1. From this it can be concluded that, if a closed collection of only three members exists, the collection must contain a subclass of $\{1,2,3,4\}$. Two such collections are listed in Table 6.3-7d. These were determined by a procedure which is essentially one of trial and error. The minimum-state flow table which can be formed from collection (i) is shown in Table 6.3-7e.

At the present time, no satisfactory general procedure is known for obtaining closed collections of compatibility classes containing a minimum number of members: all existing procedures require excessive enumeration.

However, there is one very important type of flow table for which it has been shown that only maximum compatibility classes need be considered in forming the minimum-state flow table. This restriction to maximum compatibility classes allows a simple procedure to be formulated.

Type A Flow Tables

The most common source of unspecified entries in a flow table is the restriction on multiple-input changes for fundamental-mode operation. For a large class of such flow tables it is possible to formulate a simple procedure for obtaining a minimum-state flow table.

Definition. A flow table is of type A if and only if:
1. It is a flow table for fundamental-mode operation.
2. It is a primitive flow table.
3. Each unstable next-state entry refers to an internal state which is stable for the corresponding input state.
4. The only unspecified entries are those which occur because of a restriction on the input states which can directly follow each possible input state.

It has been shown [9] that for flow tables which satisfy these conditions any collection of maximum compatibility classes which contains each state in at least one maximum compatibility class will automatically be closed. It follows from this that it is always possible to obtain a minimum-state table by considering only maximum compatibility classes.

6.4 FORMATION OF TRANSITION AND EXCITATION TABLES

Once a minimum-state flow table has been obtained, the next step in designing a sequential circuit is the formation of a transition table. This requires that a unique combination of values of the internal variables be assigned to each state of the flow table. For a flow table having r internal states, at least $S_0 = \langle \log_2 r \rangle$ internal variables are necessary, since the number of different combinations of values which can be assigned to m variables is 2^m. It is always possible to realize a pulse-mode circuit which uses only S_0 internal variables; for a fundamental-mode circuit it may be necessary to use more than S_0 internal variables in order to avoid critical races. The formation of transition and excitation tables for pulse-mode circuits will be considered first, and then a discussion of the problems peculiar to fundamental-mode operation will be presented.

The flow table of Example 6.1-1 (Table 6.1-1) is repeated in Table 6.4-1a, and the corresponding transition table which results from arbitrarily assigning $y_1 y_2$ values 00,01,11,10 to states A,B,C,D, respectively,

Sequential-circuit Synthesis

Table 6.4-1. Transition Table for Table 6.1-1

(a) Flow table

s	00	01	11	10	z
A	A	C	–	B	0
B	B	C	–	B	1
C	C	D	–	A	0
D	D	D	–	A	1

x_1x_2

S

(b) Transition table

y_1y_2	00	01	11	10	z
A– 00	00	11	dd	01	0
B– 01	01	11	dd	01	1
C– 11	11	10	dd	00	0
D– 10	10	10	dd	00	1

x_1x_2

Y_1Y_2

(c) Maps for Y_1 and Y_2

y_1y_2	00	01	11	10
00	0	1	d	0
01	0	1	d	0
11	1	1	d	0
10	1	1	d	0

x_1x_2

Y_1

$$Y_1 = x_1'y_1 + x_2$$

y_1y_2	00	01	11	10
00	0	1	d	1
01	1	1	d	1
11	1	0	d	0
10	0	0	d	0

x_1x_2

Y_2

$$Y_2 = x_1'x_2'y_2 + x_2y_1' + x_1y_1'$$

is shown in Table 6.4-1b. Two internal variables are necessary since S_0 equals two for a four-state table. Note that the dash entries of the flow table become don't cares in the transition table since any values of internal variables are permitted in total states for which the next-state is unspecified. The formation of the transition table is straightforward since it involves a direct replacement of the same binary number for each appearance of a state symbol of the flow table.

If amplifiers or relays are to be used for the circuit realization, the excitation table is identical with the transition table. The functional expressions for Y_1, Y_2, etc., can be obtained directly from the transition table by the appropriate minimization procedure. Table 6.4-1c shows the maps used in forming the minimal sums for Y_1 and Y_2 of Table 6.4-1b. If a circuit is to be constructed using flip-flops, the excitation table differs from the transition table and a procedure for the formation of the excitation table is necessary.

Formation of Flip-flop Excitation Tables

For a flip-flop, the required signals at its input terminals are determined directly by the corresponding values of y and Y. Table 6.4-2 shows the

6.4 formation of transition and excitation tables

*Table 6.4-2. Flip-flop Application Table
for a Set-Reset Flip-flop*

y	Y	S	R
0	0	0	d
0	1	1	0
1	0	0	1
1	1	d	0

dependence of S and R upon y and Y for a set-reset flip-flop. Whenever the flip-flop is to change state (y and Y are different), a signal must be applied to the appropriate terminal to cause the change. When the flip-flop is to remain in the same state, there is a don't-care entry in the table to allow the possibility of applying a signal to the terminal which causes the flip-flop to enter the state which it is already in. Thus, if the flip-flop is in the 0, or reset, state and is to remain in this state, it is permissible either to apply no input signal or to apply an input signal to the reset R input. A table such as Table 6.4-2 which shows the relations among y, Y, and the flip-flop inputs is called a *flip-flop application table*. The excitation table for set-reset flip-flops that is obtained from the transition table of Table 6.4-1b is shown in Table 6.4-3. This table is formed by making use of the rules embodied in Table 6.4-2.

For set-reset flip-flops it is possible to obtain a more compact excitation table by making use of the special encoding which is shown in Table 6.4-4. This possibility arises because only five of the nine combinations of values of S and R can ever occur and because S and R are never required to both equal 1 in the same state. The minimal sums for S and R can be determined directly from maps of the coded values. The minimal sum for S is formed by including all fundamental products encoded with an S and as many fundamental products encoded with s or d as are helpful. In the minimal sum for R it is necessary to include all fundamental products encoded with an R and as many fundamental products encoded with r or d as are helpful. The compact form of the excitation table and the formation of the minimal sums are illustrated in Table 6.4-5.

Internal Variable Assignments–Symmetries

For pulse-mode operation, any internal variable assignment for which a unique combination of values of the internal variables is assigned to each internal state will lead to a legitimate sequential circuit. However, the choice of a particular assignment can have a considerable effect on the economy of the final circuit. It would therefore be desirable to have a method of choosing that assignment which would result in the most

Sequential-circuit Synthesis

Table 6.4-3. *Use of Set-Reset Flip-flops for Table 6.4-1b*

(a) Excitation table

y_1y_2	00	01	11	10	z
00	0d 0d	10,10	dd,dd	0d,10	0
01	0d d0	10,d0	dd,dd	0d,d0	1
11	d0,d0	d0,01	dd,dd	01,01	0
10	d0,0d	d0,0d	dd,dd	01,0d	1

S_1R_1, S_2R_2

(b) Formation of minimal sums

x_1x_2

y_1y_2	00	01	11	10
00	0	1	d	0
01	0	1	d	0
11	d	d	d	0
10	d	d	d	0

$S_1 = x_2$

x_1x_2

y_1y_2	00	01	11	10
00	d	0	d	d
01	d	0	d	d
11	0	0	d	1
10	0	0	d	1

$R_1 = x_1$

x_1x_2

y_1y_2	00	01	11	10
00	0	1	d	1
01	d	d	d	d
11	d	0	d	0
10	0	0	d	0

$S_2 = y_1'x_2 + y_1'x_1$

x_1x_2

y_1y_2	00	01	11	10
00	d	0	d	0
01	0	0	d	0
11	0	1	d	1
10	d	d	d	d

$R_2 = y_1x_1 + y_1x_2$

$z = y_1y_2' + y_1'y_2$

economical circuit. Of course this would depend on the criteria of economy which are used, and different assignments could result for different criteria. One possibility would be to form the excitation table for each possible assignment and then to choose the most economical of the corresponding circuits. Before advocating such a procedure it would be wise to determine the number of such assignments which would have to be considered [11].

It is not true that the most economical sequential circuit for a given flow table will always contain only S_0 internal variables. It may be possible to decrease the total number of elements in the circuit by using more

6.4 formation of transition and excitation tables

Table 6.4-4. Encoding of Excitation-table Entries for Set-Reset Flip-flops

y	Y	S	R	Coded entry
0	0	0	d	r
1	0	0	1	R
0	1	1	0	S
1	1	d	0	s
–	d	d	d	d

Table 6.4-5. Compact Form of Table 6.4-3

(a) Excitation table

y_1y_2 \ x_1x_2	00	01	11	10
00	r,r	S,S	d,d	r,S
01	r,s	S,s	d,d	r,s
11	s,s	s,R	d,d	R,R
10	s,r	s,r	d,d	R,r

S_1R_1, S_2R_2

(b) Minimal sums

y_1y_2 \ x_1x_2	00	01	11	10
00	r	S	d	r
01	r	S	d	r
11	s	s	d	R
10	s	s	d	R

$S_1 = x_2,\ R_1 = x_1$

y_1y_2 \ x_1x_2	00	01	11	10
00	r	S	d	S
01	s	s	d	s
11	s	R	d	R
10	r	r	d	r

$S_2 = y_1'x_2 + y_1'x_1$
$R_2 = y_1x_2 + y_1x_1$

than the minimum number of flip-flops or feedback loops. In counting the number of assignments, only those involving S_0 internal variables will be considered. This results in a conservative estimate of the amount of work involved in enumeration but simplifies the discussion. It will not affect the final conclusions. For a flow table with r rows, the number of

Sequential-circuit Synthesis

Table 6.4-6. Permutation and Complementation of Internal Variables

(a) Assignments

Internal states	Assignment I		Assignment II		Assignment III	
	y_1	y_2	y_1	y_2	y_1	y_2
A	0	0	0	0	0	1
B	0	1	1	0	0	0
C	1	1	1	1	1	0
D	1	0	0	1	1	1

(b) Excitation functions

Assignment I
$Y_1 = f_1(y_1, y_2, x_1, x_2)$
$Y_2 = f_2(y_1, y_2, x_1, x_2)$
$Z = g(y_1, y_2, x_1, x_2)$
$S_1 = h_1(y_1, y_2, x_1, x_2)$
$R_1 = h_2(y_1, y_2, x_1, x_2)$
$S_2 = k_1(y_1, y_2, x_1, x_2)$
$R_2 = k_2(y_1, y_2, x_1, x_2)$

Assignment III
$Y_1 = f_1(y_1, y_2', x_1, x_2)$
$Y_2 = [f_2(y_1, y_2', x_1, x_2)]'$
$Z = g(y_1, y_2', x_1, x_2)$
$S_1 = h_1(y_1, y_2', x_1, x_2)$
$R_1 = h_2(y_1, y_2', x_1, x_2)$
$S_2 = k_2(y_1, y_2', x_1, x_2)$
$R_2 = k_1(y_1, y_2', x_1, x_2)$

different assignments of S_0 variables is

$$\frac{2^{S_0}!}{(2^{S_0} - r)!}$$

This is a rapidly growing function, equaling 24 for $r = 3$ and 6,720 for $r = 5$. However, it is not necessary to consider each of these assignments individually in order to determine the most economical circuit.

If a circuit were designed for a flow table using assignment I of Table 6.4-6a, then a circuit corresponding to assignment II could be obtained merely by relabeling the appropriate leads in the circuit for assignment I. Since assignment II involves a permutation of the variables of assignment I, there can be no gain in economy by using assignment II rather than assignment I. Thus, assignments which are permutations of the variables of other assignments need not be considered explicitly, and only one representative from each permutation class need be studied.

Assignment III of Table 6.4-6a is obtained from assignment I by complementing y_2. It is possible to obtain the excitation functions for assignment III directly from those for assignment I, as shown in Table 6.4-6b. In a flip-flop circuit the change from assignment I to assignment

Table 6.4-7. Number of Distinct Assignments of S_0 Variables to r States

r	S_0	Number of distinct assignments
1	0	1
2	1	1
3	2	3
4	2	3
5	3	140
6	3	420
7	3	840
8	3	840
9	4	10,810,800

III might involve some rewiring (depending on the type of flip-flop) but would not involve the addition of any components. In a circuit using feedback loops there is a possibility of a change in the number of required inverters. In any case, it is not necessary to construct new excitation tables for assignments which differ from an already studied assignment only in some complemented variables. Two assignments are said to be distinct if it is not possible to obtain one assignment from the other by complementing and permuting variables; the number of distinct assignments of S_0 variables to r states is

$$\frac{(2^{S_0} - 1)!}{(2^{S_0} - r)! S_0!}$$

The derivation of this formula is given in [11] and will not be discussed here. Table 6.4-7 lists the values given by this formula for values of r from 1 to 9. It is clear from these values that enumeration by hand is feasible for values of r up to 4 and that for values of r greater than 9 even use of a high-speed digital computer would be highly questionable. Three distinct assignments for four states are shown in Table 6.4-8. Research into techniques for obtaining assignments for economical circuits without resorting to enumeration has been reported in the literature [12,13,14,15]. The details of this work will not be presented here since they are quite specialized.

Sequential-circuit Synthesis

Table 6.4-8. Three Distinct Assignments of Two Variables to Four States

Assignment I	Assignment II	Assignment III
y_1 y_2	y_1 y_2	y_1 y_2
0 0	0 0	0 0
0 1	0 1	1 1
1 1	1 0	0 1
1 0	1 1	1 0

Critical Races [1]

A fundamental-mode flow table is shown in Table 6.4-9a, and a transition table for this flow table is shown in Table 6.4-9b. The three entries of Table 6.4-9b which are marked with an asterisk correspond to races, for both the internal variables are required to change. The race in the $x_1x_2 = 11$ column is noncritical because the stable 11 state will eventually be reached independently of the order in which the internal variables change. Both the races in the $x_1x_2 = 10$ column are critical. Thus it must be concluded that this particular assignment of internal variables does not lead to a circuit free of critical races. Permuting or complementing internal variables has no effect on the situation with respect to races. It is therefore reasonable to search for an assignment corresponding to a transition table without critical races by examining the other two distinct assignments of two variables. It is easily verified that each of the two resulting transition tables will also contain critical races. This investigation shows that it is not possible to design a circuit for the given flow table which contains only two internal variables and is free of critical races.†

A state table which corresponds to the flow table of Table 6.4-9a is shown in Table 6.4-9c. For this state table it is possible to form a transition table which does not involve any critical races. Such a table is shown in Table 6.4-9d. The introduction of two additional states into the state table and the specification of multiple transitions permit this elimination of critical races. Of course, one additional internal variable is required.

Row A of the flow table of Table 6.4-9a shows that it must be possible to move from state A to state B by changing one internal variable and also that it must be possible to go from state A to state C by changing only

---

† This statement is true for the design techniques being considered here. If more general techniques involving the insertion of controlled delays into the combinational circuitry are used, the discussion given here is no longer applicable.

6.4 formation of transition and excitation tables

Table 6.4-9. Elimination of Critical Races

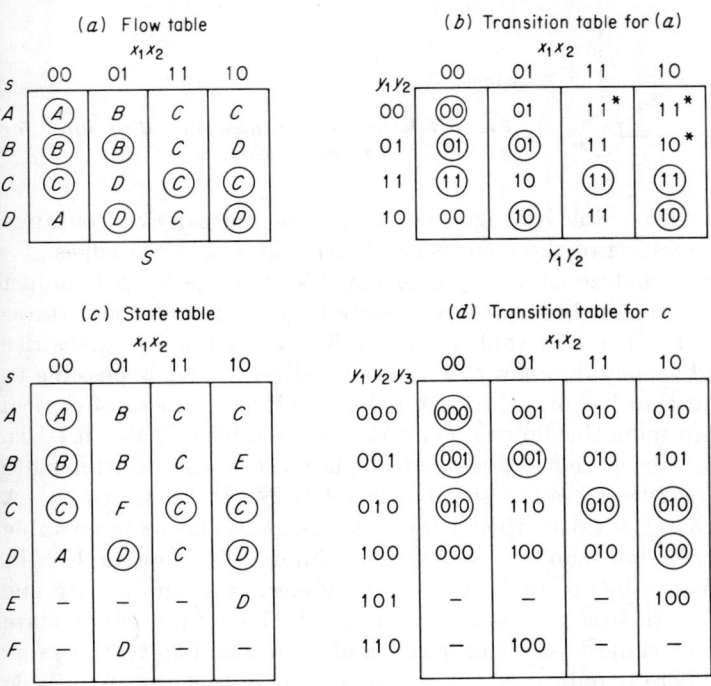

one internal variable. Thus state A must differ from state B in only one internal variable and must also differ from state C in some other single variable. Row D of the table shows that states A and D must also differ in only one variable. Clearly, with only two internal variables it is not possible for state A to differ from each of the remaining states in a different single internal variable. These relations are illustrated in Fig. 6.4-1. In this figure each internal state is represented by a node, and two nodes are joined by an edge only if the corresponding states must differ in a single internal variable. A diagram like this will be called a *state adjacency diagram*. Such a diagram is similar to the n-cubes discussed in Chap. 2 in that each edge represents a change in a single variable (two nodes connected directly by an edge must differ in only one variable). It is possible to obtain a transition table which corresponds directly to a given flow table (not to some equivalent state table) and is free of critical races if and only if it is possible to label the nodes of an n-cube with the states of the flow table so that every pair of states which are connected by an edge on the state adjacency diagram are also connected by an edge on the n-cube.

Sequential-circuit Synthesis

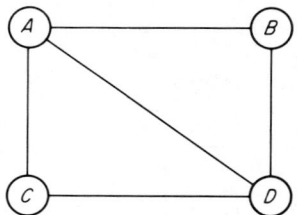

Fig. 6.4-1. State adjacency diagram for Table 6.4-9a.

Clearly this is not possible for Fig. 6.4-1 since no n-cube can ever contain a closed path consisting of three edges or of any odd number of edges.

The failure of a state adjacency diagram to satisfy the conditions just given does not necessarily mean that a state table with additional states must be formed. The flow table shown in Table 6.4-10a also gives rise to the state adjacency diagram of Fig. 6.4-1. However, it is possible to obtain a circuit that has only two internal variables and is free of critical races by transforming this flow table into the state table of Table 6.4-10b. This can be done because it is possible to replace the transition from state A to state D by successive transitions from A to B and thence to D. A possibility such as this exists whenever the same state occurs as an unstable next-state entry more than once in a single column of the flow table. In such cases it is possible to replace the state adjacency diagram with one or more weak state adjacency diagrams, each of which represents a state table having the same number of states and corresponding to the same flow table. There is only one weak state adjacency diagram for Table 6.4-10a, and it is shown in Fig. 6.4-2.

Table 6.4-10. Weak Adjacency

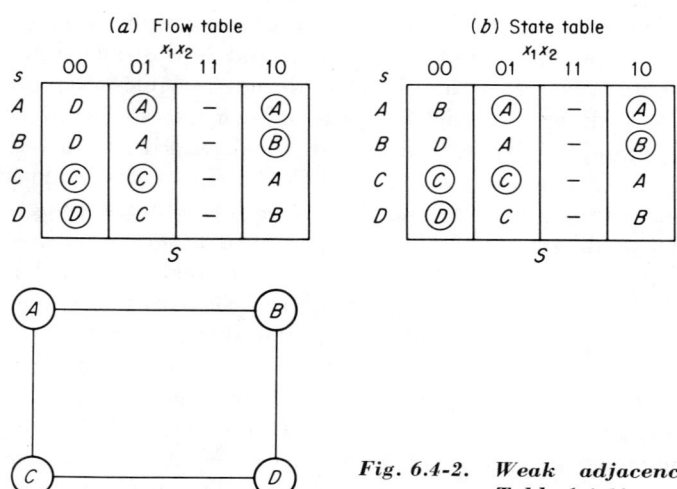

Fig. 6.4-2. Weak adjacency diagram for Table 6.4-10.

6.4 formation of transition and excitation tables

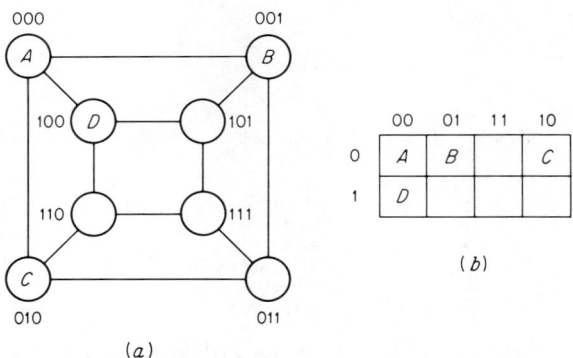

Fig. 6.4-3. Labeling of the 3-cube for the diagram of Fig. 6.4-1. (a) 3-cube; (b) 3-cube map.

It is possible to give general conditions for assignments leading to circuits free of critical races in terms of the weak adjacency diagrams and n-cubes. Any assignment which corresponds to a circuit having no critical races can be represented by a labeling of an n-cube such that corresponding to each edge of the weak adjacency diagram there is a path between the appropriate nodes of the n-cube which does not pass through any other nodes of the n-cube with different labels.† For Table 6.4-9a, the weak state adjacency diagram is the same as the state adjacency diagram. A 3-cube labeling that corresponds to Table 6.4-9c and satisfies Fig. 6.4-1 is shown in Fig. 6.4-3.

It is possible to show that any four-row flow table can be realized with a circuit free of critical races with at most three internal variables. This is done by considering a "worst case" in which all pairs of states are required to be adjacent, as in the state adjacency diagram of Fig. 6.4-4. There are several labelings of the 3-cube which satisfy the requirements of this diagram and therefore of all other diagrams involving four states. One such labeling of particular interest is shown in Fig. 6.4-5. This scheme is peculiar in that there are two nodes for each state. It is always possible

† This is a necessary but not a sufficient condition, for it is possible to have the paths "interfere" so that some of the intermediate unlabeled nodes would have to satisfy conflicting requirements.

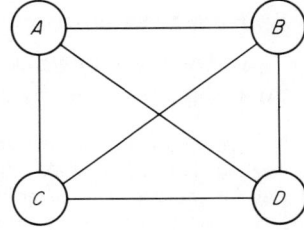

Fig. 6.4-4. State adjacency diagram for "worst-case" situation involving four states.

Sequential-circuit Synthesis

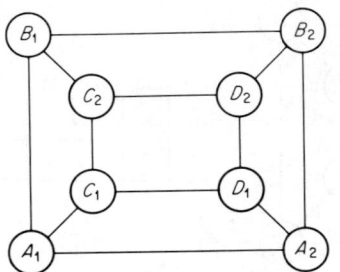

Fig. 6.4-5. *Labeling of a 3-cube to satisfy "worst-case" requirements for a 4-state table.*

Table 6.4-11. *A State Table Corresponding to Table 6.4-9a and Using the Labeling of Fig. 6.4-5*

(a) Flow table

s	$x_1 x_2$ 00	01	11	10
A	(A)	B	C	C
B	(B)	(B)	C	D
C	(C)	D	(C)	(C)
D	A	(D)	C	(D)

S

(b) State table

s	$x_1 x_2$ 00	01	11	10
A_1	(A_1)*	B_1	C_1	C_1
A_2	(A_2)	B_2	A_1	A_1
B_1	(B_1)	(B_1)	C_2	B_2
B_2	(B_2)	(B_2)	B_1	D_2
C_1	(C_1)	D_1	(C_1)	(C_1)
C_2	(C_2)	D_2	(C_2)	(C_2)
D_1	A_2	(D_1)	C_1	(D_1)
D_2	D_1	(D_2)	C_2	(D_2)

S

to go from any node to a node labeled with any arbitrary other state either directly or by passing through a node with the same label as the first node. Thus nodes B_1 and C_1 can be reached directly from node A_1, and node D_1 can be reached from node A_1 by passing through node A_2. Table 6.4-11 shows how this labeling of the 3-cube would be used to form a state table corresponding to the flow table of Table 6.4-9a.†

Standard Assignments for General Flow Tables

In the preceding discussion it has been shown that any four-state flow table can be realized by a circuit requiring at most three internal variables.

† In Table 6.4-11b, the stable A_1 entry in the total state for $x_1 x_2 = 00$, $s = A_1$ is marked with an asterisk. This entry could be replaced by a dash since there is no way for the circuit to enter this total state: there are no other stable states in the row and there is no other A_1 entry in the column.

6.4 formation of transition and excitation tables

A question still remains as to how many internal variables are required to realize any flow table having r states. Indeed, it is not obvious that an arbitrary flow table can always be realized by means of any circuit which is free of critical races. It has been shown [1,16] that for any flow table with r states, and thus requiring a minimum of S_0 internal variables, it is always possible to obtain a circuit which has at most $2S_0 - 1$ internal variables and does not contain any critical races. Moreover, for any value of S_0, there will be some flow tables which require exactly $2S_0 - 1$ variables. The details of this demonstration are somewhat involved and specialized. They will not be presented here. Instead a technique whereby $2S_0 + 1$ variables can be used for any flow table will be discussed.

If $2S_0 + 1$ variables are used as internal variables for a flow table having 2^{S_0} states, then 2^{2S_0+1} combinations of values of the internal variables must be assigned to the 2^{S_0} states. This is done by assigning $2^{2S_0+1}/2^{S_0} = 2^{S_0+1}$ combinations to each state. In other words a state table is formed in which each of the states of the flow table is replaced by 2^{S_0+1} states. The manner in which the 2^{S_0+1} states are assigned is illustrated in Table 6.4-12 for $S_0 = 2$ and is carried out as follows:

1. An arbitrary correspondence between the 2^{S_0} states of the flow table and S_0 internal variables, $y_1, y_2, \ldots, y_{S_0}$, is chosen. In Table 6.4-12 this correspondence is shown in the y_1 and y_2 columns.

2. The $2S_0 + 1$ variables to be used in forming the transition table are designated $y_{\alpha_1}, y_{\alpha_2}, \ldots, y_{\alpha_{S_0}}, y_{\beta_1}, y_{\beta_2}, \ldots, y_{\beta_{S_0}}$, and y_0. For all combinations of values of these variables for which $y_0 = 0$, the corresponding state is determined only by the values of the y_{α_i} variables—the values of the y_{β_j} variables have no effect on the assignment. Specifically, when $y_0 = 0$, the combination of values is assigned to that state for which the $y_1, y_2, \ldots, y_{S_0}$ values agree with the $y_{\alpha_1}, y_{\alpha_2}, \ldots, y_{\alpha_{S_0}}$ values of the combination. For example, in Table 6.4-12 the combination

$$y_{\alpha_1} y_{\alpha_2} y_0 y_{\beta_1} y_{\beta_2} = 00011$$

is assigned to state A because $y_{\alpha_1} y_{\alpha_2} = 00$ and $y_1 y_2 = 00$ for this state and combination.

When $y_0 = 1$, the state is determined by the values of the y_{β_j} variables in the same fashion. Thus the combination $y_{\alpha_1} y_{\alpha_2} y_0 y_{\beta_1} y_{\beta_2} = 00111$ of Table 6.4-12 is assigned to state C since $y_1 y_2 = 11$ for this state. Table 6.4-13 lists those combinations of values which would be assigned to a single state for $S_0 = 3$. This technique can be directly extended for arbitrary values of S_0.

It is now necessary to show that an assignment such as just described can be used for a flow table in which transitions between all pairs of states are required. The labeling of a 5-cube map corresponding to the assignment of Table 6.4-12 is shown in Fig. 6.4-6. This figure shows that every

Table 6.4-12. Assignment of $2S_0 + 1$ Variables to 2^{S_0} States for $S_0 = 2$

States	y_1y_2	$y_{\alpha_1}y_{\alpha_2}$	y_0	$y_{\beta_1}y_{\beta_2}$
A	0 0	0 0	0	0 0
		0 0	0	0 1
		0 0	0	1 1
		0 0	0	1 0
		0 0	1	0 0
		0 1	1	0 0
		1 1	1	0 0
		1 0	1	0 0
B	0 1	0 1	0	0 0
		0 1	0	0 1
		0 1	0	1 1
		0 1	0	1 0
		0 0	1	0 1
		0 1	1	0 1
		1 1	1	0 1
		1 0	1	0 1
C	1 1	1 1	0	0 0
		1 1	0	0 1
		1 1	0	1 1
		1 1	0	1 0
		0 0	1	1 1
		0 1	1	1 1
		1 1	1	1 1
		1 0	1	1 1
D	1 0	1 0	0	0 0
		1 0	0	0 1
		1 0	0	1 1
		1 0	0	1 0
		0 0	1	1 0
		0 1	1	1 0
		1 1	1	1 0
		1 0	1	1 0

6.4 formation of transition and excitation tables

Table 6.4-13. *Assignment of Variables to a Single State for* $S_0 = 3$

State	$y_1y_2y_3$	$y_{\alpha_1}y_{\alpha_2}y_{\alpha_3}$	y_0	$y_{\beta_1}y_{\beta_2}y_{\beta_3}$
		0 1 1	0	0 0 0
		0 1 1	0	0 0 1
		0 1 1	0	0 1 0
		0 1 1	0	0 1 1
		0 1 1	0	1 0 0
		0 1 1	0	1 0 1
		0 1 1	0	1 1 0
A	0 1 1	0 1 1	0	1 1 1
		0 0 0	1	0 1 1
		0 0 1	1	0 1 1
		0 1 0	1	0 1 1
		0 1 1	1	0 1 1
		1 0 0	1	0 1 1
		1 0 1	1	0 1 1
		1 1 0	1	0 1 1
		1 1 1	1	0 1 1

A state is adjacent to some B state and to some D state but that only the A states marked with an asterisk are adjacent to a C state. Thus, in order to make the transition from a specific A state to a C state, it may first be necessary to enter one of the A states adjacent to a C state before making the transition to a C state. If $y_0 = 0$ for the original A state, this is done by changing the $y_{\beta_1}y_{\beta_2}$ variables until they become equal to the values of the y_1y_2 variables for the state C. Then the value of y_0 is changed

$y_0 = 0$

$y_{\beta_1}y_{\beta_2}$ \ $y_{\alpha_1}y_{\alpha_2}$	00	01	11	10
00	A	B	C	D
01	A	B	C	D
11	A*	B	C	D
10	A	B	C	D

$y_0 = 1$

$y_{\beta_1}y_{\beta_2}$ \ $y_{\alpha_1}y_{\alpha_2}$	00	01	11	10
00	A	A	A*	A
01	B	B	B	B
11	C	C	C	C
10	D	D	D	D

Fig. 6.4-6. *The 5-cube labeling corresponding to the assignment of Table 6.4-12.*

Table 6.4-14. A Fragment of a Flow Table Illustrating Use of a $2S_0 + 1$ Assignment for $S_0 = 2$

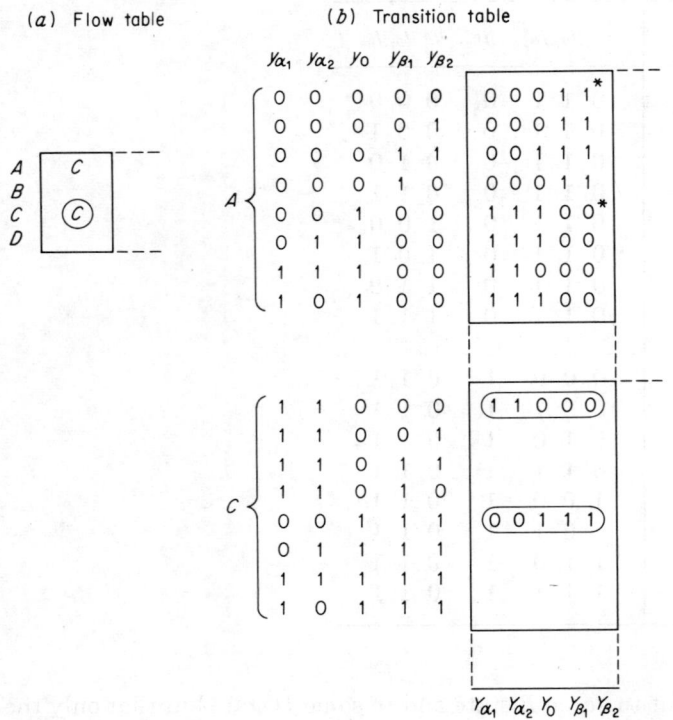

(a) Flow table   (b) Transition table

so that the state is determined by the y_{β_1}, y_{β_2} variables. An analogous technique is used when $y_0 = 1$ for the original state. This is illustrated in the fragment of a transition table shown in Table 6.4-14. The corresponding transition diagram is shown in Fig. 6.4-7.

Two of the entries in Table 6.4-14b are marked with an asterisk because they involve races. These races are not critical, for the relative order in which the two unstable variables change will not affect the final stable state reached. The time required for any transition in a circuit using this type of assignment cannot be greater than twice the longest time required to change any internal variable. This is because any transition involves passing through at most two unstable entries in the state table. A noncritical race involves only one reaction time since both variables are changing at the same time. It is possible by means of a different technique to realize any flow table by a circuit which requires only one reaction time for any transition [1,16]. However, this technique requires $2^{S_0} - 1$ internal variables.

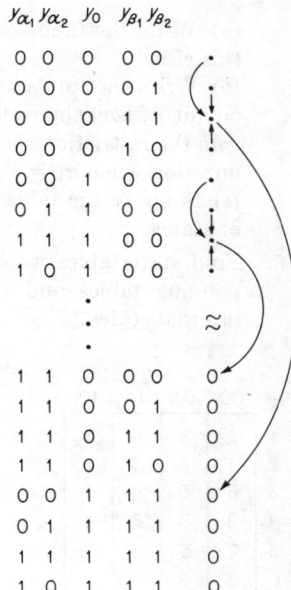

Fig. 6.4-7. Transition diagram for Table 6.4-14.

This presentation shows that any flow table can be realized by using at most $2S_0 + 1$ variables. The feasibility of using only $2S_0 - 1$ variables is demonstrated in the references cited earlier. The problem of determining the minimum number of internal variables required to realize a given flow table by means of a circuit free of critical races is still unsolved.

Problems

1. A circuit is to be designed having two pulse inputs x_1 and x_2 and one d-c output z. Whenever an x_1 pulse is received, the output is to become equal to 1, provided that there have been exactly two x_2 pulses after the last previous x_1 pulse. Otherwise the output is to remain equal to 0. Once the output becomes equal to 1, it is to remain equal to 1 until the next x_2 pulse. Whenever an x_2 pulse is received, the output is to become equal to 0.

   Write a (pulse-mode) flow table for this circuit.

2. A circuit is to be designed having two pulse inputs x_1 and x_2 and two d-c outputs Z_1 and Z_2. The inputs are restricted so that x_1 and x_2 are never simultaneously equal to 1. When either x_1 or x_2 is equal to 1, the corresponding output Z_1 or Z_2 is to be equal to 1. When x_1 and x_2 are both equal to 0, Z_1 is to be equal to 1 if x_1 was the last input equal to 1 and Z_2 is to be equal to 1 if x_2 was the last input equal to 1. Z_1 and Z_2 are never both equal to 1.

Sequential-circuit Synthesis

(a) Write the fundamental-mode primitive flow table and output table for this circuit.

(b) Write the fundamental-mode flow table and output table for the same circuit *without* the restriction that x_1 and x_2 are never both equal to 1, but *with* the restriction that the x_1 pulses and x_2 pulses both have the same fixed duration when $x_1 = x_2 = 1$, $Z_1 = Z_2 = 1$.

(c) Simplify the tables of (a) and (b) to tables having a minimum number of states.

3. Find state tables which specify the same external behavior as the accompanying tables and which also have the minimum possible number of internal states.

Table I

S	x_1x_2 00	01	11	10		S	x_1x_2 00	01	11	10
1	2	5	4	1		1	0	1	1	0
2	1	8	3	5		2	1	0	1	1
3	6	5	4	1		3	0	1	1	0
4	2	5	3	4		4	0	1	1	0
5	2	5	3	7		5	1	1	0	0
6	3	8	1	5		6	1	0	1	1
7	1	6	4	5		7	1	0	1	1
8	4	2	1	4		8	1	0	1	1
	S						Z			

Table II

S	x_1x_2 00	01	11	10	z_1z_2
1	1	–	7	4	00
2	2	3	8	–	00
3	6	–	2	3	01
4	6	6	1	–	01
5	–	A	–	4	00
6	2	6	8	4	01
7	8	4	B	–	10
8	7	3	5	2	10
	S				

(a) Let A be—
  B be—
(b) Let A be 6
  B be—
(c) Let A be 6
  B be 1

4. A circuit is to be designed in which two push buttons A and B control the lighting of two lamps G and R. Whenever both push buttons are released, neither lamp is to be lit. Starting with both buttons released, the operation

of either button causes lamp G to light. Operation of the other button, with the first button still held down, causes lamp R to light. Henceforth, as long as either button remains operated, the button which first caused lamp R to light controls lamp R—causing it to extinguish when the button is released and to light when the button is operated. The other button controls lamp G in the same fashion. It is not possible to operate or release both buttons simultaneously.

(a) Form the primitive state table for the circuit just described.
(b) Reduce the number of states if possible.
(c) Assign secondary variables so that no critical races occur.

5. The accompanying flow table specifies the behavior of a circuit in which the duration of the input pulses is controlled so as not to exceed a fixed time interval τ. Draw a *primitive* flow table for a circuit which has the same behavior when the restriction on the length of the pulses is removed. Assume that no double changes of input occur and that x_1 and x_2 are never both equal to 1.

S	x_1x_2 00	01	11	10
1	1	2	–	1
2	2	3	–	1
3	3	4	–	2
4	4	4	–	3

S

S	x_1x_2 00	01	11	10
1	10	10	10	10
2	00	00	00	00
3	00	00	00	00
4	01	01	01	01

z_1z_2

6. For the accompanying flow table:

S	cx_1x_2 000	001	011	010	100	101	111	110
1	1	1	1	1	2	3	4	3
2	2	2	2	2	3	4	1	4
3	3	3	3	3	4	1	2	1
4	4	4	4	4	1	2	3	2

S

(a) Write the transition table when the following assignment of internal variables is used:

s_j	y_1	y_2
1	0	0
2	0	1
3	1	1
4	1	0

(b) Write the excitation table when (set-reset) flip-flops are used for memory devices, and derive the expressions for S_1, R_1, S_2, R_2.

(c) Write the excitation table when flip-flops are used which go to the set state ($y = 1$) when the set lead J is pulsed, which go to the reset state

Sequential-circuit Synthesis

($y = 0$) when the reset lead K is pulsed, and which change state when both set and reset leads are pulsed.

7. Simplify the accompanying flow table if possible.

S	$x_1 x_2$ 00	01	11	10		S	$x_1 x_2$ 00	01	11	10
1	①	4	4	2		1	0	0	0	1
2	1	4	3	②		2	0	0	0	1
3	1	4	③	4		3	0	0	1	1
4	1	④	④	④		4	0	0	0	0
	S						Z			

8. An electronic sequential circuit is to be designed using flip-flops and diode gates. The two circuit inputs X and C are pulses which never occur simultaneously. The C pulse occurs periodically, as shown in Fig. P6-8. The X pulse can appear (if it does appear) only singly and midway between two successive C pulses. The single *level* output Z is high in the interval between two successive C pulses if and only if the preceding interval contained an X pulse.
(a) Derive a minimum-row flow table and output table for the circuit specified above.
(b) Derive an excitation table.
(c) Derive an economical circuit using flip-flops and diode gates.

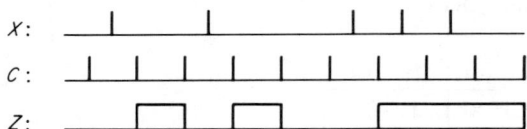

9. A sequential relay circuit is controlled by two keys K_1 and K_2 and has a single output Z. Either key, when depressed, remains depressed for a fixed interval of time, d. (You may assume that no double changes of input state occur.) The output Z changes state only when either K_1 or K_2 is depressed (no output change occurs when a key is released). The closing of K_1 assures that Z changes to (or remains at) the 0 state. The closing of K_2 assures that Z changes to (or remains at) the 1 state.
(a) Draw a primitive flow table for this circuit.
(b) Draw a minimum-row flow table—two rows are sufficient.
(c) Design an economical relay circuit. A total of two coils and nine springs is sufficient. (No isolation is necessary between inputs and output or between the secondary-relay control network and the output.)

10. For the accompanying primitive flow table, draw a diagram showing which rows can be merged. Draw all possible merged flow tables which require no more than two secondary relays.

S	$x_1 x_2$				Z
	00	01	11	10	
1	①	2	3	–	0
2	1	②	–	5	0
3	1	–	③	4	0
4	–	2	6	④	1
5	–	2	6	⑤	0
6	1	–	⑥	5	0

11. A sequential circuit is designed using the assignment of internal variables shown in Table I, and the resulting equations are

$$Y_1 = f_1(y_1, y_2, y_3, x_1, x_2, \ldots, x_r)$$
$$Y_2 = f_2(y_1, y_2, y_3, x_1, x_2, \ldots, x_r)$$
$$Y_3 = f_3(y_1, y_2, y_3, x_1, x_2, \ldots, x_r)$$
$$Z_1 = g_1(y_1, y_2, y_3, x_1, x_2, \ldots, x_r)$$
$$Z_2 = g_2(y_1, y_2, y_3, x_1, x_2, \ldots, x_r)$$

If the assignment of Table II is used for the same flow table, write expressions for Y_1, Y_2, Y_3, Z_1, and Z_2 in terms of f_1, f_2, f_3, g_1, and g_2.

Table I

S	y_1 y_2 y_3
1	0 0 0
2	0 0 1
3	0 1 0
4	0 1 1
5	1 0 0
6	1 0 1
7	1 1 0
8	1 1 1

Table II

S	y_1 y_2 y_3
1	1 0 0
2	1 0 1
3	1 1 0
4	1 1 1
5	0 0 0
6	0 0 1
7	0 1 0
8	0 1 1

12. A sequential circuit (serial adder) having four inputs—c, v, w, x—is to be designed. The input c represents a clock pulse, and the inputs v, w, and x represent three binary numbers. There is to be a single output z, which represents the arithmetic sum of the three inputs.
   One bit of the sum is to occur as an output pulse on either the z or the z' leads whenever a pulse occurs on the input lead.
   (a) Draw a pulse-mode flow table for this circuit—include only those columns of the flow table which correspond to $c = 1$.
   (b) Draw a flow table for a circuit which has the same performance as (a) except that the output is a level which remains on the z lead until the next clock pulse occurs.

Sequential-circuit Synthesis

13. The accompanying flow tables describe fundamental-mode sequential circuits. You are to assign combinations of internal variables $(y_1 y_2 \cdots y_s)$ to the internal states, so as to avoid critical races, and form an excitation table for $Y_1 Y_2 \cdots Y_s$. Assign the all-0 combination to state 1. Do not reorder the rows of the table. Additional rows may be added if necessary. Use as few internal variables as possible.

(a)

s	$x_1 x_2$ 00	01	11	10
1	2	3	–	(1)
2	(2)	5	–	(2)
3	6	(3)	–	5
4	(4)	6	–	2
5	4	(5)	–	(5)
6	(6)	(6)	–	1

S

(b)

s	$x_1 x_2$ 00	01	11	10
1	3	(1)	(1)	4
2	4	(2)	(2)	2
3	(3)	2	1	(3)
4	(4)	1	2	(4)

S

14. Find the minimum-row flow table which has the same terminal behavior as the accompanying flow table:

s	$x_1 x_2$ 00	01	11	10
1	3	5	–	–
2	3	5	–	–
3	2	3	–	1
4	2	3	–	5
5	–	5	–	1

S

s	$x_1 x_2$ 00	01	11	10
1	0	1	–	–
2	0	–	–	–
3	–	0	–	–
4	0	–	–	–
5	–	0	–	–

Z

REFERENCES

1. Caldwell, S. H.: "Switching Circuits and Logical Design," John Wiley & Sons, Inc., New York, 1958.
2. Runyon, J. P.: Derivation of Completely and Partially Specified State Tables, in E. J. McCluskey, Jr. and T. C. Bartee (eds.), "A Survey of Switching Circuit Theory," chap. 8, pp. 121–144, McGraw-Hill Book Company, New York, 1962.
3. Rabin, M. O., and D. Scott: Finite Automata and Their Decision Problems, *IBM J. Research Development*, vol. 3, no. 2, pp. 114–125, April, 1959.
4. Kleene, S. C.: Representation of Events in Nerve Nets and Finite Automata, in C. E. Shannon and J. McCarthy (eds.), "Automata Studies," pp. 3–41, Princeton University Press, Princeton, N.J., 1956.

5. Moore, E. F.: Gedanken-experiments on Sequential Machines, in C. E. Shannon and J. McCarthy (eds.), "Automata Studies," pp. 129–153, Princeton University Press, Princeton, N.J., 1956.
6. Birkhoff, G., and S. Maclane: "A Survey of Modern Algebra," rev. ed., The Macmillan Company, New York, 1953.
7. Paull, M. C., and S. H. Unger: Minimizing the Number of States in Incompletely Specified Sequential Switching Functions, *IRE Trans. on Electronic Computers*, vol. EC-8, no. 3, pp. 356–367, September, 1959.
8. Unger, S. H.: Simplification of State Tables in E. J. McCluskey, Jr. and T. C. Bartee (eds.), "A Survey of Switching Circuit Theory," chap. 9, pp. 145–170, McGraw-Hill Book Company, New York, 1962.
9. McCluskey, E. J.: Minimum-state Sequential Circuits for a Restricted Class of Incompletely Specified Flow Tables, *Bell System Tech. J.*, vol. 40, no. 6, pp. 1759–1768, November, 1962.
10. Narasmhan, R.: Minimizing Incompletely Specified Sequential Switching Functions, *IRE Trans. on Electronic Computers*, vol. EC-10, no. 3, pp. 531–532, September, 1961.
11. McCluskey, E. J., Jr., and S. H. Unger: A Note on the Number of Internal Assignments for Sequential Switching Circuits, *IRE Trans. on Electronic Computers*, vol. EC-8, no. 4, pp. 439–440, December, 1959.
12. Armstrong, D. B.: On the Efficient Assignment of Internal Codes to Sequential Machines, *IRE Trans. on Electronic Computers*, vol. EC-11, no. 5, pp. 611–622, October, 1962.
13. Dolotta, T. A., and E. J. McCluskey, The Coding of Internal States of Sequential Circuits, *IEEE Trans. on Electronic Computers*, vol. EC-13, no. 5, pp. 549–562, October, 1964.
14. Stearns, R. E., and J. Hartmanis: On the State Assignment Problem for Sequential Machines, II, *IRE Trans. on Electronic Computers*, vol. EC-10, no. 4, pp. 593–603, December, 1961.
15. Karp, R. M.: Some Techniques of State Assignment for Synchronous Sequential Machines, *IEEE Trans. on Electronic Computers*, vol. EC-13, no. 5, pp. 507–518, October, 1964.
16. Huffman, D. A.: A Study of the Memory Requirements of Sequential Switching Circuits, *Research Lab. Electronics MIT Tech. Rept.* 293, April, 1955.

# TRANSIENT BEHAVIOR
OF SWITCHING CIRCUITS

In the preceding chapters certain idealizing assumptions regarding the performance of the elements of switching networks have been made, and the effects of delays in responding to signal changes or in propagating signal changes have been ignored. It is possible for the spurious delays in a network to affect the network performance so that the behavior is changed significantly. This can be true in both combinational and sequential networks, although the effects are most serious in sequential networks, either because of effects inherent in the sequential nature of the network or because of the results of these effects in the combinational portion of the sequential network. The purpose of this chapter is to investigate the effects of spurious delays in switching networks and to develop techniques for controlling these effects.

7.1 COMBINATIONAL NETWORKS

In Chap. 3, a switching algebra was developed for representing combinational circuits. The assumptions made in this development apply specifically to steady-state performance of combinational circuits and are not all valid when circuit inputs are being changed. During an input change it is possible for a circuit to have an output different from that predicted by its switching-algebra representation. A simple situation in which this can occur is illustrated in Fig. 7.1-1. For this circuit the switching algebra predicts that S_2 will equal 0 whenever both w and x equal 0, irrespective of the value of y. However, many types of set-reset flip-flops have the property that when the flip-flop is changing state there is a short period of time during which both the y and y' outputs are at a high voltage level. Thus it is possible for y and y' momentarily *both* to be equal to 1. If the delays through the

7.2 analysis

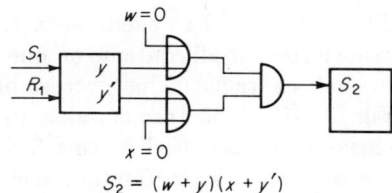

Fig. 7.1-1. Circuit with possible spurious output.

$S_2 = (w+y)(x+y')$

remaining gates are sufficiently uniform, it is possible for both inputs to the AND gate momentarily to be equal to 1. If this occurs, a short pulse will appear at the S_2 input. Whether this pulse has any effect on flip-flop 2 depends on the specific electrical characteristics of the flip-flop. The possibility does exist that the flip-flop will be switched to the set state by the pulse.

There are several techniques for correcting a situation such as that shown in Fig. 7.1-1. One technique is to design the flip-flops so that they will not respond to short input pulses. This usually results in slower-acting flip-flops and can slow down the response of the entire circuit. A circuit designed in this fashion is sensitive to changes in circuit parameters and is in this sense somewhat unreliable.

Another possibility is that of redesigning the combinational circuit so that it is not possible for spurious output pulses to occur. For the circuit of Fig. 7.1-1 this can be done by using the expression $(w+y)(x+y')(w+x)$ for S_2 rather than $(w+y)(x+y')$. The introduction of the gate corresponding to $(w+x)$ prevents the generation of the spurious output pulse.

In this chapter systematic procedures will be developed for analyzing a circuit to determine whether or not there is any possibility of spurious outputs being generated. Synthesis procedures for designing circuits which cannot produce spurious outputs will also be developed.

7.2 ANALYSIS

In developing switching algebra, it was implicitly assumed that there were no delays in combinational circuits. All make contacts on the same relay were represented by the same variable, implying that they all close or open simultaneously. In gate networks, the propagation time of a signal along a wire and the delay in a gate responding to an input change were both ignored. In order to develop a technique for treating transient performance, it is necessary explicitly to include the possibility of such delays.

Contact Networks

In contact networks the possibility of nonsimultaneous action of the contacts on a single relay can be explicitly accounted for by assigning a different symbol to each contact. This can easily be done by using the same letter for all contacts on one relay and attaching a different subscript to each contact. This process of subscripting variables is illustrated in Fig. 7.2-1. The transmission function for a contact network with the variables subscripted in this fashion is called the *transient transmission function* T_{tr}. This function accurately represents the network performance even during input changes. The performance can also be represented by means of the P sets and S sets described in Chap. 3 if subscripted variables are used. The P sets and S sets can be obtained directly from the circuit diagram by tracing paths. They can also be obtained from the transient transmission function T_{tr}. If T_{tr} is "multiplied out" into sum-of-products form by using the theorems $X(Y + Z) = XY + XZ$, $X \cdot X = X$, $X + X = X$, and $X + XY = X$, each product term in the resulting expression will correspond to one P set. Thus, for the circuit of Fig. 7.2-1

$$T_{tr} = a_1(b_3 + b_2'c_1) + b_1(c_1 + b_2'b_3)$$
$$= a_1b_3 + a_1b_2'c_1 + b_1c_1 + b_1b_2'b_3$$

and the P sets are

$\{a_1,b_3\}$
$\{a_1,b_2',c_1\}$
$\{b_1,c_1\}$
$\{b_1,b_2',b_3\}$

The S sets can be obtained by "adding out" the transient transmission function into product-of-sums form by using the theorems $X + YZ = (X + Y)(X + Z)$, $X + X = X$, $X \cdot X = X$, $X(X + Y) = X$.

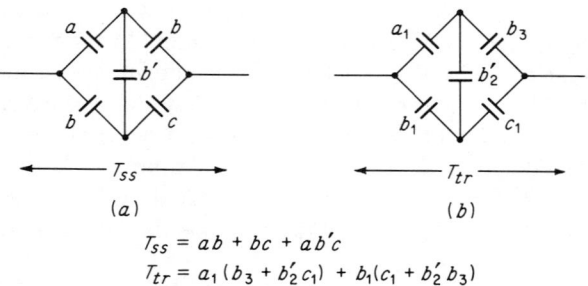

$T_{ss} = ab + bc + ab'c$
$T_{tr} = a_1(b_3 + b_2'c_1) + b_1(c_1 + b_2'b_3)$

Fig. 7.2-1. *Subscripting of variables in a contact network.* (a) *Unsubscripted variables;* (b) *subscripted variables.*

Each sum factor in the resulting expression corresponds to an S set. Thus

$$\begin{aligned}
T_{tr} &= a_1(b_3 + b_2'c_1) + b_1(c_1 + b_2'b_3) \\
&= a_1(b_3 + b_2')(b_3 + c_1) + b_1(c_1 + b_2')(c_1 + b_3) \\
&= (b_3 + c_1)[a_1(b_2' + b_3) + b_1(b_2' + c_1)] \\
&= (b_3 + c_1)(a_1 + b_1)(a_1 + b_2' + c_1)(b_1 + b_2' + b_3)(b_2' + b_3 + c_1) \\
&= (b_3 + c_1)(a_1 + b_1)(a_1 + b_2' + c_1)(b_1 + b_2' + b_3)
\end{aligned}$$

and the S sets are

$\{b_3, c_1\}$
$\{a_1, b_1\}$
$\{a_1, b_2', c_1\}$
$\{b_1, b_2', b_3\}$

These P sets and S sets should be verified by path tracing on the circuit diagram.

Gate Networks

In gate networks it is necessary to account explicitly for the delay in propagating a signal along a wire. This can be done by associating a direction of propagation† and a delay with each wire in the circuit. Each wire is assigned an arbitrary number as in Fig. 7.2-2. If the lead is labeled i, it is assumed that the delay associated with the lead is d_i. There are two ends for each wire—end I, which is connected either to a circuit input or to a gate output, and end II, which is connected to a gate input or to a circuit output. The direction of signal propagation is from end I to end II. If the signal occurring at end I of a lead labeled i is $a(t)$, then the signal present at end II of this lead will be $a(t - d_i)$. This relationship will be indicated by labeling end II with the symbol a_i, representing $a(t - d_i)$ as shown in Fig. 7.2-3. Similarly, if end I of lead i is labeled a_{jk}, representing $a(t - d_j - d_k)$, then end II will be labeled a_{jki}, representing

† The following development is valid only when a unique direction of propagation can be associated with each lead in a circuit. This is not an important restriction, for it is normally possible to do this for gate networks.

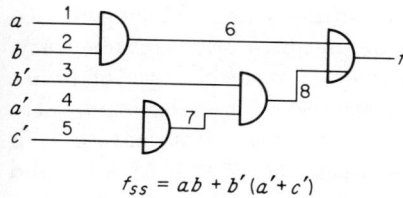

Fig. 7.2-2. Labeling of leads in a GATE network.

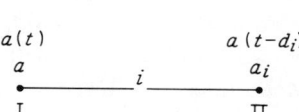

Fig. 7.2-3. Labeling of leads.

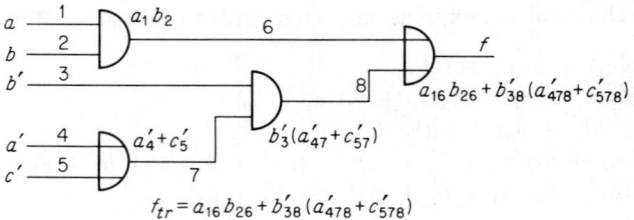

Fig. 7.2-4. *Formation of transient output function.*

$a(t - d_i - d_j - d_k)$; and if end I is labeled $a_j(b'_r + c_s)$, then end II is labeled $a_{ji}(b'_{ri} + c_{si})$. The analysis of a gate network is carried out by following these labeling rules and by making use of the logical properties of the gates to derive an output expression from the symbols for the input signals. This is illustrated in Fig. 7.2-4 for the circuit of Fig. 7.2-2. The resulting output expression is called the *transient output function* and represents the output correctly even when the inputs are changing provided that the subscripted variables are properly interpreted.

The transient performance of gate networks can also be represented by means of P sets and S sets. The extension of the Chap. 3 techniques to the transient situation is analogous to the extension for contact networks. Thus, for Fig. 7.2-4 the P sets and S sets are as shown in the tabulation.

P sets	S sets
$\{a_{16}, b_{26}\}$	$\{a_{16}, b'_{38}\}$
$\{a'_{478}, b'_{38}\}$	$\{b_{26}, b'_{38}\}$
$\{b'_{38}, c'_{578}\}$	$\{a_{16}, a'_{478}, c'_{578}\}$
	$\{a'_{478}, b_{26}, c'_{578}\}$

7.3 STATIC HAZARDS

In the preceding section, analysis techniques which are valid for combinational circuits during transient conditions were developed. Of particular interest is the question of whether or not spurious, momentary, false outputs can occur in a specific network during changes of input. Before developing techniques for testing networks for such outputs, the network of Fig. 7.1-1 will be analyzed by means of the transient-analysis techniques to illustrate the relations between the P sets, S sets, and spurious outputs.

The transient analysis of the circuit of Fig. 7.1-1 with the flip-flops removed is shown in Fig. 7.3-1. In the previous discussion of this circuit

7.3 static hazards

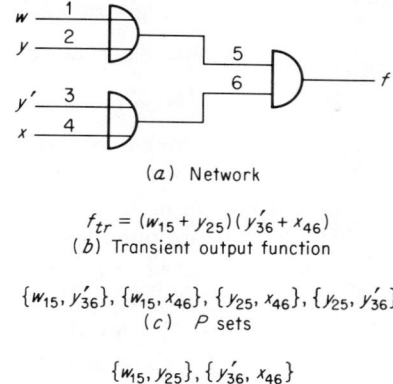

(a) Network

$$f_{tr} = (w_{15} + y_{25})(y'_{36} + x_{46})$$
(b) Transient output function

$$\{w_{15}, y'_{36}\}, \{w_{15}, x_{46}\}, \{y_{25}, x_{46}\}, \{y_{25}, y'_{36}\}$$
(c) P sets

$$\{w_{15}, y_{25}\}, \{y'_{36}, x_{46}\}$$
(d) S sets

Fig. 7.3-1. Transient analysis for Fig. 7.1-1.

it was concluded that a spurious 1 output could occur when $w = x = 0$ and both y and y' become equal to 1. This possibility is explicitly indicated by the P set $\{y_{25}, y'_{36}\}$. It will be shown that the possibility of a spurious output is always indicated by the presence of a pair of complementary literals, such as y_{25} and y'_{36}, in a single S set or P set.

The following discussion will be restricted to single-variable changes at the input, since it is not generally possible to prevent spurious outputs for multiple-variable changes. As in the earlier chapters an input state is an assignment of a value to each of the input variables, and two input states are adjacent if they differ only in the value assigned to one of the input variables. Transitions between pairs of input states which both produce the same steady-state output will be considered first. If it is possible for a spurious output to be produced during such a transition, the transition is said to correspond to a *static hazard*.

Definition. A *static 1 hazard* is a transition between a pair of adjacent input states which both produce a 1 output, during which transition it is possible for a momentary 0 output to occur.

Definition. A *static 0 hazard* is a transition between a pair of adjacent input states which both produce a 0 output, during which transition it is possible for a momentary 1 output to occur.

Thus the behavior discussed in connection with Fig. 7.3-1 corresponds to a static 0 hazard. The presence of a static hazard in a network does not mean that the corresponding transition will actually produce a spurious output. Whether or not the spurious output occurs will depend on the specific values of the delays in the circuit. The precise relation between static hazards and P sets or S sets having pairs of complementary literals is given by the following theorem:

Theorem 7.3-1. A static 0 hazard exists in a network if and only if the following two conditions are satisfied:

1. There is a P set of the network,

$$K_1 = \{a_1^*, b_2^*, \ldots, x_i, x_j', \ldots, z_k^*\}$$

in which the same variable may appear with different (multiple) subscripts but exactly one variable (x) appears both complemented and uncomplemented.

2. There is at least one pair of (adjacent) input states of the network satisfying the following:

(a) Both input states produce 0 outputs.

(b) The variable x is equal to 0 for one of the states and equal to 1 for the other state of the pair.

(c) Each other (non-x) *literal* of K_1 is equal to 1 for both input states.

PROOF. First it will be shown that if the conditions of the theorem are satisfied there will be a static 0 hazard present (sufficiency of the theorem). Condition 2(a) guarantees that there is a pair of adjacent input states which both produce 0 outputs. If the network is placed in the one of these input states for which $x = 0$, all the subscripted literals of K_1 except x_i will become equal to 1 before any further input changes. This is true because of condition 2 and because it is assumed that an input variable is not changed until all previous input changes have propagated through the network (the network has "settled down"). If x is now changed to 1, it is possible for x_i to become equal to 1 before x_j' becomes equal to 0. If this happens, all the literals of K_1 are equal to 1 and a 1 output must be produced. Eventually x_j' will become equal to 0, and the network output will return to 0. Thus, if the conditions of the theorem are satisfied, it is possible to have a spurious 1 output depending on the sequence in which x_i and x_j' change.

In order to prove the necessity of the theorem, it will be assumed that a static 0 hazard exists. Thus there must be two adjacent input states, which both produce 0 outputs, such that during a transition between these two states a spurious 1 output may be produced. Let the variable in which these input states differ be x.

The spurious 1 output requires that the network contain a P set with the following properties: (1) Each variable which is equal to 0 for the pair of input states must either be absent from the P set or must appear only complemented. (2) Each variable which is equal to 1 for the pair of input states must either be absent from the P set or must appear only uncomplemented. If a variable which is equal to 0 (1) for both input states appeared uncomplemented (complemented) in the P set, it would not be possible for all literals in the P set to be equal to 1 during the transition between the two input states and consequently the P set could not produce a 1 output during this transition. (3) If only the literals specified by conditions (1) and (2) appeared in the P set, all literals of the P set would be equal to 1 for both input states. This cannot be, because it was assumed that both input states produce 0 outputs. If x or x' appeared in the P set, all literals of the P set would equal 1 for one of the input states. The only way to have at least one literal of the P set equal to 0 for each input state of the pair of adjacent input states and still satisfy conditions (1) and (2) is to have *both* x and x' appear in the P set. If another variable such as y appeared in the P set both complemented and uncomplemented, it would not be possible to have all literals of the P set equal to 1 during a single input-variable change. When only one variable is changing, only one pair of complementary literals can both equal 1. Conditions (1), (2), and (3) are the same conditions as are given for the P set K_1 of the theorem. Therefore the necessity of the theorem is proved.

It should be pointed out that there are two assumptions inherent in the proof just given, and the theorem is valid only when these assumptions are legitimate. The first assumption is that the input is changed only after the network has settled down; i.e., all changes from the previous input change have propagated throughout the network. The second assumption is that no "bounce" signals are present—once an input changes, it retains its new value until the next input change. This second assumption is not always valid for contact networks, since a contact may close, then bounce open, and then finally close and remain closed. It is clear that this phenomenon can cause a spurious output, but this type of behavior will not be discussed here.

There is a corresponding theorem for static 1 hazards. Since the proof of this theorem is directly analogous to the proof just given, it will be omitted.

Theorem 7.3-2. A static 1 hazard exists in a network if and only if the following two conditions are satisfied:

1. There is an S set of the network,

$$K_0 = \{a_1^*, b_2^*, \ldots, x_i, x_j', \ldots, z_k^*\}$$

where the same variable may appear with different subscripts but exactly one variable (x) appears both complemented and uncomplemented.

2. There is at least one pair of (adjacent) input states satisfying the following:
    (a) Both input states produce 1 outputs.
    (b) The variable x is equal to 0 for one of the states and equal to 1 for the other state of the pair.
    (c) Each other (non-x) *literal* of K_0 is equal to 0 for both input states.

Removal of Subscripts

In the two preceding theorems no conditions are placed on the subscripts of the variables referred to. Thus it is possible to omit the subscripts and use only the original unsubscripted variables. Whether a variable such as z appears only once as z_k^* in the P set K_1 or appears several times as z_k^*, z_m^*, ..., the conditions of part 2 of the theorem remain unchanged. The important thing is that if z_k^* equals z_k then z_m^* must equal z_m, etc., or if z_k^* equals z_k', then z_m^* must equal z_m', etc. Identifying the same literals with different subscripts will not change any of the conditions of the theorems. This can be done by simply omitting the subscripts. It is, however, necessary to preserve carefully the difference between a variable and its complement. In particular the possibility of both a variable and its complement appearing in the same P set or S set must be preserved, since this is the critical characteristic of P sets or S sets which give rise to static hazards.

    Definition. The set which results when the subscripts are removed from the literals of a P set (S set) and then repetitions of the same literals are removed will be called a 1 set (0 set).

Since it is usually easier to work with the unsubscripted variables, Theorems 7.3-1 and 7.3-2 will be restated in terms of 1 sets and 0 sets.

    Theorem 7.3-3. A static 0 hazard (1 hazard) exists in a network if and only if the following two conditions are satisfied:

    1. There is a 1 set (0 set) of the network,

$$L = \{a^*, b^*, \ldots, x, x', \ldots, z^*\}$$

where exactly one variable (x) appears both complemented and uncomplemented.

    2. There is at least one pair of (adjacent) input states of the network satisfying the following:
    (a) Both input states produce 0 (1) outputs.
    (b) The variable x is equal to 0 for one of the input states and equal to 1 for the other state of the pair.
    (c) Each other (non-x) *literal* of L is equal to 1 (0) for both input states.

7.3 static hazards

It would be possible to obtain the P sets and S sets with subscripted variables and then to remove the subscripts. This process is not particularly attractive in that it would in no way reduce the complexity of the analysis procedure. Much more desirable would be a procedure that never required the introduction of subscripted variables. The subscripts need never be introduced provided that any pairs of complementary literals such as x and x' are treated as if they had different subscripts. That is, x and x' must be treated as distinct literals rather than as complements. Removal of the subscripts allows the identification of the same literals and use of the corresponding theorems, such as $X + X = X$, $X \cdot X = X$, $X + XY = X$, $X(X + Y) = X$. Pairs of complementary literals must be treated as if they corresponded to different variables, and theorems such as $X + X' = 1$, $X \cdot X' = 0$, $X + X'Y = X + Y$, $X(X' + Y) = XY$, $XY + X'Z + YZ = XY + X'Z$, and $(X + Y)(X' + Z)(Y + Z) = (X + Y)(X' + Z)$ must *not* be used.

DeMorgan's theorem

$$(X + Y + \cdots + Z)' = X'Y' \cdots Z'$$

and

$$(XY \cdots Z)' = X' + Y' + \cdots Z$$

involves no "cancellation" of complementary literals and *may still be used*.

An Example of the Use of Theorem 7.3-3

In Fig. 7.3-2 a network is shown which will be used to illustrate Theorem 7.3-3. Specifically, this network will be analyzed to determine which static hazards are present. The 1 sets are listed in Table 7.3-1 and can be obtained as follows:

$$f = wxy + (w + z)(w' + y')$$
$$f = wxy + w(w' + y') + z(w' + y')$$
$$f = wxy + ww' + wy' + w'z + y'z$$

Table 7.3-1 also lists the 0 sets, which can be obtained as follows:

$$f = wxy + (w + z)(w' + y')$$
$$f = (wxy + w + z)(wxy + w' + y')$$
$$f = (w + z)(wxy + w' + y')$$
$$f = (w + z)(w + w' + y')(x + w' + y')(y + w' + y')$$

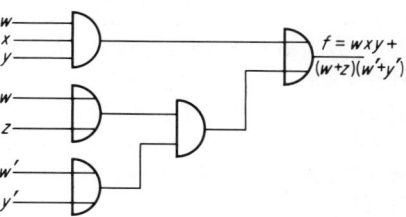

Fig. 7.3-2. Network to illustrate Theorem 7.3-3.

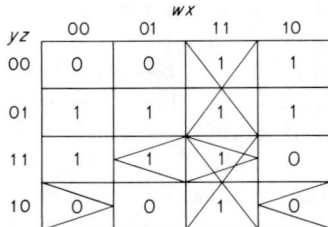

Fig. 7.3-3. *Map for the output function for the network of Fig. 7.3-2. Each static hazard of the network is indicated by a rhombus on the map.*

Table 7.3-1. *The 1 Sets and 0 Sets for Fig. 7.3-2*

1 sets	0 sets
$\{w,x,y\}$	$\{w,z\}$
$\{w,y'\}$	$\{w',x,y'\}$
$\{w',z\}$	$\{w,w',y'\}$
$\{y',z\}$	$\{w',y,y'\}$
$\{w,w'\}$	

There are two 0 sets and one 1 set which satisfy the conditions given in Theorem 7.3-3 for the set L. The 0 set $\{w',y,y'\}$ will be considered first. This 0 set will give rise to a hazard if and only if there is a pair of adjacent input states of the network which:

1. Both produce 1 outputs
2. Have $y = 1$ for one of the input states and $y = 0$ for the other input state
3. Have $w = 1$ for both input states

The map for the output function of this circuit is shown in Fig. 7.3-3. Examination of the map shows that the pair of states for which $w = 1$, $x = 1$, $z = 1$ satisfies these conditions as does also the pair of states for which $w = 1$, $x = 1$, $z = 0$. Thus, the 0 set $\{w',y,y'\}$ gives rise to two static 1 hazards.

By a similar process it can be determined that the 0 set $\{w,w',y'\}$ gives rise to a static 1 hazard corresponding to the pair of input states for which $x = 1$, $y = 1$, $z = 1$. The 1 set $\{w,w'\}$ will give rise to a static 0 hazard if there are two adjacent input states which both produce a 0 output and differ only in the value of the w variable. Inspection of Fig. 7.3-3 shows that the pair of input states for which $x = 0$, $y = 1$, $z = 0$ satisfies these conditions and therefore corresponds to a static 0 hazard. Each pair of input states which corresponds to a static hazard is enclosed in a rhombus in Fig. 7.3-3.

7.3 static hazards

A Test Procedure

In using Theorem 7.3-3 to test for the presence of hazards in a network it is inconvenient to have to search pairs of states on a map in order to determine whether condition (2) of the theorem is satisfied. For functions of a large number of variables such a search can be extremely tedious, if not impossible. It is possible to restate part (2) of Theorem 7.3-3 in various forms which are more convenient for computation. These are rather straightforward and are described in the literature [1].

Alternative Conditions for Static Hazards

The Theorem 7.3-3 technique for analyzing networks requires that both the 0 sets and 1 sets be formed. For many networks it is relatively easy to obtain the 1 sets but comparatively difficult to form the 0 sets, or vice versa. Also, any synthesis procedure for hazard-free networks based on Theorem 7.3-3 would require that both the 0 sets and 1 sets of the network be "controlled" in the procedure. Usually, in designing networks it is possible to specify only the 1 sets (or 0 sets). By developing the theory of hazards further it is possible to specify techniques which require knowledge of the 0 sets or 1 sets, but not both, and which permit a network to be analyzed for the presence of hazards or a hazard-free network to be designed. These techniques depend on the theorems to be proved next.

In the previous theorems, the relationship between 0 sets and 1 hazards (1 sets and 0 hazards) was studied. The effects of 1 sets on 1 hazards (0 sets on 0 hazards) must be considered next.

Consider two adjacent input states which differ only in the value of x, and suppose that there is a 1 set which has all its literals equal to 1 for *both* the input states. For Fig. 7.3-2 such a pair of states might be those with $w = 0$, $y = 0$, $z = 1$, and the appropriate 1 set would be $\{w',z\}$. In a transition between such a pair of input states there can be no spurious 0 output, since the 1 set has all its literals equal to 1 throughout the transition (the 1 set cannot contain x or x', since all its literals equal 1 for both input states). The converse of this statement is also true. If there is a pair of adjacent input states which both produce a 1 output and there is not a 1 set having all its literals equal to 1 for both input states, the pair of input states will correspond to a hazard.

Definition. A 1 set is said to cover an input state if all the literals of the 1 set are equal to 1 for the input state.

The 1 set $\{w,x,y\}$ covers the input states $w = 1$, $x = 1$, $y = 1$, $z = 0$ and $w = 1$, $x = 1$, $y = 1$, $z = 1$ but does not cover the input state $w = 0$, $x = 1$, $y = 1$, $z = 1$.

Definition. A 0 set is said to cover an input state if all the literals of the 0 set are equal to 0 for the input state.

The 0 set $\{w,z\}$ covers the input states for which $w = 0$ and $z = 0$ but does not cover any input states having $w = 1$ or $z = 1$.

A 1 set or 0 set which contains one or more pairs of complementary literals cannot cover any input states. In the following it will be useful to treat such 1 sets and 0 sets specially.

Definition. A 0 set (1 set) which does not contain any pair of complementary literals will be called a *stable 0 set (stable 1 set)*.

Definition. A 0 set (1 set) which contains at least one pair of complementary literals will be called an *unstable 0 set (unstable 1 set)*.

Theorem 7.3-4.† A static 1 hazard (0 hazard) exists in a network if and only if:

1. There is a pair of adjacent input states which both produce 1 outputs (0 outputs).
2. There is no 1 set (0 set) of the network which covers both the input states of the pair.

PROOF. The necessity of the theorem was demonstrated in the previous discussion. To prove sufficiency, assume that a pair of adjacent input states exists, differing only in one variable (say ω), which both produce 1 outputs and that no 1 set covers both input states. There are two types of 1 sets: those which do not cover either input state of the pair and those which cover one of the input states of the pair. The 1 sets which cover neither input state have at least one literal equal to 0 for both input states. The 1 sets which cover one input state must include either an ω or ω' since this is the only variable in which the input states differ. If both ω and ω' are set equal to 0, there will be no 1 set with all its literals equal to 1. Thus the network output must be 0 for this situation, and there must be a corresponding 0 set which includes ω, ω', and perhaps some additional literals (complemented if the variable is equal to 1 for the input states, and uncomplemented if the variable is equal to 0 for the input states). Thus the conditions of Theorem 7.3-3 must be satisfied, and the pair of states must correspond to a static 1 hazard.

Since unstable 1 sets (0 sets) cannot cover any input states, condition (2) could be stated for stable 1 sets (0 sets) only without changing the proof of the theorem. For functions of a small number of variables it is easy to determine whether or not the conditions of Theorem 7.3-4 are satisfied by plotting the stable 1 sets (or 0 sets) on a map. The function for Fig.

† This theorem was proved by D. A. Huffman [2] for contact networks.

7.3 static hazards

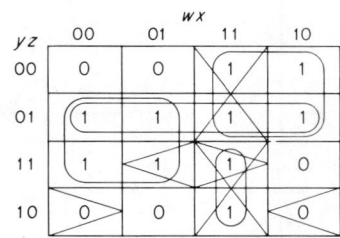

Fig. 7.3-4. *Map for the output function for the network of Fig. 7.3-2. The 1 sets are shown encircled. Each static hazard is indicated by a rhombus.*

7.3-2 is shown in Fig. 7.3-4. Examination of this map will show that each pair of adjacent states that both produce 1 outputs and that are not covered by the same 1 set correspond to a hazard. Functions which involve so many variables that use of a map is inconvenient can easily be analyzed for hazards by means of the following theorem:

Theorem 7.3-5. A static 1 hazard (0 hazard) exists in a network if and only if there is a pair of stable 1 sets (0 sets) $\{\alpha^*, \ldots, \beta^*, \omega\}$, $\{\lambda^*, \ldots, \theta^*, \omega'\}$ such that:

1. There is only one variable which is complemented in one of the 1 sets and uncomplemented in the other 1 set (any number of *literals* may appear in both 1 sets).
2. Each of the other stable 1 sets (0 sets) contains ω, ω', or the complement of one of the other literals included in either of the original pair of 1 sets (0 sets), or a variable which does not appear in either of the original pair of 1 sets (0 sets).

PROOF. If two adjacent input states both produce 1 outputs and are not both covered by the same 1 set, they must each be covered individually by 1 sets. These 1 sets must therefore be of the form $\{\alpha^*, \ldots, \beta^*, \omega\}$, $\{\lambda^*, \ldots, \theta^*, \omega'\}$, where α^* may equal λ^* but α^* may not equal $(\lambda^*)'$. Since both input states are not included in a single 1 set, there must not be a 1 set which contains only $\{\alpha^*, \ldots, \beta^*, \ldots, \theta^*\}$ or some subset thereof.

Analysis-procedure Example

It is possible to discover whether a network contains any static hazards by determining only the 1 sets or the 0 sets and then applying Theorems 7.3-3 and 7.3-4. This procedure will be illustrated by means of the network of Fig. 7.3-5. The 1 sets for this network are $\{w',x,z\}$, $\{w,x',z\}$, $\{w,x,y'\}$, and $\{x,y',z\}$. They are plotted in Fig. 7.3-6 on a map of the function of the network. Since there are no unstable 1 sets, there can be no 0 hazard (Theorem 7.3-3). There is a static 1 hazard between the two states for which $w = 1$, $y = 0$, $z = 1$ since both states correspond to 1 outputs and are not both covered by any single 1 set.

Transient Behavior of Switching Circuits

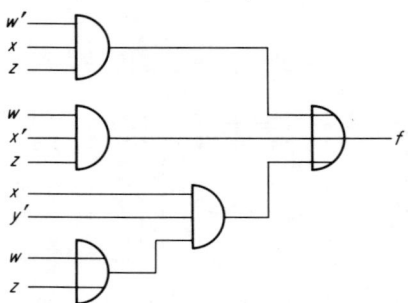

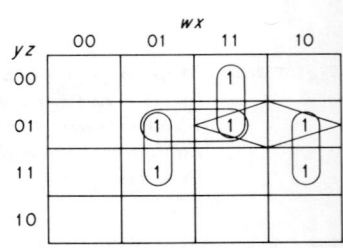

Fig. 7.3-5. A network to illustrate the analysis procedure for static hazards.

Fig. 7.3-6. Map of the function of the network of Fig. 7.3-5 showing 1 sets.

It is also possible to arrive at the same conclusion by using only the 0 sets. The 0 sets for the network of Fig. 7.3-5 are

{w,x}   {w,w',y'}
{w',x',y'}   {x,x'}
{x,z}
{y',z}
{w,z}

Since there are two unstable 0 sets, the possibility of a 1 hazard must be checked by applying Theorem 7.3-3. For the set $\{w,w',y'\}$ to produce a 1 hazard, there must be a pair of input states which differ only in w and for both of which $y' = 0$. Inspection of the map of Fig. 7.3-7 shows that no such pair exists. The set $\{x,x'\}$ does give rise to a 1 hazard since the states for which $w = 1$, $y = 0$, $z = 1$ differ only in x and both have 1 outputs. Inspection of Fig. 7.3-7 shows that there is no pair of adjacent input states which both produce 0 outputs and which are not included in a single 0 set. Therefore there are no static 0 hazards. Although use of the 0 sets leads to the same conclusions reached on the basis of the 1 sets,

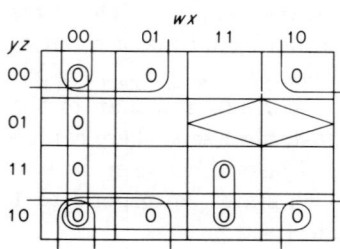

Fig. 7.3-7. Map of the function of the network of Fig. 7.3-5 showing 0 sets.

7.4 dynamic hazards

there is substantially more work involved in obtaining and using the 0 sets for this network.

7.4 DYNAMIC HAZARDS

Even in the absence of any static hazards it is still possible for a network to produce spurious transient outputs. The preceding discussion of static hazards treats network behavior only for input changes which produce no change in the steady-state network output. Spurious outputs can also be developed during input changes which do affect the steady-state output. This type of spurious output can be developed in the network of Fig. 7.4-1. The map for the output function of this network is shown in Fig. 7.4-1b. Since there are no pairs of adjacent input states that both produce the same output, there can be no static hazards in this network. However, false outputs can appear during the transition from $a = 0$, $b = 1$, $c = 0$ to $a = 0$, $b = 1$, $c = 1$. Figure 7.4-2a shows the structure which the network assumes when relay A is released and relay B is operated. If relay C is released, there will be no conduction through the network. Assume that relay C is energized and that the contacts respond in the order given by their subscripts. When contact c_1 closes, the network will become a closed circuit. The opening of the c_2' contact will open this circuit, and the subsequent closure of contact c_3 will again close the path through the network. Finally, the opening of contact c_4' will have no effect on the closed path through the network. These actions are illustrated in the timing diagram of Fig. 7.4-2b. During this transition from an open circuit to a closed circuit, a momentary closing and a momentary opening of the network can occur if the contacts operate in the manner prescribed.

The present discussion is concerned with transitions between pairs of adjacent input states which produce changes in the network output. If

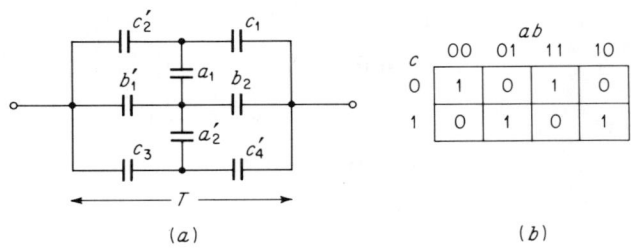

Fig. 7.4-1. A network with a dynamic hazard but no static hazards. (a) Network; (b) map of transmission function.

Transient Behavior of Switching Circuits

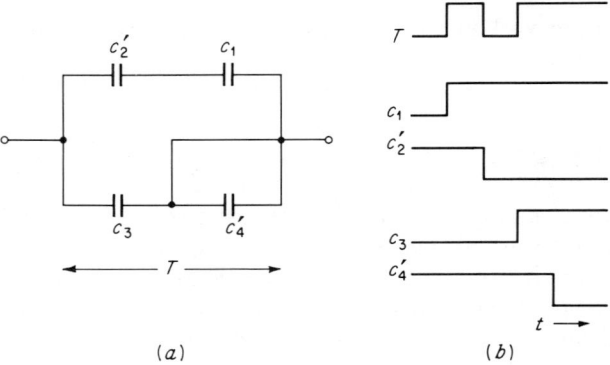

Fig. 7.4-2. Network of Fig. 7.4-1 with $A = 0$, $B = 1$, and C changing from 0 to 1. (a) Network; (b) timing diagram.

momentary spurious outputs can occur during such a transition, the transition is said to correspond to a *dynamic hazard*.

Definition. A *dynamic hazard* is a transition between a pair of adjacent input states, one of which produces a 1 output and the other of which produces a 0 output, during which transition it is possible for a momentary 0 output and a momentary 1 output to occur.

Theorem 7.4-1. A dynamic hazard exists in a network if and only if the following conditions are all satisfied:

1. The following P sets and S sets exist in the network:
  (a) An unstable P set of the network†

$$K_1 = \{\cdots a_\mu^* \cdots, x_i^+, (x_j^+)', (x_k^+)'\}$$

  (b) An unstable S set of the network,

$$K_0 = \{\cdots b_\nu^* \cdots, (x_i^+)', x_k^+, (x_j^+)'\}$$

  (c) A stable P set of the network,

$$L_1 = \{\cdots c_\sigma^* \cdots, x_i^+, x_j^+, x_k^+\}$$

  (d) A stable S set of the network,

$$L_0 = \{\cdots d_\tau^* \cdots, x_i^+, x_j^+, x_k^+\}$$

† The symbols a^*, b^*, etc., represent literals. The symbol x_i^+ also represents a literal, but the $+$ symbol is consistent in referring to a prime or to no prime. That is, if $x_i^+ = x_i$, then $x_k^+ = x_k$ and $(x_j^+)' = x_j'$; or if $x_i^+ = x_i'$, then $x_k^+ = x_k'$, $(x_j^+)' = (x_j')' = x_j$, etc.

7.4 dynamic hazards

2. These P sets and S sets satisfy the following conditions:

   (a) None of these P sets and S sets contain any pairs of complementary literals other than those explicitly shown.

   (b) The underlined literals must be present. The presence or absence of the literals that are not underlined will have no effect on the existence of a dynamic hazard.

   (c) The two P sets can have literals in common, and the S sets can have literals in common, but no P set can have any literals in common with any S set except for the common literals shown explicitly.

   (d) No literal can occur complemented in one P set (S set) and uncomplemented in the other P set (S set) except for those shown explicitly.

   (e) Any literal may occur complemented in one or both of the P sets and uncomplemented in one or both of the S sets, or vice versa.

PROOF. First sufficiency will be demonstrated by showing that a dynamic hazard is present if the conditions of the theorem are satisfied. The network is placed in the input state for which $x^+ = 1$, all the remaining (non-x) literals of the P sets, K_1 and L_1, are equal to 1, and all the remaining (non-x) literals of the S sets, K_0 and L_0, are equal to 0. The circuit output must equal 1 for this input state because all the literals of the P set L_1 are equal to 1. The input x is now changed to 0, and it is assumed that the contacts or signals change in the following sequence: $x_k^+ \to 0$, $x_j^+ \to 0$, $x_i^+ \to 0$. When x_k^+ becomes equal to 0, the P set L_1 is no longer effective, but the S set K_0 has all its literals equal to 0, and therefore the circuit output becomes equal to 0. The S set K_0 becomes ineffective when x_j^+ changes to 0, but the P set K_1 then becomes effective, causing the circuit output to equal 1. Finally, x_i^+ changing to 0 makes K_1 ineffective, the S set L_0 effective, and the circuit output equal to 0. Thus, if the contacts or signals change in the order specified above, two spurious outputs will be developed during this transition.

To show the necessity of the theorem, it will be assumed that a dynamic hazard exists for a pair of adjacent states differing only in the value of the x variable. The existence of a dynamic hazard requires that the network output change for this transition: there must be a P set such as L_1 and an S set such as L_0 to control the network output for the two input states of the transition. Since two momentary outputs are developed, there must be an unstable P set such as K_1 and an unstable S set such as K_0. The conditions (a to e) of part 2 of the theorem follow directly from the fact

Table 7.4-1. Sequence in Which P Sets and S Sets Become Effective for a Dynamic Hazard

f	L_0			K_1			K_0			L_1			Sequence of states		
	x_i^+	x_j^+	x_k^+	x_i^+	$x_j^{+\prime}$	$x_k^{+\prime}$	$x_i^{+\prime}$	x_j^+	x_k^+	x_i^+	x_j^+	x_k^+	x_i^+	x_j^+	x_k^+
0	(0	0	0)	0	1	1	1	1	0	0	0	0	0	0	0
1	1	0	0	(1	1	1)	0	1	0	1	0	0	1	0	0
0	1	1	0	1	0	1	(0	0	0)	1	1	0	1	1	0
1	1	1	1	1	0	0	0	0	1	(1	1	1)	1	1	1

that this is a transition between a pair of adjacent input states, and all the P sets and S sets must be effective at some time during the transition. The order in which the x variables change and the P and S sets become effective is shown in Table 7.4-1.

Because the conditions of this theorem do involve the subscripts on the variables, it is not possible to restate the theorem in terms of 1 sets and 0 sets. However, it is possible to state, in terms of 1 sets and 0 sets, a necessary condition for the existence of a dynamic hazard. This condition, as stated in the following corollary, is useful in the synthesis of hazard-free networks:

Corollary. Any network which contains a dynamic hazard must contain at least one unstable 1 set and one unstable 0 set. The same variable must occur both complemented and uncomplemented in both the 1 set and the 0 set.

Analysis Example

Table 7.4-2 lists the P sets and S sets for the network of Fig. 7.4-1. These will be employed to illustrate the use of Theorem 7.4-1 to test whether or not a dynamic hazard is present in a network. Since two of the unstable S sets contain two pairs of complementary variables, these S sets cannot satisfy the conditions for K_0 in Theorem 7.4-1. The remaining two unstable S sets both contain the c variable primed and unprimed. Therefore only the first two unstable P sets can possibly satisfy the Theorem 7.4-1 conditions for K_1. There are two possible choices for K_0 and two possible choices for K_1, or four possible combinations for K_0 and K_1. Each combination must be tested further since they each have a literal shared between K_0 and K_1, as is required by the theorem—$(x_j^+)'$. Only one of the choices—$K_0 = \{b_2,c_1,c_4'\}$, $K_1 = \{c_1,c_2'\}$—will be tested in detail here.

Since b_2 occurs in K_0, no P set containing a b can qualify as L_1 and no S set containing a b' can qualify as L_0. Further, L_1 must contain c_4',

Table 7.4-2. P Sets and S Sets for Fig. 7.4-1

	S sets		P sets
	(unstable)		(unstable)
(K_0)	$b_2\, c_1\, c_4'$	(K_1)	$c_1\, c_2'$
	$b_1'\, c_2'\, c_3$		$c_3\, c_4'$
X	$a_1\, a_2'\, b_2\, c_2'\, c_3$	X	$a_1\, a_2'\, c_2'\, c_4'$
X	$a_1\, a_2'\, b_1'\, c_1\, c_4'$	X	$b_1'\, b_2$
		X	$a_1\, a_2'\, c_1\, c_3$
	(stable)		(stable)
	$a_1\, b_2\, c_2'\, c_4'$		$a_2'\, b_1'\, c_4'$
	$a_2'\, b_2\, c_1\, c_3$		$a_1\, b_1'\, c_1$
X	$a_1\, b_1'\, c_1\, c_3$	X	$a_2'\, b_2\, c_3$
X	$a_2'\, b_1'\, c_2'\, c_4'$	X	$a_1\, b_2\, c_2'$

and L_0 must contain c_2'. Thus the only possibilities are $L_1 = \{a_2', b_1', c_4'\}$ and $L_0 = \{a_1, b_2, c_2', c_4'\}$. These choices do satisfy the conditions of Theorem 7.4-1 and thus correspond to a dynamic hazard for $a = 0, b = 0$, and c changing (initially $c_2' = c_4' = 0, c_1 = 1$; then $c_2' \to 1, c_1 \to 0, c_4' \to 1$). The other three possible choices for K_0 and K_1 satisfy the conditions of the theorem and thus correspond to a different dynamic hazard. It is perhaps worth noting that the appearance of c_4' in L_0 is permitted but not required by the theorem.

7.5 SYNTHESIS OF HAZARD-FREE COMBINATIONAL NETWORKS

The design techniques presented in Chap. 4 all involve some method for controlling the 1 sets or 0 sets of the network being designed. By making use of the theorems presented in Secs. 7.3 and 7.4, it is possible to prove a theorem which gives sufficient conditions on the 1 sets of a network so that no hazards will be present. These are not necessary conditions, for it is possible to have a hazard-free network which does not satisfy them. These conditions usually lead to reasonable networks, and more general design methods are not known at the present time. It should be emphasized that this is a technique of using additional switching elements in a network in order to avoid hazards. Techniques using additional delay elements have been discussed elsewhere [3]. The important characteristic

Transient Behavior of Switching Circuits

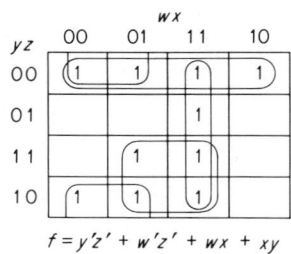

Fig. 7.5-1. *Design of a hazard-free network using a map.*

$f = y'z' + w'z' + wx + xy$

of the conditions to be presented here is that they require control of only the 1 sets *or* the 0 sets, but not both.

Theorem 7.5-1. A network whose 1 sets (0 sets) satisfy the following conditions will not contain any static or dynamic hazards:

1. For each pair of adjacent input states that both produce a 1 output (0 output), there is at least one 1 set (0 set) that includes both input states of the pair.
2. There are no 1 sets (0 sets) that contain exactly one pair of complementary literals.

PROOF. By Theorem 7.3-4 there will be no static 1 hazards (0 hazards) if condition 1 is satisfied. If condition 2 is satisfied, there will be no static 0 hazards (1 hazards), because of Theorem 7.3-3; and because of the corollary to Theorem 7.4-1, there will be no dynamic hazards.

The techniques given in Chap. 4 for designing combinational networks start by finding a minimal sum for the output function. The network is then designed so that each 1 set of the network corresponds to one of the product terms in the minimal sum. In order to design a hazard-free network, it is necessary only to modify the procedures for obtaining a minimal sum so that the 1 sets derived from the resulting product terms satisfy the conditions of Theorem 7.5-1. If a map is used to form the sum-of-products expression, the procedure of choosing the prime implicants to include in the expression must be modified as follows: A sufficient number of prime implicants must be picked so that each pair of adjacent input states which both produce 1 outputs is included in a single prime implicant. This is illustrated in the map shown in Fig. 7.5-1.

The tabular method for generating a minimal sum which is presented in Secs. 4.4 and 4.5 can be modified in a similar fashion. No change is required in the method of generating the prime implicants as described in Sec. 4.4. The process of selecting those prime implicants which are included in the final expression must be modified as follows: A column is added to the prime implicant table for each pair of adjacent input states

Table 7.5-1. *A Tabular Method for Obtaining a Hazard-free Network*

(a) Determination of the prime implicants

	w x y z			w x y z
0	0 0 0 0 ✓		(0,2)	0 0 – 0 ✓
2	0 0 1 0 ✓		(0,4)	0 – 0 0 ✓
4	0 1 0 0 ✓		(0,8)	– 0 0 0 ✓
8	1 0 0 0 ✓		(2,6)	0 – 1 0 ✓
6	0 1 1 0 ✓		(4,6)	0 1 – 0 ✓
12	1 1 0 0 ✓		(4,12)	– 1 0 0 ✓
			(8,12)	1 – 0 0 ✓
7	0 1 1 1 ✓		(6,7)	0 1 1 – ✓
13	1 1 0 1 ✓		(6,14)	– 1 1 0 ✓
14	1 1 1 0 ✓		(12,13)	1 1 0 – ✓
			(12,14)	1 1 – 0 ✓
15	1 1 1 1 ✓		(7,15)	– 1 1 1 ✓
			(13,15)	1 1 – 1 ✓
			(14,15)	1 1 1 – ✓

	w x y z
(0,2,4,6)	0 – – 0
(0,4,8,12)	– – 0 0
(4,6,12,14)	– 1 – 0
(6,7,14,15)	– 1 1 –
(12,13,14,15)	1 1 – –

(b) Selection of prime implicants to be included in sum expression

	0 2	0 4	0 8	2 6	4 6	4 12	8 7	6 14	12 13	12 14	7 15	13 15	14 15	
$w'z'$	⊗	×	⊗	×										*
$y'z'$		×	⊗		×	⊗								*
xz'				×	×			×		×				
xy							⊗	×			⊗		×	*
wx								⊗	×			⊗	×	*

$$f = w'z' + y'z' + xy + wx$$

Transient Behavior of Switching Circuits

which both produce 1 outputs. The standard techniques for selecting rows from the prime implicant table can then be used. This procedure is illustrated in Table 7.5-1. Note that the columns added to the prime implicant table correspond to the rows in the second step in the process for generating the prime implicants. For the function of Table 7.5-1 it is not necessary to retain in the prime implicant table any of the columns corresponding to the fundamental products, since each input state that corresponds to a 1 output is adjacent to another input state that also corresponds to a 1 output.

A network can be designed directly from the sum-of-products expression. Often a more desirable network results if the expression is first modified by means of the theorems of Boolean algebra. In designing hazard-free networks, only the theorems which do not modify the 1 sets and 0 sets of the network can be generally used. These are the theorems described in the discussion following Theorem 7.3-3. The theorem most commonly used is the factoring theorem: $XY + XZ = X(Y + Z)$, $(X + Y)(X + Z) = X + YZ$.

7.6 ESSENTIAL HAZARDS

In the preceding sections the effects of delays on the performance of combinational circuits have been studied. It was demonstrated that spurious temporary outputs could be developed in networks containing hazards but that it was always possible to design a hazard-free network for any output-function specification. If a combinational network containing a hazard is used in designing a sequential circuit, the resulting sequential circuit can enter an incorrect internal state due to a hazard pulse. The sequential circuit can thus have a false output which is not momentary but which lasts as long as the circuit remains in the incorrect internal state. It follows from this discussion that, in order to ensure the proper functioning of a sequential circuit in spite of variations in stray delays, hazard-free networks should be used in the combinational portions of the circuit. Although this is a necessary condition for obtaining a circuit in which the terminal behavior is independent of the stray delays, it is not sufficient. It has already been pointed out that circuits which are operated in pulse mode must always have the internal delays controlled. This requirement is brought about by the relationship between the input pulse width and the delay in transmitting changes in the y_i variables to the combinational circuitry. No similar requirement exists for fundamental-mode operation. However, if there are stray delays in the combinational circuitry of a sequential circuit being operated in fundamental mode, it is possible for the circuit to malfunction even if the combinational circuits are hazard-free. This will be demonstrated by means of the following example:

7.6 essential hazards

Example. In the sequential circuit of Fig. 7.6-1 there are no hazards in the combinational circuits. However, the performance of this circuit is directly affected by the stray delays present. In order to examine this dependence on the stray delays, it will be assumed that the circuit is placed in the condition for which $x = y_1 = y_2 = 0$, and then the response of a change in x will be studied. Labels have been placed on the leads in Fig. 7.6-1 to correspond to this condition.

When x changes from 0 to 1, there are two paths by which gate B can be affected, as shown by the heavier lines on the figure. If the delay along the lower path (through the inverter) is shorter than the delay along the upper path (through gate A and flip-flop 1), then the bottom input to gate B will change to 0 before the top input changes to 1 and the output of gate B will remain equal to 0. The circuit will thus remain in the condition where $x = 1$, $y_1 = 1$, $y_2 = 0$, as predicted by the transition table for this circuit as shown in Table 7.6-1.

On the other hand, if the delay along the lower path is longer than along the upper path, the top input to gate B will change to 1 before the bottom input changes to 0. The output of gate B will thus equal 1 from the time when its top input changes to 1 until the time when the bottom input changes to 0. A pulse will thus occur on the set input of flip-flop 2. The duration of this pulse is equal to the difference in delay along the two paths from x to gate B. If the pulse is sufficiently long, it will cause flip-flop 2 to become set $y_1 = 1$, which will in turn cause a signal to appear on the reset lead of flip-flop 1 via gate C. The circuit will finally enter the condition in which $x = 1$, $y_1 = 0$, $y_2 = 1$. This illustrates a situation in which the circuit's internal state and output depend directly on the difference in delay along two paths. When the delay is greater for the lower path through the inverter than the delay for the path through flip-flop 1, the combinational circuit generating

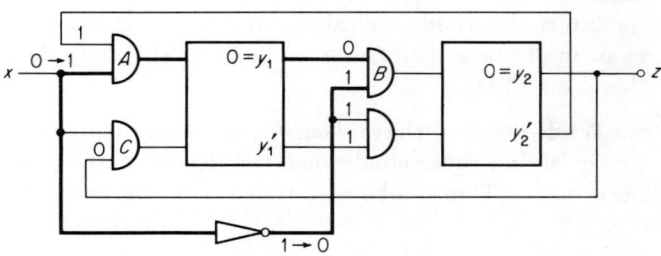

Fig. 7.6-1. Sequential circuit containing an essential hazard (fundamental mode.)

Table 7.6-1. Analysis of the Circuit of Fig. 7.6-1

(a) Excitation functions

$$S_1 = xy_2' \quad R_1 = xy_2$$
$$S_2 = x'y_1 \quad R_2 = x'y_1' \quad Z = y_2$$

(b) Excitation table

$y_1 y_2$	$x=0$	$x=1$
00	0 0 0 1	1 0 0 0
01	0 0 1 0	1 0 0 0
11	0 0 1 0	0 1 0 0
01	0 0 0 1	0 1 0 0

S_1R_1, S_2R_2

(c) Transition table

$y_1 y_2$	$x=0$	$x=1$	Z
00	(0 0)	1 0	0
10	1 1	(1 0)	0
11	(1 1)	0 1	1
01	0 0	(0 1)	1

$Y_1 Y_2$

the excitation for flip-flop 2 receives the change in the state of flip-flop 1 before it receives the change in the circuit input. Thus flip-flop 2 reacts as if the circuit were in the state for which $x = 0$, $y_1 = 1$, $y_2 = 0$, rather than the state with $x = 1$, $y_1 = 0$, $y_2 = 0$ or $x = 1$, $y_1 = 1$, $y_2 = 0$. For the circuit of Fig. 7.6-1, the insertion of delays in the four leads which connect y_1, y_1', y_2, y_2' from the flip-flop outputs to the input circuitry will prevent the occurrence of any malfunctions of the type just described.

It has been shown [4] that any sequential circuit operated in fundamental mode will have its performance unaffected by variations in stray delays in the combinational circuitry if these conditions are satisfied:

1. There are no races.
2. The combinational circuitry is hazard-free.
3. Only one input signal is changed at a time.
4. Each y_i signal has a delay of appropriate magnitude in the path whereby it is connected to the combinational circuitry which generates the internal variable excitation functions. It must be assumed that there is an upper bound to the magnitude of the stray delays in the combinational circuitry.

While it is true that it is always possible to guarantee proper operation by satisfying these conditions, the delays in the y_i leads are not always necessary. This can be seen by considering the accompanying fragment

	$x=0$	$x=1$
A	(A)	B
B	*	(B)

7.6 essential hazards

Table 7.6-2. Possible Sequential-circuit Responses to a Single Change in Input (Fundamental Mode)

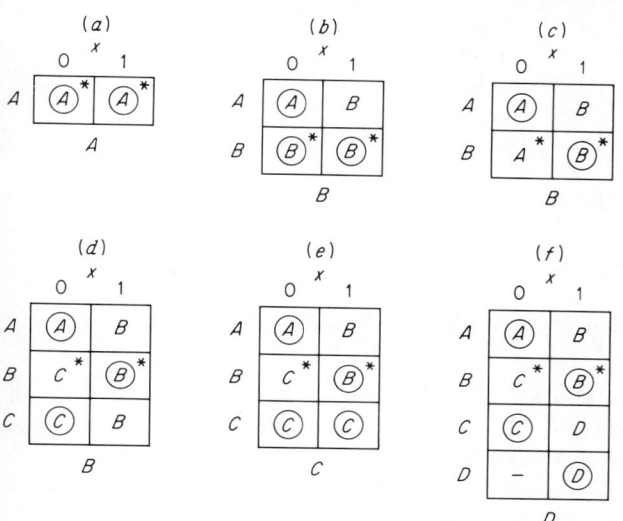

of a flow table. The circuit is initially stable in state A with $x = 0$. When x changes to 1, the circuit goes to state B. Because of stray delays some of the excitation circuitry may receive the signal corresponding to the change to state B before the change in x is received. Therefore the circuit may respond as if it were in the total state for which $x = 0$ and the internal state is B. Whether or not delays are required in the y_i leads depends on whether this response leaves the circuit in a different final state from what would otherwise be the case. Since the x change will eventually appear at the excitation circuitry, this final state will be determined by the action which occurs when the circuit is initially in state B with $x = 0$ and then x changes to 1.

In Table 7.6-2, six flow-table fragments are shown which illustrate all possible relevant circuit actions. In each flow table two total states are marked with an asterisk—one with $x = 1$ that corresponds to the correct total state after x changes from 0 to 1 and another with $x = 0$ that corresponds to the total state when the x change is delayed. By following the circuit response for the $x = 0$ asterisk state—the circuit being allowed to settle down and then x to change to 1—it is possible to determine the state which the circuit will finally settle down in when x is delayed. This final state is listed below each table. If this final state is the same state as the $x = 1$ asterisk state, as in Table 7.6-2a, b, c, and d, then the input delay has no effect on the circuit performance. In Table 7.6-2e and f the final state is different from the $x = 1$ asterisk state, and thus the circuit performance does depend on the input delay. It has been shown by Unger [4] that any flow table containing a situation such as that illustrated in Table 7.6-2e or f will always require y_i delays to ensure proper operation when

Table 7.6-3. A Flow Table with One Essential Hazard

s \ x_1x_2	00	01	11	10
1	①	2	–	4
2	3	②	3	–
3	③	2	③	4
4	1	④*	④	④

stray combinational delays are present. A more formal statement of his results follows:

Definition. A total state S_j and an input variable x_i represent an *essential hazard* for a flow table T if and only if, when the table is initially in state S_j, the state reached after one change in x_i is different from the state reached after three changes in x_i when fundamental-mode operation is assumed.

The flow table shown in Table 7.6-3 has only one essential hazard, that with total state 4, $x_1 = 0$, $x_2 = 1$, and variable x_2. One change in x_2 leads to state 1, $x_1 = 0$, $x_2 = 0$, two changes lead to state 2, $x_1 = 0$, $x_2 = 1$, and three changes lead to state 3, $x_1 = 0$, $x_2 = 0$.

Theorem 7.6-1. A sequential circuit for fundamental-mode operation that is not affected by stray delays in the combinational circuitry can be constructed without delays in the y_i leads if and only if the corresponding flow table does not contain any essential hazards [4].

Unger [4] has also shown that if essential hazards are present it is always possible to construct a sequential circuit with only one delay element. This is done by using a combinational circuit to switch the single delay element into the appropriate y_i branch during each input change.

A technique for eliminating the effects of both essential hazards and critical races by inserting delay elements in the input variable leads has been developed by Eichelberger [5]. This technique very often also allows a reduction in the number of required internal variables.

Problems

1. (a) Write the P sets and S sets for the networks of Fig. P7-1.
   (b) Write the 1 sets and 0 sets for the networks of Fig. P7-1.

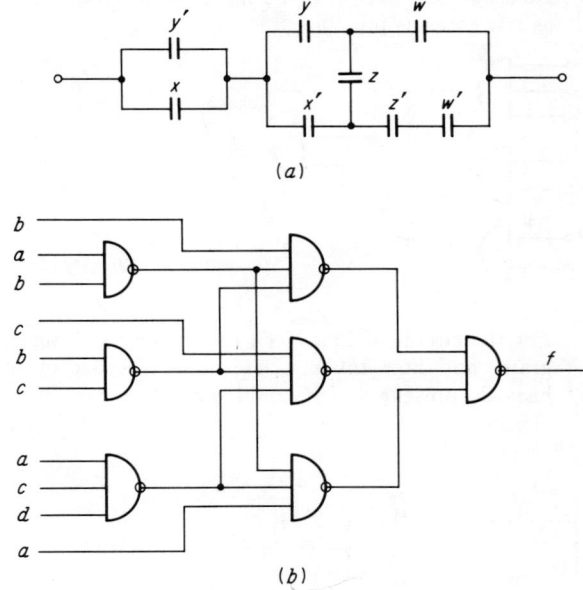

Fig. P7-1

2. For the circuit of Fig. P7-2:
   (a) Determine the 1 sets.
   (b) Analyze the circuit for static hazards using only the 1-sets.
   (c) Determine the 0 sets.
   (d) Analyze the circuit for static hazards using only the 0 sets.

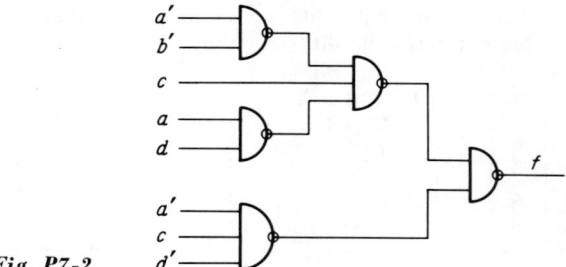

Fig. P7-2

3. Analyze the circuit of Fig. P7-3 for both static and dynamic hazards.

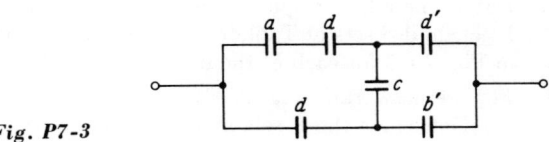

Fig. P7-3

4. Redesign the network of Fig. P7-4 to eliminate all static hazards. Use as few gates as possible.

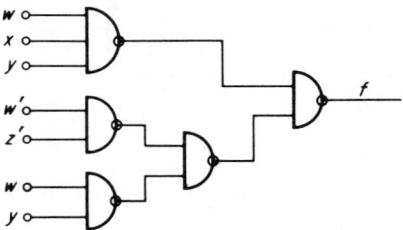

Fig. P7-4

5. For the circuit of Fig. P7-5, determine the transition table, transition diagram, and flow table. Determine whether or not there are any static hazards present which could *actually* occur during the operation of this circuit.

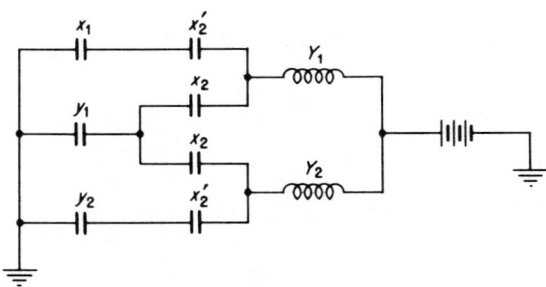

Fig. P7-5

6. Analyze the sequential circuit shown in Fig. P7-6. Are there any hazards present in this circuit? Do they have any effect on the circuit performance?

S	$x_1 x_2$ 00	01	11	10
1	2	1	—	3
2	3	2	4	2
3	4	—	4	—
4	1	4	—	—

S

S	$x_1 x_2$ 00	01	11	10
1	0	—	0	1
2	—	0	0	—
3	1	—	0	—
4	—	1	—	1

z

Fig. P7-6

7. Design a two-stage hazard-free network which realizes the function
$f(u,v,w,x,y,z) = \Sigma(0,1,2,3,5,7,9,11,12,13,14,15,17,21,23,25,27,29,35,39,43,47,52,58,60)$.

8. Determine all four dynamic hazards for the network of Fig. 7.4-1 (use the P sets and S sets of Table 7.4-1). Draw networks and diagrams like those in Fig. 7.4-2 for each of the dynamic hazards.

9. For the network of Fig. P7-9:
   (a) Determine the P sets and S sets of Y.

(b) Determine whether or not there are any dynamic hazards in this network.

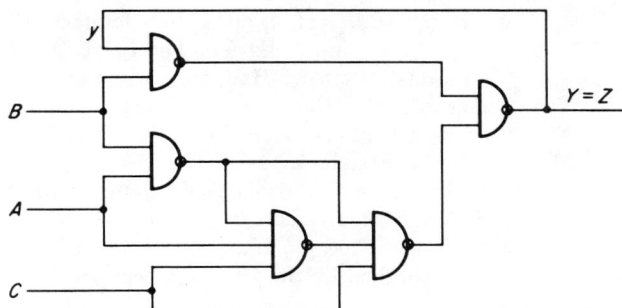

Fig. P7-9

10. Are there any static hazards in the network of Fig. 7.4-1? Explain your answer.
11. For the following flow table:
    (a) Form an excitation table, using the assignment shown.

s	$y_1 y_2$
1	00
2	01
3	11
4	10

(b) Write expressions for Y_1, Y_2, and Z; and draw a circuit using OR-NOT gates.

There are to be no incorrect transient outputs (hazards) in Y_1, Y_2, or Z corresponding to allowed transitions. Use as few gates as possible.

	$x_1 x_2$			
s	00	01	11	10
1	①	4	4	2
2	1	4	3	②
3	1	4	③	4
4	1	④	④	④

S

	$x_1 x_2$			
s	00	01	11	10
1	0	0	0	1
2	0	0	0	1
3	0	0	1	1
4	0	0	0	0

Z

REFERENCES

1. McCluskey, E. J.: Transients in Combinational Logic Circuits, in R. H. Wilcox and W. C. Mann (eds.), "Redundancy Techniques for

Computing Systems," pp. 9–46, Spartan Books, Washington, D.C., 1962.
2. Huffman, D. A.: The Design and Use of Hazard-free Switching Networks, *J. ACM*, vol. 4, pp. 47–62, January, 1957.
3. Maley, G. A., and J. Earle: "The Logic Design of Transistor Digital Computers," Prentice-Hall, Inc., Englewood Cliffs, N.J., 1963.
4. Unger, S. H.: Hazards and Delays in Asynchronous Sequential Switching Circuits, *IRE Trans. on Circuit Theory*, vol. CT-6, no. 1, pp. 12–25, March, 1959.
5. Eichelberger, E. B.: Sequential Circuit Synthesis Using Input Delays, Switching Circuit Theory and Logical Design, *Proc. Fourth Ann. Symposium*, Chicago, Ill., Oct. 28–30, 1963, pp. 105–116, September, 1963, published by the IEEE, *Special Publ.* S-156.

ADDITIONAL REFERENCES

Caldwell, S. H.: "Switching Circuits and Logical Design," John Wiley & Sons, Inc., New York, 1958.
Eichelberger, E. B.: Hazard Detection in Combinational and Sequential Switching Circuits, Switching Circuit Theory and Logical Design, *Proc. Fifth Ann. Symposium*, Princeton, N.J., Nov. 11–13, 1964, pp. 111–120, October, 1964, published by the IEEE, *Special Publ.* S-164.
Keister, W., A. E. Ritchie, and S. H. Washburn: "The Design of Switching Circuits," D. Van Nostrand Company, Inc., Princeton, N.J., 1951.
Kliman, M., and O. Lowenschuss: Asynchronous Electronic Switching Circuits, 1959 *IRE Nat. Conv. Record*, pt. 4, pp. 267–274.
Moisil, Gr. C.: Sur l'application des logiques à trois valeurs à l'etude des schémas à contacts et relais, *Actes proc. congr. intern. automatique*, 1956, p. 48.
Montgomerie, G. A.: Sketch for an Algebra of Relay and Contactor Circuits, *J. IEE*, vol. 95, pt. II, no. 36, July, 1948.
Muller, D. E.: Treatment of Transition Signals in Electronic Switching Circuits by Algebraic Methods, *IRE Trans. on Electronic Computers*, vol. EC-8, no. 3, p. 401, September, 1959.
Naslin, P.: Les Aleas de continuité dans les circuits de commutation a séquences, *Automatisme*, vol. 4, no. 6, pp. 220–225, June, 1959.
Roginskii, V. N.: The Operation of Relay Networks in Transitional Periods, *Avtomatika i Telemekhanika*, vol. 20, no. 10, pp. 1408–1416, October, 1959.
Stone, H. E.: Minimization of Hazard-free Switching Networks, B.S.E. thesis, Department of Electrical Engineering, Princeton University, Princeton, N.J., May, 1960.
Yoeli, M., and S. Rino: Application of Ternary Algebra to the Study of Static Hazards, *J. ACM*, vol. 11, no. 1, pp. 84–97, June, 1964.

INDEX

Adder, serial, 212–213
Addition, binary, 47
  logical, 69n.
  in switching algebra, 69
Algebra, Boolean, 97–104
  of sets, 97, 103–104
  switching (*see* Switching algebra)
Algorithm, 44
Analysis of combinational circuits, 70–74
  AND-NOT networks, 93–97
  contact networks, 70–73
  gate networks, 74–75
  OR-NOT networks, 93–97
AND-NOT networks, 93–97
Arithmetic, addition, 47
  binary, 46–53
  division, 51–53
  multiplication, 51
  subtraction, 47–50

Base, conversion of, 43–46
  number, 42
Binary arithmetic, 46–53
Binary-coded-decimal numbers, 53–55
  biquinary, 55
  excess-3, 54
  self-complementing, 54
  two-out-of-five, 55
  weighted, 54
Binary codes, 53–62
Binary counter, 5
Binary numbers, 43, 46
  arithmetic, 46–53
  geometric representation, 55–59
Boolean algebra, 97–104
Boolean ring, 104
Branching method, 154–155

Canonical expressions, 78–81
  product, 79–81
  sum, 79–81
Carry, end-around, 49
Characteristic function, amplifier, 192
  set-reset flip-flops, 189

Clock pulses, 210
Clocked sequential circuits, 210–216
Codes, binary, 53–62
  error-correcting, 62
  error-detecting, 62
  Gray, 61–62, 77, 236
  unit distance, 59–62
Compatibility class, 254–261
  closed, 254–261
Compatibility table, 256–260
Compatible states, 253–254
Complement(s), 48–50, 100
  diminished radix, 50
  radix, 50
Complete sets, 89
Complete sum, 123, 167–171
Conjunction, 100
Connective, logical, 100
Consensus operation, 166–167
Critical race, 196
Cryotrons, 22–28
  crossed-film, 26–28
  flip-flop, 23–26
  wire-wound, 23–26
Current-switching gates, 20
Cyclic prime implicant tables, 152–157

Decimal specification of switching functions, 76
DeMorgan's theorems, 87
Denial, 100
Diode circuits, 8–11
Direct-coupled transistor logic, 19
Disjunction, 101
Distance on n cube, 57–59
Distinguishable states, 246
Distinguished fundamental product, 137
Distinguished 1 cell, 125
Division, 51–53
  nonrestoring, 53
  restoring, 52
Dominance, 147–152
Don't-care conditions, 77n.
Duality, principle of, 70
Dynamic hazards, 299–303

Equivalence relation, 244
Error-detecting and -correcting codes, 62
Essential hazards, 306
Essential multiple-output prime implicants, 136–139
Essential prime implicant, 125–126
Essential rows, 146–147
Excitation functions, 188
Excitation table, 188–189, 261–263

Flip-flop, 4–5, 18, 23–26
   application table, 263
   characteristic function, 189
   cryotron, 23–26
   excitation table, 262–263
   symmetrical-triggering, 5
   vacuum tube, 4
Flow table, 199, 224–235
   covering, 252–253
   formation, 224–235
   incompletely specified, 248–261
   minimum state, 244–246
   primitive form, 230–233
   simplification, 237–261
Fundamental mode, 183–215
   output specifications, 233–235
Fundamental products, 78–79
Fundamental sums, 78–79

Gates, AND, 9, 69
   AND-NOT, 17, 91
   current-switching, 20
   DCTL, 19–20
   EXCLUSIVE OR, 91–93
   NAND, 18
   NOR, 18
   OR, 9, 69
   OR-NOT, 19, 91
   sum modulo two, 91–93
Gray code, 61–62, 77, 236

Hazard, dynamic, 299–303
   essential, 306–310
   static, 288–299
Hazard-free networks, 303–306
Huffman, D. A., 183, 296n.

Inaccessible states, 237–239
Inclusion, 103

Incompatibility class, 260
Incompletely specified flow table, 248–261
Incompletely specified functions, 126
Indistinguishability classes, 243–248
Indistinguishable circuits, 239–241
Indistinguishable states, 241–243
Induction, perfect, 84
Initial states, 235
Input state, 186
Intersection, 103
Irredundant sum, 124
Iterative consensus, 165–174

Keys, 1–2

Literal, 68
Logic, double-rail, 89
   propositional, 99–103
Logical connective, 100

Magnetic cores, 28–35
   mirror symbols, 32–33
   output connections, 33–35
Map, n cube, 57, 116–127
Material implication, 102
Maxterm, 78–79
Memory in sequential circuits, 190–192
Minimal product, 127–131
Minimal sum, 114–115, 123–131
Minimum-state flow tables, 244–246
Minterm, 78–79
Mirror symbols, 32–33
Moore model circuits, 205n.
Multiple-output minimal sums, 135–139
Multiple-output networks, 131–139, 157–165
Multiple-output prime implicant tables, 160–165
Multiple-output prime implicants, 132–135
Multiplication, binary, 51
   in switching algebra, 69

n cube, 56–57, 61
   distance, 57–59
   map, 57, 116–127
   symmetries, 61–62

NAND gate, 18
Next-state variables, 186
NOR gate, 18
Number of functions, 83
Number systems, base, 42
 binary, 43, 46–53
 conversion, 43–46
 positional notation, 42–43
 radix, 42

Octal numbers, 46
 use in forming prime implicants, 143–146
Operating point, 187, 207
OR-NOT networks, 93–97
Output specifications, fundamental mode, 233–235
 pulse mode, 227–228

P set, 73, 286
Partial ordering, 103
Petrick's method, 155n.
 (*See also* Prime implicant function)
Prime implicant function, 155–157
Prime implicant tables, 146–156
 branching method, 154–155
 cyclic, 152–157
 dominance in, 147–152
 essential rows, 146–147
 multiple-output, 160–165
 prime implicant function, 155–157
Prime implicants, 117–122
 determination using octal numbers, 143–146
 essential, 125
 by iterative consensus, 171–174
 multiple-output, 132, 157–160
 tabular determination of, 140–146, 157–160, 171–174
 theorem, 129–131
Prime implicates, 127
Primitive-form flow tables, 230–233
Propositional logic, 99
Pulse-mode operation, 199–210
 output specifications, 227–228

Quine, W. V., 119n.
Quine-McCluskey method, 140n.

Races in sequential circuits, 195–199, 208–210, 268–272
 eliminating, 268–272
 fundamental mode, 195–199
 pulse mode, 208–210
Radix, 42
Radix point, 42
Regular expression, 218
Relays, 3–4
 series control, 68

S set, 73, 287
Sequential circuits, AND-NOT gates, 194–195
 clocked, 210–216
 diode gates, 192–194
 flow table, 199
 excitation functions, 188
 excitation table, 188–189, 261–263
 fundamental-mode operation, 183
 initial states, 235
 input state, 186
 internal state, 186
 internal variable assignments, 263–277
 internal variables, 186
 Mealy model, 205n.
 memory, 190–192
 Moore model, 205n.
 operating point, 187–207
 pulse-mode operation, 199–210
 races, 195–199, 208–210, 268–272
 realizability, 235–237
 regular expression, 218
 state diagram, 216–218
 synchronous, 211n.
 total state, 184
 transition diagram, 187
 transition table, 186, 261–277
Serial adder, 212–213
Set, 97
Shifting, 50–51
State adjacency diagram, 269–272
State diagram, 216–218
State table, 188
Static hazards, 288–299
Subset, 103
Subtraction, 47–50
Sum, canonical, 79–81
 complete, 123, 167–171
 irredundant, 124
 minimal, 114–115, 123–131
Switches, 1–2

Switching algebra, 66–97
  addition, 69
  multiplication in, 69
  postulates, 67–70
  theorems, 84–89
Symbols, AND gate, 36
  AND-NOT gate, 38
  gate, 35–38
  inverter, 36
  mirror, 32–33
  OR gate, 36
  OR-NOT gate, 37
  relay, 3
Symmetries of the n cube, 61–62
Synchronous sequential circuits, $211n$.

Table of combinations, 75–76
Tabular determination of prime
    implicants, 140
Total state, 184
Transient analysis, 285–288

Transient transmission function, 286–287
Transistors, 11–21
  characteristics, 11–13
  current-switching gates, 20–21
  diode logic, 16–18
  direct-coupled logic (DCTL), 19–20
  resistor logic, 18–19
Transition diagram, 187
Transition table, 186, 261–277
Transmission, network, 68
Truth table, $76n$.
Two-out-of-five code, 55
Two-stage circuit, 115
Type A flow tables, 261

Union, 103
Unit-distance codes, 59–62

Vacuum-tube circuits, 4–8
Venn diagram, 104